# Elliptic Theory on Singular Manifolds

# Differential and Integral Equations and Their Applications

A series edited by:
A.D. Polyanin
*Institute for Problems in Mechanics, Moscow, Russia*

---

**Volume 1**
Handbook of First Order Partial Differential Equations
*A.D. Polyanin, V.F. Zaitsev and A. Moussiaux*

**Volume 2**
Group-Theoretic Methods in Mechanics and Applied Mathematics
*D.M. Klimov and V. Ph. Zhuravlev*

**Volume 3**
Quantization Methods in Differential Equations
*V.E. Nazaikinskii, B.-W. Schulze and B. Yu. Sternin*

**Volume 4**
Hypersingular Integral Equations and Their Applications
*I.K. Lifanov, L.N. Poltavskii and G.M. Vainikko*

**Volume 5**
Equations of Mathematical Diffraction Theory
*Mezhlum A. Sumbatyan and Antonio Scalia*

**Volume 6**
High-Precision Methods in Eigenvalue Problems and Their Applications
*Leonid D. Akulenko and Sergei V. Nesterov*

**Volume 7**
Elliptic Theory on Singular Manifolds
*Vladimir E. Nazaikinskii, Anton Yu. Savin, Bert-Wolfgang Schulze and Boris Yu. Sternin*

# Elliptic Theory on Singular Manifolds

**Vladimir E. Nazaikinskii**
**Anton Yu. Savin**
**Bert-Wolfgang Schulze**
**Boris Yu. Sternin**

CRC Press
Taylor & Francis Group
Boca Raton London New York

CRC Press is an imprint of the
Taylor & Francis Group, an **informa** business
A CHAPMAN & HALL BOOK

CRC Press
Taylor & Francis Group
6000 Broken Sound Parkway NW, Suite 300
Boca Raton, FL 33487-2742

First issued in paperback 2019



ISBN-13: 978-1-58488-520-7 (hbk)
ISBN-13: 978-0-367-39229-1 (pbk)



**Library of Congress Cataloging-in-Publication Data**

Elliptic theory on singular manifolds / Vladimir E. Nazaikinskii ... [et al.].
p. cm. -- (Differential and integral equations and their applications ; v. 7)
Includes bibliographical references and index.
ISBN 1-58488-520-3
1. Elliptic operators. 2. Manifolds (Mathematics) 3. Singularities (Mathematics) I. Nazaæikinskiæi, V. E. II. Series.

QA329.42.E55 2005
515'.7242--dc22 2005049376

**Visit the Taylor & Francis Web site at**
**http://www.taylorandfrancis.com**

**and the CRC Press Web site at**
**http://www.crcpress.com**

# Contents

## Preface

Index theory of elliptic operators, which arose in the 1960s in the work of Atiyah, Singer, and Bott, establishes important connections between analysis and topology, most remarkably reflected in the Atiyah–Singer index theorem and subsequent fundamental results. At first, the theory dealt mainly with elliptic operators on smooth compact manifolds without boundary and boundary value problems. However, later it became clear that elliptic theory on manifolds with singularities is also of great theoretical and practical interest, and the development of index theory for manifolds with singularities became topical. Note that analysis and topology of elliptic operators on manifolds with singularities is much more complicated than in the smooth case: specific problems arising here have no analogs in the case of smooth manifolds and require completely new mathematical notions and theories. The first attempts to clarify the situation in the case of the simplest (isolated) singularities were made relatively long ago, and mathematicians and physicists, mainly in the American school, obtained results having important applications in geometry. After these preliminary results, the general index problem (i.e., the index problem for general elliptic operators on manifolds with singularities) became the challenge. This problem was attacked by American, French, German, Russian, and Scandinavian mathematicians. Recently, there has been dramatic progress in the field, and in a number of cases (e.g., isolated singularities) definitive results were obtained. Moreover, very recently the authors have managed to solve the index problem for manifolds with nonisolated singularities of the first level of complexity, so-called edge singularities.

Unfortunately, many of the above-mentioned results are scattered over various journals and preprints. From this viewpoint, it seems that there is an urgent need to give a systematic exposition in the form of a book clearly displaying the problems and main results.

The present monograph is an attempt to fill the gap. It is of interest to a wide readership including specialists in differential equations, topology, and related fields. We also hope that the book will be of interest to physicists, since the corresponding index theory in the smooth case has strong connections with theoretical physics, in particular, with quantum field theory (Atiyah, Witten, and others). The book is also easily accessible to graduate, postgraduate, and postdoctoral students specializing in the above-mentioned fields.

The results included in the book were obtained with partial support of the Deutsche Forschungsgemeinschaft and of Russian Foundation for Basic Research grants nos. 02-01-00118 and 05-01-00982.

*The Authors*
*Moscow–Potsdam*

## About the Authors

**Vladimir E. Nazaikinskii, Ph.D.,** is an actively working mathematician specializing in partial differential equations, mathematical physics, and noncommutative analysis. He was born in 1955 in Moscow, graduated from Moscow Institute of Electronic Engineering in 1977, defended his Ph.D. in 1980, and worked at the Institute for Automated Control Systems, Moscow Institute of Electronic Engineering, and Moscow State University. Currently he is a senior researcher at the Institute for Problems in Mechanics, Russian Academy of Sciences. He is the author of five monographs and more than 40 papers on various aspects of noncommutative analysis, asymptotic problems, and elliptic theory.
**Address:** Institute for Problems in Mechanics, Russian Academy of Sciences, 119526, pr. Vernadskogo 101-1, Moscow, Russia
**E-mail:** nazaikinskii@yandex.ru

**Anton Yu. Savin, Ph.D.,** is a mathematician specializing in partial differential equations, topology, and noncommutative geometry. He was born in Moscow in 1975, graduated from the Department of Computational Mathematics and Cybernetics, Moscow State University in 1997, defended his Ph.D. in 2000, and worked at Moscow State University. Currently he is a researcher at the Independent University of Moscow. He is the author of more than 30 papers on differential equations, topology and mathematical physics.
**Address:** Independent University of Moscow, 119002, Bolshoy Vlasyevskiy Pereulok 11, Moscow, Russia
**E-mail:** antonsavin@mail.ru

**Professor Bert-Wolfgang Schulze, D.Sc.,** is Professor of Analysis at the University of Potsdam, Germany. He was born in 1944 in Erfurt and studied geophysics and mathematics in Leipzig between 1963-1968. In 1970 he became active in potential theory and boundary value problems at the Academy of Sciences in (East) Berlin; his Habilitation followed in 1974 at the University of Rostock, and in 1980 he was named Professor of Analysis at the Karl Weierstrass Institute in Berlin. He was head of the research group "Partial Differential Equations" of the Max Planck Society in the years 1992-1996, and in 1994 he was appointed to the Professorship in Analysis at the University of Potsdam. He is active in the fields of global analysis, index theory of boundary value problems, operator algebras on manifolds with singularities, asymptotics of solutions to elliptic equations, and parabolicity in algebras of Volterra pseudodifferential operators. His publications include ten monographs and over 160 papers.
**Address:** Universität Potsdam, Institut für Mathematik, Postfach 601553, D-14415 Potsdam, Germany
**E-mail:** schulze@math.uni-potsdam.de

**Professor Boris Yu. Sternin, D.Sc.,** is an internationally renowned scientist specializing in partial differential equations, complex analysis, differential geometry, topology, and mathematical physics. He was born in Moscow in 1939, graduated from the Department of Mechanics and Mathematics, Moscow State University in 1962, defended his Ph.D. in 1965 and D.Sc. in 1984 and worked at the Institute for Problems in Mechanics, Moscow Institute of Electronic Engineering, and Moscow Civil Aviation Institute. He was a professor at Moscow State University for ten years. Currently he is a professor at the Independent University of Moscow. He is the author of eleven monographs published by Springer-Verlag, Kluwer Academic Publishers, Walter de Gruyter, CRC Press, etc. and of more than 300 papers in various fields of pure and applied mathematics.
**Address:** Independent University of Moscow, 119002, Bolshoy Vlasyevskiy Pereulok 11, Moscow, Russia
**E-mail:** sternin@mail.ru

## Notation

The notation used throughout the book is largely standard, and we mention only a few specific notation conventions which might cause some difficulties for the reader. Note that here we do not give any *definitions*; these are to be found in the relevant chapters.

***Manifolds and coordinates.*** The letter $\Omega$ is used to denote a smooth closed manifold playing the role of the base of the *infinite model cone*

$$K_\Omega = \{\Omega \times [0,\infty)\}/\{\Omega \times \{0\}\}.$$

Points of (and coordinates on) $\Omega$ are denoted by $\omega$ ( $= (\omega_1, \dots, \omega_k)$, $k = \dim \Omega$), and the variable $r \in [0, \infty)$ is referred to as the *radial variable*. By $K_\Omega^\circ$ we denote the open subset of $K_\Omega$ obtained by deleting the vertex. The change of variables $r = e^{-t}$ takes $K_\Omega^\circ$ to the infinite cylinder

$$C_\Omega = \Omega \times (-\infty, \infty),$$

and the variable $t$ is called the *cylindrical variable*.

Manifolds with singularities are denoted by the letter $\mathcal{M}$ (possibly with subscripts). The *stretched manifold* corresponding to $\mathcal{M}$ is denoted by $M$. This book deals only with manifolds with conical points or edges, and in both cases $M$ is a smooth manifold with boundary $\partial M$. Specifically,

• if $\mathcal{M}$ is a manifold with conical singularities, then $\partial M$ is the disjoint union of bases $\Omega_j$ of model cones at the singular points

• if $\mathcal{M}$ is a manifold with an edge, then the edge is denoted by $X$. It is a smooth manifold, and one has a smooth locally trivial bundle

$$\pi : \partial M \xrightarrow{\Omega} X$$

with fiber the base $\Omega$ of the model cone. The points of (and coordinates on) the edge are denoted by $x$ ( $= (x_1, \dots, x_n)$, $n = \dim X$). Thus local coordinates on $\partial M$ are $(x, \omega)$.

By $W$ we denote the *infinite wedge* associated with $\mathcal{M}$, and by $\mathcal{U}$ a collar neighborhood of the edge in $\mathcal{M}$ (and hence in $W$). The corresponding neighborhood in $M$ is denoted by $U$. Local coordinates in $U$ are $(x, \omega, r)$.

By $T^*\mathcal{M}$ we denote the *stretched cotangent bundle* of $\mathcal{M}$. This is a vector bundle over $M$; canonical local coordinates on $T^*\mathcal{M}$ in a neighborhood of $\partial M$ are denoted by $(x, \omega, r, \xi, q, p)$, where $\xi, q, p$ are the momentum variables dual to $x, \omega, r$, respectively. (The variables $x, \xi$ are missing if $\mathcal{M}$ is a manifold with conical singularities.) In particular, $(x, \xi)$ are coordinates on $T^*X$. The variable $p$ is called the *conormal variable*.

By $\mathcal{M}^\circ = M^\circ$ we denote the interior of $M$ (equal to $\mathcal{M} \setminus X$).

***Function spaces.*** The most commonly used spaces that do not yet have standard notation in the literature are

- the weighted Sobolev spaces $H^{s,\gamma}(K_\Omega)$ and $H^{s,\gamma}(\mathcal{M})$ on the infinite cone and on a manifold with conical singularities
- the weighted Sobolev space $\mathcal{K}^{s,\gamma}(K_\Omega)$ on the infinite cone, differing from the preceding in the behavior at infinity
- the weighted Sobolev space $\mathcal{W}^{s,\gamma}(\mathcal{M})$ on a manifold $\mathcal{M}$ with edges

Here $s$ and $\gamma$ are real parameters, called the *smoothness exponent* and the *weight exponent*, respectively.

***Symbols and symbol spaces.*** For a (pseudo)differential operator $D$ on a singular manifold $\mathcal{M}$,

- $\sigma(D)$ is the *interior symbol* of $D$, and $\sigma_\partial(D)$ is the boundary symbol of $D$, i.e., the restriction of $\sigma(D)$ to the boundary of the stretched cotangent bundle
- $\sigma_c(D)$ is the *conormal symbol* of $D$ (if $\mathcal{M}$ is a manifold with conical singularities)
- $\sigma_\wedge(D)$ is the *edge symbol* of $D$ (if $\mathcal{M}$ is a manifold with edges)
- $\sigma(\sigma_\wedge(D))$ and $\sigma_c(\sigma_\wedge(D))$ are the *interior* and *conormal symbols* of $\sigma_\wedge(D)$
- the pair $(\sigma(D), \sigma_c(D))$ or $(\sigma(D), \sigma_\wedge(D))$ is the *principal symbol* of $D$ if $\mathcal{M}$ is a manifold with conical singularities or edges, respectively

Apart from the usual symbol spaces of Hörmander type for scalar symbols, we use the spaces $S^0_{CV}(T^*X, \mathfrak{H}_1, \mathfrak{H}_2)$ of symbols of compact fiber variation for operator-valued symbols.

***Miscellanea.*** Finally, we mention some notation that does not fall within any of the preceding.

By $\mathcal{L}_\gamma$ we denote the weight line corresponding to the weight exponent $\gamma$ in the complex plane.

By $\varkappa_\lambda$ we denote the dilation group acting on functions of the variable $r$ as follows:

$$\varkappa_\lambda f(r) = \lambda^{(k+1)/2} f(\lambda r).$$

Here $k = \dim \Omega$.

# Part I

# Singular Manifolds and Differential Operators

# Chapter 1

# Geometry of Singularities

## 1.1. Preliminaries

The subject of interest in this book is singular manifolds with differential operators defined on them. For an arbitrary "singular" topological space, it is obviously not an easy task even to define reasonably what differential operators are, let alone study their properties. Hence we do not deal with general singular spaces; on the contrary, we restrict our attention to spaces that possess rich additional structures permitting one to define differential operators in a natural way. Most generally speaking, we consider spaces $\mathcal{M}$ with a given $C^\infty$ structure on a dense open subset $\mathcal{M}^\circ$. Then the algebra $\operatorname{Diff}(\mathcal{M}^\circ)$ of all differential operators with smooth coefficients on $\mathcal{M}^\circ$ is well defined, and we single out some subalgebra $\mathcal{D} \subset \operatorname{Diff}(\mathcal{M}^\circ)$ by imposing conditions on the behavior of operators near the singularity set $\mathcal{M} \setminus \mathcal{M}^\circ$. For example, the operators should "degenerate" there or, on the opposite, their coefficients are allowed to go to infinity at a specified rate. It is the pair $(\mathcal{M}^\circ, \mathcal{D})$ rather than $\mathcal{M}$ alone that is of interest to us. From the viewpoint of differential equations, two pairs $(\mathcal{M}^\circ, \mathcal{D})$ and $(\mathcal{M}^\circ, \widetilde{\mathcal{D}})$, where $\mathcal{D} \neq \widetilde{\mathcal{D}}$, represent *different* manifolds with singularities and certainly should be distinguished in the study. The idea of defining manifolds with singularities as ringed spaces (understood as pairs of this sort) goes back to Schulze, Sternin and Shatalov (1997, 1998*a*). Note that in the $C^\infty$ case (where the singularity set $\mathcal{M} \setminus \mathcal{M}^\circ$ is empty) the introduction of the ring of differential operators is unnecessary, since it can be reconstructed from $C^\infty(\mathcal{M})$: all differentiations are just derivations of $C^\infty(\mathcal{M})$. In the singular case this is, however, meaningful since we are not interested in all differential operators but deal with some subalgebra.

The above-mentioned class of singular manifolds bearing differential operators is still too wide and general if we wish to obtain really informative results, and so we shall single out a comprehensible subclass that satisfies all the practical needs and at the same time enjoys a very simple and clear construction of the corresponding classes of differential operators.

We start from some simple examples in which singular manifolds and the corresponding degenerate or "singular" differential operators arise.

### 1.1.1. Examples

***Operators on the cone.*** In the space $\mathbb{R}^3$ with coordinates $(x, y, z)$, consider the right circular cone $K_\Theta$ given by the relations

$$x^2 + y^2 = z^2 \sin^2 \Theta, \quad z \geq 0 \tag{1.1}$$

(see Fig. 1.1). Here $\Theta$ is the angle between the axis and the generator of the cone.

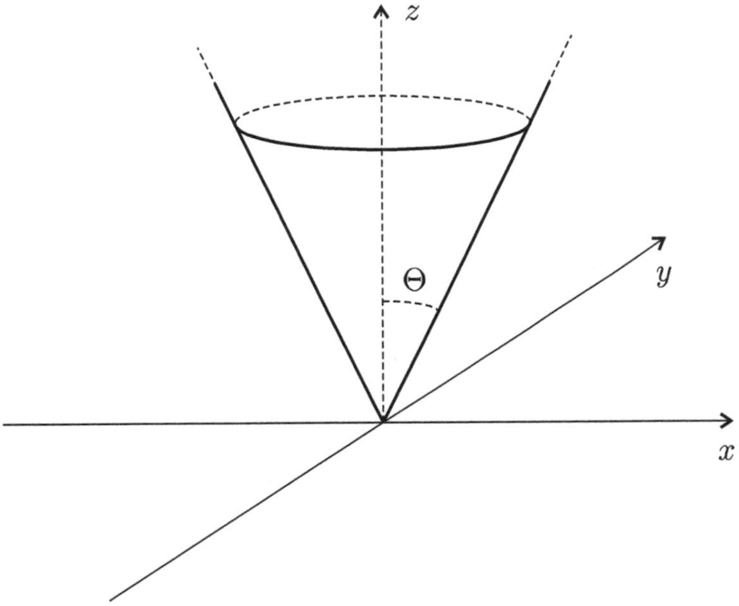


Figure 1.1. A circular cone.

The dense open subset $K_\Theta^\circ \subset K_\Theta$ obtained by deleting the vertex bears the natural $C^\infty$ structure inherited from the ambient space $\mathbb{R}^3$.

Let us see how we can naturally define an algebra of differential operators on $K_\Theta^\circ$. The restriction of the Euclidean metric $dx^2+dy^2+dz^2$ in $\mathbb{R}^3$ to $K_\Theta^\circ$ is a natural metric on $K_\Theta^\circ$. Let us express it in the coordinates $(r, \varphi)$, where $r$ is the distance from the vertex along the generator and $\varphi$ is the polar angle corresponding to the projection of a point of the cone on the $xy$-plane. Thus

$$r^2 = x^2 + y^2 + z^2 = (x^2 + y^2)(1 + \sin^2 \Theta), \quad \sin \varphi = \frac{y}{\sqrt{(x^2 + y^2)}}.$$

The metric has the form

$$ds^2 = dr^2 + r^2 \sin^2 \Theta \, d\varphi^2. \tag{1.2}$$

Once a metric is given on a manifold, one can naturally construct the associated geometric operators, e.g., the Euler operator or the Beltrami–Laplace operator. Consider, say, the corresponding Beltrami–Laplace operator

$$\triangle_c = \frac{1}{r^2}\left[\left(r\frac{\partial}{\partial r}\right)^2 + c^2\frac{\partial^2}{\partial\varphi^2}\right], \tag{1.3}$$

where the positive parameter $c^2 = (\sin\Theta)^{-2}$ is determined by the angle $\Theta$ at the cone vertex.

◀ The formula (1.3) follows from the general formula

$$\triangle = \frac{1}{\sqrt{\det g}}\frac{\partial}{\partial\alpha^i}\left(g^{ij}\sqrt{\det g}\frac{\partial}{\partial\alpha^j}\right),$$

which expresses the Beltrami–Laplace operator $\triangle$ corresponding to a metric $g$ in local coordinates $\{\alpha_j\}$ via the components and the determinant of the metric tensor $g = (g_{ij})$. ▶

Consider also the extreme case $\Theta = \pi/2$, in which the cone coincides with the $xy$-plane. In this case, the operator (1.3) becomes

$$\triangle = \frac{1}{r^2}\left[\left(r\frac{\partial}{\partial r}\right)^2 + \frac{\partial^2}{\partial\varphi^2}\right],$$

which is the well-known expression for the Laplacian

$$\triangle = \frac{\partial^2}{\partial x^2} + \frac{\partial^2}{\partial y^2}$$

on the plane in the polar coordinates related to the Cartesian coordinates by the formulas

$$x = r\cos\varphi, \quad y = r\sin\varphi.$$

Thus in this case the singularity is *fictitious*: we have just taken a usual differential operator with smooth coefficients in Cartesian coordinates and proceeded to polar coordinates. However, the operator $\triangle_c$, which can be viewed as a perturbation of $\triangle \equiv \triangle_1$, has singularities in the coefficients even in the Cartesian coordinates; note that one can hardly tell the qualitative difference between these operators by examining their expressions in polar coordinates.

EXERCISE 1.1. Write out the expression of $\triangle_c$ in the Cartesian coordinates.

Modulo the factor $r^{-2}$, the operator (1.3) is a polynomial in $r\partial/\partial r$, i.e., a Fuchs degenerate operator at $r = 0$. We see that a natural geometric operator on the cone, the Beltrami–Laplace operator, is a differential operator with Fuchsian degeneration at the vertex. It turns out that the same is true for other geometric operators.

Indeed, a straightforward computation shows that both the exterior differentiation operator and the Euler operator on the cone are operators with Fuchsian degeneration.

So, what natural candidate could we take for an algebra of differential operators on the cone? The answer, in a sense, suggests itself: one should take the algebra $\mathcal{D}(K_\Theta)$ of operators of the form

$$D = r^{-m} \sum_{j+k\leq m} a_{jk}(r,\varphi)\left(r\frac{\partial}{\partial r}\right)^j \left(\frac{\partial}{\partial \varphi}\right)^k, \tag{1.4}$$

where the coefficients $a_{jk}(r,\varphi)$ are *smooth functions up to* $r = 0$. The elements of our algebra are Fuchs degenerate operators multiplied by $r^{-m}$, where $m$ is the order of the operator. One can readily check that operators of the form (1.4) with various $m$ indeed form an algebra.

The expression (1.4) seems slightly artificial owing to the factor $r^{-m}$ multiplying the sum. We can proceed to a different expression, which is free of this disadvantage. Namely, the following assertion is true.

PROPOSITION 1.2. (1) *The operator* (1.4) *can be represented in the form*

$$D = \sum_{j+k+l\leq m} b_{jkl}(r,\varphi)\left(\frac{1}{r}\right)^l \left(\frac{\partial}{\partial r}\right)^j \left(\frac{1}{r}\frac{\partial}{\partial \varphi}\right)^k, \tag{1.5}$$

*where the coefficients* $b_{jkl}(r,\varphi)$ *are smooth functions up to* $r = 0$.

(2) *The algebra* $\mathcal{D}(K_\Theta)$ *is generated by operators of multiplication by functions smooth up to* $r = 0$ *and the operators*

$$\frac{\partial}{\partial r}, \quad \frac{1}{r}\frac{\partial}{\partial \varphi}. \tag{1.6}$$

*Remark* 1.3. The most interesting fact is that the operator $r^{-1}$ is *not* on the list in (1.6). However, it arises in the commutation of the operators already present there with smooth functions. This will be discussed in more detail in Section 1.2 dealing with general cone-degenerate operators.

Thus we see that, remarkably, the algebra $\mathcal{D}(K_\Theta)$ of differential operators on the cone $K_\Theta$ is generated by the operators (1.6), which are just *vector fields of length of the order of unity* in the metric (1.2) and the operators of multiplication by functions smooth up to $r = 0$, that is, $C^\infty$ functions on the half-cylinder

$$K_\Theta^\wedge = \overline{\mathbb{R}}_+ \times \mathbb{S}^1,$$

which is the *manifold with boundary obtained from* $K_\Theta$ *by blowing up the vertex to a circle.*

These observations are important in that they will serve us as a pilot model when we proceed from examples to more general classes of singular manifolds.

*Remark* 1.4. We point out that even though the cone $K_\Theta$ can be rectified into a plane for any $\Theta$ (and just coincides with the plane for $\Theta = \pi/2$), the deletion of the vertex plays a very important role even if we deal with operators whose singularity is fictitious, i.e., stems from the passage to polar coordinates alone. Indeed, consider the equation

$$\triangle_c u = 0.$$

We require that this equation hold only outside the singularity, in other words, for $r > 0$ in the polar coordinates. Moreover, we assume that the function $u$ is also defined in general only for $r > 0$. Then for all $c$ the equation in question has not only the trivial solution $u = \text{const}$ *but also the solution* $u = C \ln r$, *where* $C$ *is an arbitrary constant.*

***Operators on the cuspidal cone.*** Now consider the cuspidal cone $\Upsilon_k \subset \mathbb{R}^3$ given by the relations

$$x^2 + y^2 = z^{2(k+1)}, \quad z \geq 0 \tag{1.7}$$

(see Fig. 1.2), where $k$ is a positive integer[1] known as the *order* of the cusp. Again

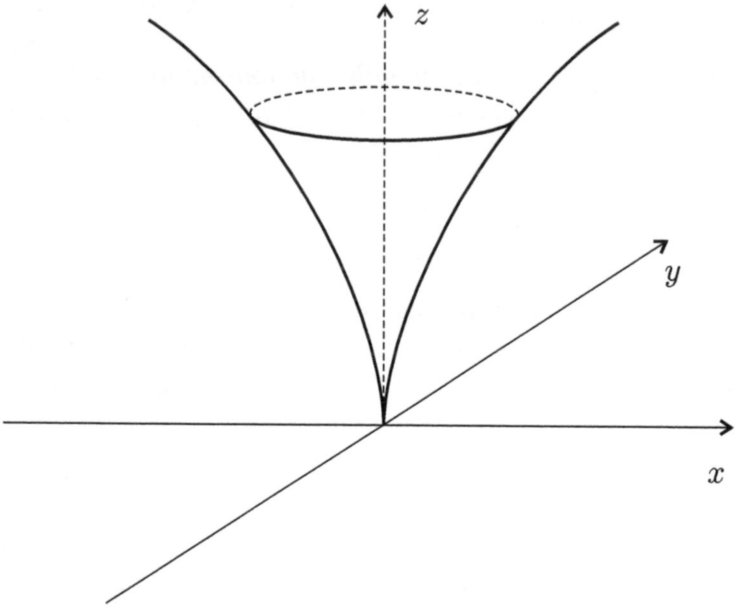


Figure 1.2. A cuspidal cone of order $k$.

the dense open subset $\Upsilon_k^\circ \subset \Upsilon_k$ obtained by the deletion of the vertex inherits a

[1]For $k = 0$, we return to the conical case. One can also consider fractional-order cusps (e.g., Schulze, Sternin and Shatalov 1998*a*), but we do not touch on this matter in the present book.

natural $C^\infty$ structure from $\mathbb{R}^3$, and topologically the cuspidal cone does not at all differ from the cone considered in the preceding example.

To write out a meaningful class of differential operators on the cuspidal cone, we again use the restriction of the Euclidean metric defined in the ambient space. To compute this restriction, we should choose some coordinates on $\Upsilon_k^\circ$; it is convenient to take them in the form $(z,\varphi)$, where $\varphi$ is the polar angle of the projection of a point of the cusp onto the $xy$-plane. Then the restriction of the Euclidean metric of $\mathbb{R}^3$ to $\Upsilon_k^\circ$ has the form

$$ds^2 = \left(1+(k+1)^2 z^{2k}\right)dz^2 + z^{2k+2}d\varphi^2, \tag{1.8}$$

and we can consider the corresponding geometric operators.

In particular, the Beltrami–Laplace operator is given by the expression

$$\triangle = z^{-2(k+1)}\left[a(z)^{-3/2}\left(z^{k+1}\frac{\partial}{\partial z}\right)a(z)^{1/2}\left(z^{k+1}\frac{\partial}{\partial z}\right)+\frac{\partial^2}{\partial\varphi^2}\right], \tag{1.9}$$

where the function $a(z) = \left(1+(k+1)^2 z^{2k}\right)$ is smooth and does not vanish.

EXERCISE 1.5. Derive the formula (1.9).

Modulo the factor $z^{-2(k+1)}$, the operator (1.9) is a polynomial in $z^{k+1}\partial/\partial z$, i.e., an operator with degeneration of order $k+1$.

One can verify that other geometric operators (like the exterior differentiation operator or the Euler operator) corresponding to the metric (1.8) on the cuspidal cone are also operators with degeneration of order $k+1$.

It is now easy to write out a natural candidate for the algebra of differential operators on the cuspidal cone. This is the algebra $\mathcal{D}(\Upsilon_k)$ of operators of the form

$$D = z^{-(k+1)m}\sum_{i+j\le m} a_{ij}(z,\varphi)\left(z^{k+1}\frac{\partial}{\partial z}\right)^i\left(\frac{\partial}{\partial\varphi}\right)^j, \tag{1.10}$$

where the coefficients $a_{ij}(z,\varphi)$ are again *smooth functions up to* $z=0$. The elements of our algebra are strongly (of order $k+1$) degenerate operators multiplied by $z^{-(k+1)m}$, where $m$ is the order of the operator.

*Remark* 1.6. Note that operators of the form (1.10) with various $m$ indeed form an algebra.

◀ A similar assertion is valid for fractional $k$, but the coefficients in this case should depend also on fractional powers of $z$. ▶

We can transform the expression (1.10) so as to get rid of the formidable factor $z^{-(k+1)m}$.

PROPOSITION 1.7. (1) *The operator* (1.10) *can be represented in the form*

$$D = \sum_{i+j+l\le m} b_{ijl}(z,\varphi)\left(\frac{1}{z}\right)^{l(k+1)}\left(\frac{\partial}{\partial z}\right)^{j}\left(\frac{1}{z^{k+1}}\frac{\partial}{\partial\varphi}\right)^{i}, \tag{1.11}$$

*where the coefficients* $b_{ijl}(z,\varphi)$ *are smooth functions up to* $z = 0$.

(2) *The algebra* $\mathcal{D}(\Upsilon_k)$ *is generated by operators of multiplication by functions smooth up to* $z = 0$ *and the operators*

$$\frac{\partial}{\partial z}, \quad \frac{1}{z^{k+1}}\frac{\partial}{\partial\varphi}. \tag{1.12}$$

Thus in this example we also see that the algebra $\mathcal{D}(\Upsilon_k)$ of differential operators on the cuspidal cone is generated by *coordinate vector fields of length of the order of unity* in the metric (1.8) and the operators of multiplication by smooth functions on the *stretched manifold*

$$\Upsilon_k^{\wedge} = \overline{\mathbb{R}}_+ \times \mathbb{S}^1$$

*of the cuspidal cone.*

***Operators on the wedge.*** Now we consider the wedge

$$W = \mathbb{R} \times K,$$

which is the direct product of the line $\mathbb{R}$ with coordinate $z$ by the cone $K$ whose base is a circle with coordinate $\varphi$. The coordinate on each of the generators of the cone is denoted by $r$.

The metric is represented in the form

$$ds^2 = dr^2 + r^2 d\varphi^2 + dz^2, \tag{1.13}$$

and accordingly, the Beltrami–Laplace operator can be written out in the coordinates $(r, \varphi, z)$ in the form

$$\triangle = \frac{1}{r^2}\left[\left(r\frac{\partial}{\partial r}\right)^2 + \frac{\partial^2}{\partial\varphi^2}\right] + \frac{\partial^2}{\partial z^2} = \frac{1}{r^2}\left[\left(r\frac{\partial}{\partial r}\right)^2 + \frac{\partial^2}{\partial\varphi^2} + \left(r\frac{\partial}{\partial z}\right)^2\right]. \tag{1.14}$$

Modulo the factor $r^{-2}$, the operator (1.14) is a polynomial in the operators

$$r\frac{\partial}{\partial r}, \quad r\frac{\partial}{\partial z}, \quad \frac{\partial}{\partial\varphi}$$

with coefficients that are smooth functions on the stretched manifold

$$W^{\wedge} = \mathbb{R}_z \times \mathbb{S}^1 \times \overline{\mathbb{R}}_+$$

of the wedge $W$. Such operators are said to be *edge-degenerate*. General edge-degenerate differential operators on $W$ have the form

$$D = r^{-m} \sum_{i+j+k \leq m} a_{ijk}(z, \varphi, r) \left( r \frac{\partial}{\partial z} \right)^k \left( r \frac{\partial}{\partial r} \right)^j \left( \frac{\partial}{\partial \varphi} \right)^i, \tag{1.15}$$

where the coefficients $a_{ijk}(z, \varphi, r)$ are smooth functions on the stretched manifold $W^\wedge$ of the wedge $W$.

*Remark* 1.8. It is obvious that *the set* $\mathcal{D}(W)$ *of operators of the form* (1.15) *with various orders* $m$ *is an algebra.*

Just as in the previous examples, operators $D \in \mathcal{D}(W)$ can be rewritten in a different form.

PROPOSITION 1.9. (1) *The operator* (1.15) *can be rewritten as*

$$D = \sum_{i+j+k+l \leq m} b_{ijkl}(z, \varphi, r) \left( \frac{1}{r} \right)^l \left( \frac{\partial}{\partial z} \right)^k \left( \frac{\partial}{\partial r} \right)^j \left( \frac{1}{r} \frac{\partial}{\partial \varphi} \right)^i, \tag{1.16}$$

*where the coefficients* $b_{ijkl}(z, \varphi, r)$ *have the same properties as* $a_{ijk}(z, \varphi, r)$.

(2) *The algebra* $\mathcal{D}(W)$ *is generated by operators of multiplication by smooth functions on* $W^\wedge$ *and the operators*

$$\frac{\partial}{\partial z}, \quad \frac{\partial}{\partial r}, \quad \frac{1}{r} \frac{\partial}{\partial \varphi}. \tag{1.17}$$

*Remark* 1.10. Here again the vector fields (1.17) are unit coordinate vector fields in the metric (1.13).

*Remark* 1.11. The wedge considered here can also be obtained as an "artificial" edge by passing to polar coordinates with respect to *part* of the variables in a space of higher dimension. Namely, in the space $\mathbb{R}^3$ with coordinates $(x, y, z)$ we pass to polar coordinates $(r, \varphi)$ in the $(x, y)$-plane. Then we obtain the decomposition

$$\mathbb{R}^3 = \mathbb{R} \times K \equiv W.$$

The Beltrami–Laplace operator on $W$ is none other than the Laplace operator

$$\Delta = \frac{\partial^2}{\partial x^2} + \frac{\partial^2}{\partial y^2} + \frac{\partial^2}{\partial z^2}$$

in the original coordinates in $\mathbb{R}^3$.

### 1.1.2. The general framework

Now let us summarize some useful information suggested by these examples.

As we have seen, the construction of a singular manifold can involve the following ingredients:

1. One starts from the stretched manifold, which is equipped with a smooth structure.

◀ In all the examples given so far, the stretched manifold is a manifold with boundary. This is only the case for the simple types of singularities dealt with in this book. For more complicated singular manifolds, the stretched manifold is also more complicated. ▶

2. A metric on the interior is given, which has a smooth continuation to the entire stretched manifold but degenerates (at least in some directions) on the boundary of the stretched manifold.

◀ In the conical and cuspidal cases, the metric is

$$ds^2 = dr^2 + r^{2k+2} d\varphi^2,$$

where $k = 0$ for the conical case. (For brevity, we have neglected smooth nonvanishing factors that do not affect our reasoning.) This metric degenerates on all directions tangent to the boundary $r = 0$ of the stretched manifold, which in this case is the half-cylinder $\overline{\mathbb{R}}_+ \times \mathbb{S}^1$. In the case of the wedge, the metric is

$$ds^2 = dx^2 + dr^2 + r^2 d\varphi^2$$

and degenerates at $r = 0$ in the directions tangent to the circle $\mathbb{S}^1$ with the coordinate $\varphi$. ▶

3. The singular manifold can be obtained from the stretched manifold by passing to the quotient with respect to the following equivalence relation: two points of the stretched manifold are equivalent if the distance between them in the metric is zero. (Clearly, this can happen only with boundary points of the stretched manifold.) With this definition, the metric descends to the quotient space and is already nondegenerate there in the sense that the distance between two arbitrary distinct points is nonzero.

◀ Indeed, in the conical or cuspidal case the distance between two arbitrary points of the boundary of the stretched manifold is zero, and so all boundary points are identified with each other to form the vertex of the cone or cuspidal cone. In the case of the wedge, the boundary of the stretched manifold is naturally fibered over the line, the fibers are circles, and the distance between two boundary points of the stretched manifold is zero if and only if both points lie in the same fiber. In the passage to the quotient, each circle is contracted into the corresponding point of the edge. ▶

4. The supply of differential operators on the singular manifold is obtained as follows: one locally takes coordinate vector fields of length one with respect to the metric. These fields, together with smooth functions on the stretched manifold, generate the algebra.

◀ This was clearly demonstrated in all examples. ▶

*Remark* 1.12. Prior to proceeding further, let us make a very important remark. Although geometric intuition (as well as topological considerations, see Part III) suggests considering singular manifolds as objects having geometric singularities (say, conical points), in analytical matters this serves purely visualization needs. Analytically, we always deal with the stretched object, or stretched manifold (which happens to be a manifold with boundary for the case of isolated singularities as well as edge singularities) equipped with an appropriate algebra of degenerate differential operators. This fruitful "ringed space" approach goes back to Schulze, Sternin and Shatalov (1996). Thus one can say that from the analytical viewpoint we study degenerate differential equations rather than singular spaces. The geometric visualization can be very helpful in understanding the material as well as relationships with applications (where degenerate equations often arise from objects having geometric singularities). That is why we pay much attention to it throughout the book and always discuss the corresponding "singular" objects. Analytically, however, they are often unnecessary and can even hamper the study. For example, one should be aware that differential operators on singular manifolds are undefined at singular points (where they usually have singularities); the natural domain where they act consists of functions on the (open) stretched manifold. Therefore, *actual analytic work is always carried out on the* (*open part of*) *the stretched manifold.*

Now we are in a position to describe the framework in which we shall operate. First, let us give the general definition of singular manifold.

DEFINITION 1.13 (cf. Schulze, Sternin and Shatalov 1998*a*). A *singular manifold* is a pair $(M^\circ, \mathcal{D})$, where $M^\circ$ is a (possibly, noncompact) $C^\infty$ manifold and $\mathcal{D}$ is an algebra of differential operators on $M^\circ$ such that the restriction of $\mathcal{D}$ to an arbitrary neighborhood $U \Subset M^\circ$ with compact closure contained in $M^\circ$ coincides with the restriction to the same neighborhood of the algebra of all differential operators with smooth coefficients on $M^\circ$.

Thus various singular manifolds with the same underlying smooth manifold $M^\circ$ differ only in the behavior of the differential operators $D \in \mathcal{D}$ at infinity in $M^\circ$, i.e., along the complements of an exhausting sequence of compact subsets of $M^\circ$.

Definition 1.13 is very general, and we naturally would like to have some tools that permit one to describe and construct singular manifolds. We shall exploit the following idea, which permits one to describe a broad class of practically important singular manifolds.

1. We assume that the algebra $\mathcal{D} = \mathcal{D}(M^\circ)$ is generated by a function space $F$ and a space $V$ of vector fields on $M^\circ$ such that

$$C_0^\infty(M^\circ) \subset F \subset C^\infty(M^\circ), \quad \mathrm{Vect}_0(M^\circ) \subset V \subset \mathrm{Vect}(M^\circ).$$

(Here $\mathrm{Vect}(M^\circ)$ is the space of smooth vector fields on $M^\circ$ and $\mathrm{Vect}_0(M^\circ)$ is the subspace of compactly supported vector fields.)

2. The space $F$ is described as follows: we take an embedding of $M^\circ$ in a compact manifold $M$ and set

$$F = C^\infty(M)\big|_{M^\circ} \equiv \{f|_{M^\circ} \mid f \in C^\infty(M)\}.$$

In particular, for cone- and edge-degenerate operators the manifold $M$ proves to be a manifold with boundary such that $M^\circ$ is the interior of $M$. For more complicated singular manifolds, $M$ proves to be a manifold with corners. We write $F = C^\infty(M)$ to simplify the notation.

3. To describe the space $V$ of vector fields, one uses the following technique. One takes a (nondegenerate) Riemannian metric $d\rho^2$ on $M^\circ$ that extends smoothly to the entire $M$ (but possibly degenerates on $\partial M = M \setminus M^\circ$). For a space $E \subset \mathrm{Vect}(M^\circ)$ of vector fields on $M^\circ$, by $E'$ we denote the dual space with respect to the $F$-valued pairing determined by the inner product $\langle \cdot, \cdot \rangle$ associated with the metric:

$$E' = \{\xi \in \mathrm{Vect}(M^\circ) \mid \langle \xi, \eta \rangle \in F \quad \text{for each } \eta \in E\}.$$

Now we determine $V$ from the following conditions:

(i) $\mathrm{Vect}(M) \subseteq V \subseteq \mathrm{Vect}(M^\circ)$

(ii) $V' = V$

In cases of interest to us, these conditions uniquely determine the space $V$. Thus we obtain a description of the algebra $\mathcal{D}$.

*Remark* 1.14. For the case in which $M^\circ$ is the interior of $M$, we often denote the singular manifold by $(M, \mathcal{D})$ or $(M, \mathcal{D}(M))$ rather than $(M^\circ, \mathcal{D})$ or $(M^\circ, \mathcal{D}(M^\circ))$, even though technically this is not precise.

4. The "geometric" singular manifold $\mathcal{M}$ associated with the pair $(M, \mathcal{D})$ is obtained from $M$ with the help of the degenerating metric $d\rho^2$.

To this metric, we can as usual associate a distance function on $M$ by setting

$$d(x, y) = \inf_\gamma \rho(\gamma),$$

where $\rho(\gamma)$ is the length of a curve $\gamma : [0, 1] \longrightarrow M$ and the infimum is taken over all continuous piecewise $C^1$ curves $\gamma$ joining $x$ and $y$:

$$\gamma : [0, 1] \longrightarrow M, \quad \gamma(0) = x, \quad \gamma(1) = y.$$

◀ This distance function obviously satisfies the triangle inequality, since if $\gamma_1$ and $\gamma_2$ are two piecewise $C^1$ curves joining $x$ with $y$ and $y$ with $z$, respectively, then their concatenation $\gamma_1 \sqcup \gamma_2$ joins $x$ with $z$, is piecewise $C^1$, and satisfies

$$\rho(\gamma_1 \sqcup \gamma_2) = \rho(\gamma_1) + \rho(\gamma_2).$$

On the other hand, the distance thus defined is not necessarily nondegenerate: the distance between two distinct points $x$ and $y$ is zero if, say, they can be joined by a curve whose tangent vector at each point lies in the null space of the degenerate metric $d\rho^2$. Of course, this can only happen if both $x$ and $y$ belong to the boundary $\partial M$. ▶

We say that two points $x, y \in M$ are *equivalent* and write $x \sim y$ if $d(x, y) = 0$.

DEFINITION 1.15. The *singular space $\mathcal{M}$ associated with the pair* $(M^\circ, \mathcal{D})$ is the quotient space

$$\mathcal{M} = M/\sim .$$

Note that $\mathcal{M}$ is equipped with the distance function $d(x, y)$ inherited from $M$ and hence $\mathcal{M}$ naturally bears the structure of a metrizable topological space. The natural projection

$$M \longrightarrow \mathcal{M}$$

is a homeomorphism on $M^\circ$, which permits us to identify $M^\circ$ with the interior $\mathcal{M}^\circ$ of $\mathcal{M}$ and equip $\mathcal{M}^\circ$ with the $C^\infty$ structure carried over from $M^\circ$. By abuse of language, we often refer to $\mathcal{M}$ itself as the singular manifold.

DEFINITION 1.16. The manifold $M$ will be called the *stretched manifold* of $\mathcal{M}$.

## 1.2. Manifolds with Conical Singularities

In this section, we introduce general manifolds with conical singularities and study the corresponding class of differential operators. Having in mind elliptic theory, we mostly deal with compact manifolds and also pay some attention to infinite cones, which play the role of model spaces in local problems.

### 1.2.1. Definition

Here we give the definition of a manifold with conical singularities. We are interested only in the case of finitely many conical points. There are important *examples* in which a manifold has several conical points, the most important of them being the so-called *spindle*, obtained from a finite cylinder with smooth base by shrinking both faces into conical points. However, when dealing with general theory, we assume that there is only one singular point so as to avoid additional subscripts. The passage to the case of finitely many singular points is essentially a trivial exercise.

Let $M$ be a smooth compact manifold with connected boundary $\Omega = \partial M$. This manifold will play the role of the stretched manifold in our construction. We equip the manifold $M$ as follows with a smooth Riemannian metric degenerating on the boundary. Consider some trivialization

$$U \simeq [0,1) \times \Omega$$

of a collar neighborhood of the boundary in $M$ (see Fig. 1.3). This trivialization will be fixed throughout the following. Thus each point of $U$ is represented by a

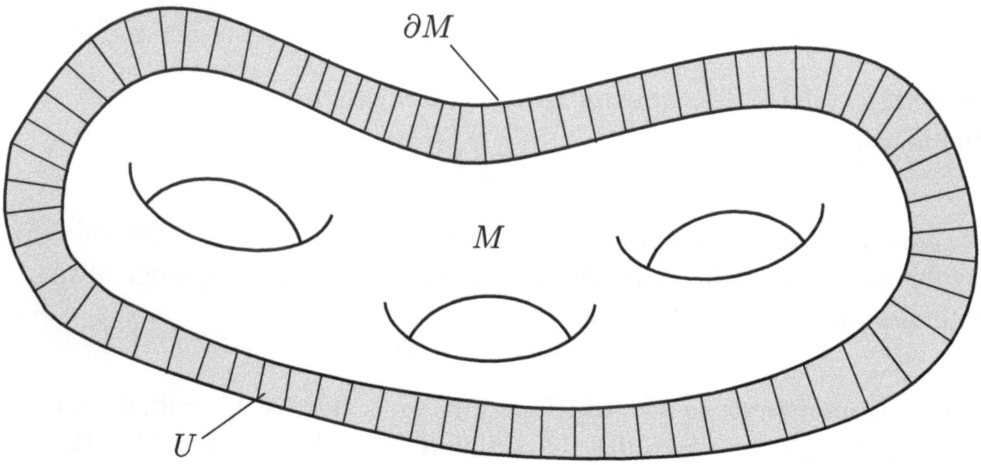


Figure 1.3. Trivialization of a collar neighborhood of the boundary.

pair $(r, \omega)$, where $r \in [0,1)$ and $\omega \in \Omega$. We define a metric in $U$ by setting

$$ds^2 = dr^2 + r^2 d\omega^2, \tag{1.18}$$

where $d\omega^2$ is an arbitrary (nondegenerate) Riemannian metric on $\Omega$, and continue it arbitrarily into $M \setminus U$ as a smooth nondegenerate metric. Thus the metric $ds^2$ degenerates only on the boundary $\partial M$ along all tangential directions.

Now, just as in the first example in Section 1.1.1, we could say that the algebra $\mathcal{D}(M)$ of cone-degenerate differential operators is generated by the operators of multiplication by smooth functions on $M$ and by coordinate vector fields of length of the order of unity with respect to the metric $ds^2$, and this indeed would be true (the computations are essentially the same as in the example of operators on the cone in Section 1.1.1), but we avoid this for the following reason. We would like to have a more direct, invariant description of the set of differential operators in terms of the metric. This can be done following the general framework introduced in the preceding section. Note that the metric $ds^2$ determines a bilinear form (pairing) on the space $\mathrm{Vect}(M)$ of vector fields on the manifold $M$ ranging in $C^\infty(M)$:

$$\begin{aligned}\mathrm{Vect}(M) \times \mathrm{Vect}(M) &\longrightarrow C^\infty(M)\\ (\xi, \eta) &\longmapsto \langle \xi, \eta \rangle,\end{aligned}$$

where $\langle \cdot, \cdot \rangle$ is the inner product corresponding to $ds^2$ (and also degenerating at the boundary). Owing to degeneracy, this pairing can be extended to wider spaces of vector fields. For a space $V \subseteq \mathrm{Vect}(M^\circ)$ of vector fields on $M^\circ$, let $V' \subseteq \mathrm{Vect}(M^\circ)$ be the space of vector fields $\eta$ such that

$$\langle \xi, \eta \rangle \in C^\infty(M)$$

for all $\xi \in V$. In general, $V \neq V'$. For example, if $V = \mathrm{Vect}(M)$, then $V'$ consists of vector fields $\xi \in \mathrm{Vect}(M^\circ)$ that have the form

$$\xi = a\frac{\partial}{\partial r} + \theta$$

in $U$, where $a \in C^\infty(U)$ and

$$r^2\theta \in C^\infty([0, 1), \mathrm{Vect}(\Omega)).$$

However, the following assertion is true.

PROPOSITION 1.17. *There exists a unique space $V \subseteq \mathrm{Vect}(M^\circ)$ of vector fields on $M^\circ$ such that*

i) $\mathrm{Vect}(M) \subseteq V$

ii) $V' = V$

*Moreover, a vector field $\xi \in \mathrm{Vect}(M^\circ)$ belongs to $V$ if and only if in the collar neighborhood $U$ of the boundary it has the form*

$$\xi = a\frac{\partial}{\partial r} + \theta, \tag{1.19}$$

*where $a \in C^\infty(U)$ and*

$$r\theta \in C^\infty([0, 1), \mathrm{Vect}(\Omega)).$$

◀ Let us sketch the proof of this assertion. Since locally $\mathrm{Vect}(M^\circ) = \mathrm{Vect}(M)$ away from the boundary, it is clear that only the behavior near the boundary should be studied, and so we argue locally, i.e., in $U$. If

$$\xi = a\frac{\partial}{\partial r} + \theta_1, \quad \eta = \alpha\frac{\partial}{\partial r} + \theta_2,$$

where the $\theta_j$, $j = 1, 2$, are vector fields on $\Omega$ depending on the parameter $r$, then

$$\langle \xi, \eta \rangle = a\alpha + r^2 d\omega^2(\theta_1, \theta_2). \tag{1.20}$$

Suppose that $V = V' \supseteq \mathrm{Vect}(M)$ and

$$\xi = a\frac{\partial}{\partial r} + \theta \in V.$$

The function (1.20) should be smooth for all $\eta \in V$ and hence for all $\eta \in \mathrm{Vect}(M)$. In view of the form of the inner product (1.20), we conclude that $a \in C^\infty(U)$ and

$$\Theta \equiv r^2\theta \in C^\infty([0, 1), \mathrm{Vect}(\Omega)). \tag{1.21}$$

Next,

$$a\frac{\partial}{\partial r} \in \mathrm{Vect}(M) \subseteq V,$$

and subtracting this from $\xi$, we find that $\theta \in V$. Consequently,

$$\langle \theta, \theta \rangle = r^2 d\omega^2(\theta, \theta) \in C^\infty(M).$$

Here we express $\theta$ from Eq. (1.21) and use the Taylor series expansion

$$\Theta = \Theta_0 + r\Theta_1 + O(r^2)$$

of $\Theta$ in powers of $r$, thus obtaining

$$\langle \theta, \theta \rangle = r^{-2} d\omega^2(\Theta_0, \Theta_0) + 2r^{-1} d\omega^2(\Theta_0, \Theta_1) + \text{a smooth function}.$$

This function should be smooth, and this is the case if and only if $\Theta_0 \equiv 0$. Hence

$$r\theta \in C^\infty([0, 1), \mathrm{Vect}(\Omega)),$$

and we have proved that the elements of $V$ have the form (1.19). ▶

DEFINITION 1.18. The algebra $\mathcal{D} \equiv \mathcal{D}(M)$ of cone-degenerate operators is generated by $C^\infty(M)$ and $V$. The pair $(M, \mathcal{D}(M))$ is called a *manifold with conical singularities*.

By this definition, each element $D \in \mathcal{D}(M)$ can be (nonuniquely) represented as a finite sum of products of the form

$$A_1 A_2 \cdots A_N, \tag{1.22}$$

where each $A_j$ is either a $C^\infty$ function on $M$ or an element of $V$:

$$D = \sum_{j=1}^{l} A_{j1} A_{j2} \cdots A_{jN_j}. \tag{1.23}$$

Define the *weight* of the product (1.22) as the number of factors that belong to $V$ and the weight of the right-hand side of (1.23) as the maximum weight of products occurring in the sum.

DEFINITION 1.19. The *order* $\operatorname{ord} D$ of the operator $D$ is the minimum possible weight of its representations in the form (1.23). The set of operators of order $\leq k$ will be denoted by $\mathcal{D}_k \equiv \mathcal{D}_k(M)$.

PROPOSITION 1.20. *The subspaces $\mathcal{D}_j$ form an increasing filtration*

$$\mathcal{D}_0 \subset \mathcal{D}_1 \subset \mathcal{D}_2 \subset \cdots \subset \mathcal{D} = \bigcup_{j=0}^{\infty} \mathcal{D}_j$$

*in the algebra $\mathcal{D}$, and*

$$\mathcal{D}_j \mathcal{D}_{j'} \subset \mathcal{D}_{j+j'}.$$

◀ The proof is an easy exercise and follows directly from the definitions. ▶

### 1.2.2. Geometric description of conical singularities

Now we describe the singular space associated with a manifold with conical singularities. We denote by $d(x, y)$ the distance between points $x, y \in M$ in the metric (1.18):

$$d(x, y) = \inf_{\gamma} l(\gamma),$$

where $l(\gamma)$ is the length of a curve $\gamma$ and the infimum is taken over all continuous piecewise smooth curves

$$\gamma : [0, 1] \longrightarrow M, \quad \gamma(0) = x, \quad \gamma(1) = y.$$

Since the metric is degenerate, it is no wonder that the distance between distinct points is not necessarily positive. Specifically, the distance $d(x, y)$, where $x \neq y$, is zero if and only if both points lie on $\partial M$. Thus we have defined the following equivalence relation on $M$:

$$x \sim y \Longleftrightarrow \begin{cases} x = y & \text{or} \\ x, y \in \partial M. \end{cases} \tag{1.24}$$

The quotient space $\mathcal{M} = M/\sim$ with respect to this equivalence relation is the space obtained from $M$ by shrinking all points of the boundary into a single point $\alpha$, the "vertex of the cone" (see Fig. 1.4). The projection $M \longrightarrow \mathcal{M}$ is one-to-one on the interior $M^{\circ} = M \setminus \partial M$ of the manifold $M$ and hence the interior $\mathcal{M}^{\circ} = \mathcal{M} \setminus \{\alpha\}$ of $\mathcal{M}$ bears a natural $C^{\infty}$ structure. A neighborhood of $\alpha$ in $\mathcal{M}$ is homeomorphic to the *cone with base* $\Omega$, which is defined as the quotient

$$K \equiv K_{\Omega} = \{[0, 1) \times \Omega\} \,/\, \{\{0\} \times \Omega\},$$

where we identify all points of the form $(0, \omega)$, $\omega \in \Omega$ (see Fig. 1.5). The point $\beta = [\{0\} \times \Omega] \in K$ is called the cone *vertex*.

The space $\mathcal{M}$ thus obtained serves as the *geometric visualization* of the would-be manifold with conical singularities.

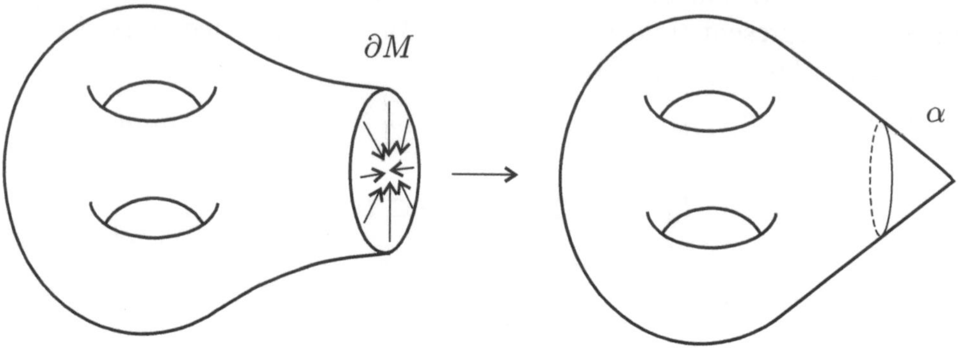


Figure 1.4. The passage to the quotient.

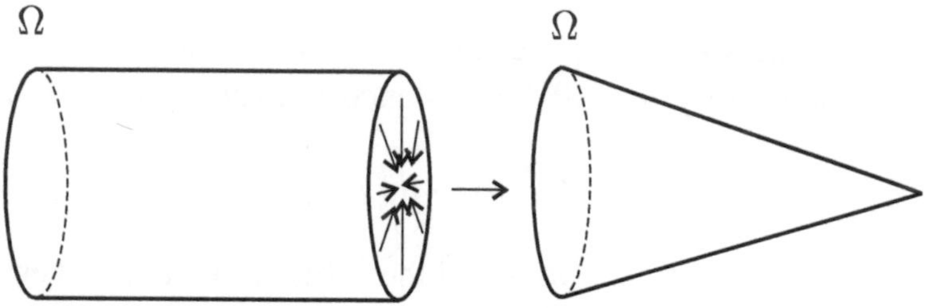


Figure 1.5. The cone with base $\Omega$.

◀ Sometimes we shall use the infinite cone $\{[0,\infty)\times\Omega\} \;/\; \{\{0\}\times\Omega\}$, which will also be denoted by $K_\Omega$. This will not lead to a misunderstanding, since the meaning will be always clear from the context. ▶

*Remark* 1.21. One can readily define a manifold with several conical points. To this end, one starts from a compact manifold with boundary consisting of several connected components and shrinks each component of the boundary into its own conical point. More complicated cases in which the bases of the cones are not necessarily connected can also be considered with the corresponding minor modifications in our definitions. In particular, the formula $d(x, y) = 0$ does not completely describe the equivalence relation $\sim$ in the case of a nonconnected base of the cone.

Formally, the equivalence relation in the case of several conical points can be described as follows. The boundary $\partial M$ is fibered over a discrete finite set $\{1, \dots, l\}$ (where $l$ is the number of conical points). The preimage $\Omega_j$ of the point $j$ is the base of the $j$th cone, $j = 1, \dots, l$. This construction anticipates the case of manifolds with edges, where the base of the bundle is no longer a finite set but a smooth manifold of positive dimension. However, apart from an abundance of indices the case of several conical points brings little new information in the analysis, and we avoid considering them. The only exception is the following example.

EXAMPLE 1.22. An important example of a manifold with several conical points is a suspension over $\Omega$ (spindle with base $\Omega$). It is obtained from the finite cylinder $C = [0,1] \times \Omega$ by shrinking both faces $\Omega \times \{0\}$ and $\Omega \times \{1\}$ into conical points $\alpha$ and $\beta$ (see Fig. 1.6).

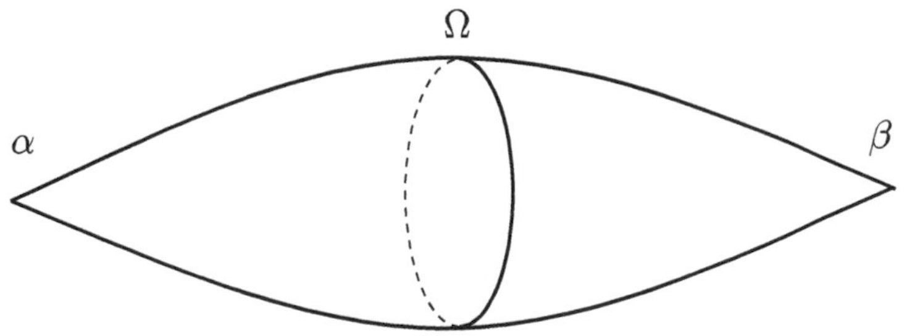


Figure 1.6. The spindle with base $\Omega$.

*Remark* 1.23. Note that the manifold $M$ with boundary can indeed be treated as a stretched manifold of $\mathcal{M}$: to obtain this stretched manifold, one cuts out an infinitesimal neighborhood of the vertex and expands the resulting smooth boundary to a "normal" size.

### 1.2.3. Local expression of cone-degenerate operators and local Hamiltonians

Differential operators on smooth manifolds can always be represented in local coordinates $y = (y_1, \dots, y_n)$ in the form

$$D = H\left(\overset{2}{y}, -i\overset{1}{\frac{\partial}{\partial y}}\right), \tag{1.25}$$

where the function $H(y,q)$ is a polynomial in the variables $q = (q_1, \dots, q_n)$ with coefficients depending on $y$,

$$\frac{\partial}{\partial y} = \left(\frac{\partial}{\partial y_1}, \dots, \frac{\partial}{\partial y_n}\right),$$

and the numbers over operators indicate the order in which they act[2]; namely, if the operator (1.25) is applied to a function, the function is first differentiated and then multiplied by the coefficients. The function $H(y,q)$ is known as the (local) *Hamiltonian* of the operator $D$. (This terminology is of course of physical origin.)

Is there any notion of Hamiltonians for cone-degenerate operators? It turns out that the answer is "yes," and we shall now describe these Hamiltonians. There

[2]Here and below, we use Maslov's ordering of noncommuting operators (e.g., Maslov 1973, Nazaikinskii, Sternin and Shatalov 1995).

are two kinds of points on a manifold with conical singularities, smooth (interior) points and singular (conical) points. It is no surprise that the notions of Hamiltonians for these two cases are different.

***Hamiltonians at interior points.*** In this case, the situation is entirely the same as for differential operators on smooth manifolds. The operator is uniquely represented in local coordinates in the form (1.25), and the function $H(y, q)$ occurring in this formula is called the *Hamiltonian* of the operator $D$.

***Hamiltonians at conical points.*** Let us study the form of a cone-degenerate operator $D \in \mathcal{D}(M)$ in the collar neighborhood of the boundary $\partial M = \Omega$. First, we shall write out expressions for elements of the algebra $\mathcal{D}(M)$ in local coordinates near some point $x_0$ on the boundary of $M$. We use local coordinates $(r, \omega)$ associated with the trivialization $U \simeq [0, 1) \times \Omega$ in a neighborhood $O$ of $x_0$. By Proposition 1.17, elements of $V$ have the form (1.19) in $U$, and we readily see that a local basis in $V$ (over $C^\infty(M)$) is formed by the vector fields

$$\frac{\partial}{\partial r}, \quad \frac{1}{r}\frac{\partial}{\partial \omega} \equiv \left(\frac{1}{r}\frac{\partial}{\partial \omega_1}, \ldots, \frac{1}{r}\frac{\partial}{\partial \omega_m}\right), \qquad m = \dim \Omega. \tag{1.26}$$

Each element $D \in \mathcal{D}(M)$ can be represented in $O$ in the form (1.23); expanding the vector fields occurring in (1.23) in the basis (1.26), we can assume without loss of generality that from the very beginning each $A_{jl}$ is either a $C^\infty$ function on $M$ or one of the operators (1.26).

The representation (1.23) is not very convenient in that it can contain terms of arbitrary length with factors repeated in various places. In the spirit of noncommutative analysis, let us try to represent cone-degenerate operators as (polynomial) functions of the *ordered* operators (1.26) with smooth coefficients depending on $x \in M$:

$$D = f\left(\overset{3}{r}, \overset{3}{\omega}, \overset{2}{\frac{1}{r}\frac{\partial}{\partial \omega}}, \overset{1}{\frac{\partial}{\partial r}}\right), \tag{1.27}$$

where $f$ is a smooth function polynomial in the last two arguments. To obtain such a representation, one should rearrange factors in the terms in (1.23) so that all differentiations will stand on the right and all coefficients on the left. Unfortunately, a moment's thought shows that the representation (1.27) is impossible in the general case.

◀ Indeed, consider, for example, the operator

$$D = \frac{1}{r}\frac{\partial}{\partial \omega}\phi(\omega) \in \mathcal{D}_1.$$

An attempt to arrange the operators in the desired order shows that

$$D = \phi(\omega)\frac{1}{r}\frac{\partial}{\partial \omega} + \frac{\phi'(\omega)}{r}.$$

Thus we see that we have gained the extra factor $r^{-1}$. The same factor occurs if we wish to commute the two differentiations:

$$\left[\frac{1}{r}\frac{\partial}{\partial\omega}, \frac{\partial}{\partial r}\right] = \frac{1}{r}\frac{1}{r}\frac{\partial}{\partial\omega}.$$

Thus we have to abandon the representation (1.27). ▶

The remedy is however simple. We supplement the set of operators (1.26) with the operator of multiplication by $r^{-1}$.

LEMMA 1.24. *Every operator $D \in \mathcal{D}_k$ can be represented in $O$ in the form*

$$D = f\left(\overset{3}{r}, \overset{3}{\omega}, \overset{2}{\frac{1}{r}}, \overset{2}{\frac{1}{r}\frac{\partial}{\partial\omega}}, \overset{1}{\frac{\partial}{\partial r}}\right), \tag{1.28}$$

*where $f(r, \omega, \theta, q, p)$ is a polynomial of degree at most $k$ in the variables $(\theta, q, p)$.*

It is probably of interest to express the product of operators $D \in \mathcal{D}$ in terms of symbols $f$ in the representation (1.28). A general recipe for this can be found in Maslov (1973) and Nazaikinskii, Sternin and Shatalov (1995). Here we give only the definitive answer.

PROPOSITION 1.25. *The left regular representation operators for the operator tuple*

$$\overset{3}{r}, \quad \overset{3}{\omega}, \quad \overset{2}{\frac{1}{r}}, \quad \overset{2}{\frac{1}{r}\frac{\partial}{\partial\omega}}, \quad \overset{1}{\frac{\partial}{\partial r}}$$

*have the form*

$$\overset{3}{r}, \quad \overset{3}{\omega}, \quad \overset{2}{\theta}, \quad q + \overset{2}{\theta}\frac{\partial}{\partial\omega}, \quad p + \frac{\partial}{\partial r} - \overset{1}{\theta}\frac{\partial}{\partial\theta} - \theta q\frac{\partial}{\partial q}.$$

Recall that this means that if $D_1, D_2 \in \mathcal{D}$ are two operators representable in the form (1.28) with $f = f_1$ and $f = f_2$, respectively, then their product $D = D_1 D_2$ is also representable in the form (1.28) with

$$f(r,\omega,\theta,q,p) = f_1\left(\overset{3}{r}, \overset{3}{\omega}, \overset{2}{\theta}, \overset{2}{\overline{q + \theta\frac{\partial}{\partial\omega}}}, \overset{1}{\overline{p + \frac{\partial}{\partial r} - \theta\frac{\partial}{\partial\theta} - \theta q\frac{\partial}{\partial q}}}\right) f_2(r,\omega,\theta,q,p).$$

*Remark* 1.26. Though very interesting, the composition formula given by this proposition is not as useful as one can expect, because it is valid only for symbols polynomial in the momentum variables and cannot be extended to any reasonably wide class of nonpolynomial symbols.

The representation (1.28) could be used for defining the Hamiltonian of the operator $D$ (which would be the function $f(r, \omega, \theta, q, p)$ in this case), but it has the disadvantage of being "too local": this representation is *not* defined in a neighborhood of the conical point but rather in a neighborhood of a boundary point of the stretched manifold $M$. Now we shall introduce another representation, free of this disadvantage.

PROPOSITION 1.27. *Every operator $D \in \mathcal{D}_k(M)$ can be uniquely represented in $U \equiv [0,1) \times \Omega$ in the form*

$$D = r^{-k} \sum_{j \le k} a_j(r) \left( ir\frac{\partial}{\partial r} \right)^j \tag{1.29}$$

*where the coefficients $a_j(r)$ are smooth functions (up to $r = 0$) ranging in the space of differential operators of order $\le k - j$ with smooth coefficients on $\Omega$.*

◀ One can readily carry out the proof in local coordinates by showing that every operator $D \in \mathcal{D}_k(M)$ can be uniquely represented in the local coordinates $(r, \omega)$ in the form

$$D = r^{-k} \sum_{j+\sigma \le k} \widetilde{a}_{j\sigma}(r,\omega) \left( ir\frac{\partial}{\partial r} \right)^j \left( -i\frac{\partial}{\partial \omega} \right)^\sigma, \tag{1.30}$$

where the coefficients $\widetilde{a}_{j\sigma}(r,\omega)$ are smooth functions (up to $r = 0$). ▶

We see that the algebra $\mathcal{D}(M)$ consists of operators with Fuchs degeneration $(r\partial/\partial r)$ with respect to the variable $r$ as $r \to 0$ (neglecting the factor $r^{-k}$). That is why it is called the algebra of *cone-degenerate* operators.

DEFINITION 1.28. The operator-valued function

$$H(r,p) = \sum_{j \le k} a_j(r) p^j$$

is called the *Hamiltonian* of the operator $D$ near the conical point.

*Remark* 1.29. The reason for eliminating the factor $r^{-k}$ is twofold. First, the Hamiltonian is nonsingular. Second, the operator $D$ is represented as a function of the operator $ir\partial/\partial r$ rather than $i\partial/\partial r$, which proves advantageous when we further define pseudodifferential operators. (Nonpolynomial functions of the operator $i\partial/\partial r$ on the half-line $\overline{\mathbb{R}}_+$ are an extremely difficult object which should be avoided.)

### 1.2.4. The cylindrical representation

There is another important representation of cone-degenerate operators, known as the *cylindrical representation*. Let us omit the factor $r^{-k}$ in (1.29) and make the change of variables

$$r = e^{-t}, \tag{1.31}$$

which maps a deleted neighborhood of the singular point onto the infinite half-cylinder $\Omega \times (0, \infty)$. In other words, $M^\circ$ is equipped with the structure of a manifold with cylindrical end $U^\circ \simeq \Omega \times (0, \infty)$. (See Fig. 1.7, where a manifold with several singular points is shown.)

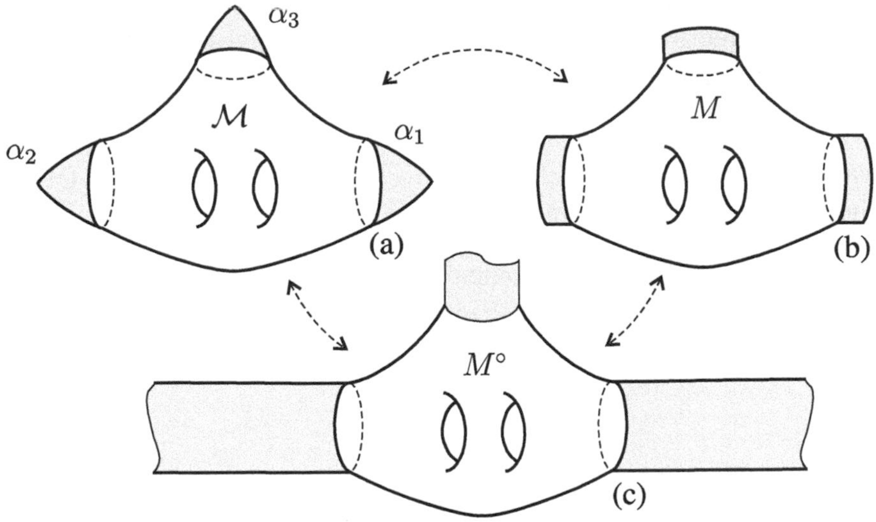


Figure 1.7. Three representations: (a) a manifold with singularities; (b) the stretched manifold (a manifold with boundary); (c) a manifold with cylindrical ends.

DEFINITION 1.30. The variables $(\omega, t)$, where $t$ is related to $r$ by the change of variables (1.31) and the variables $\omega$ are coordinates on the base $\Omega$ of the cone, are called the *cylindrical variables*. The representation of a manifold with isolated singularities as a manifold with cylindrical ends, as well as the representation of the cone as an infinite cylinder, via this change of variables is called the *cylindrical representation*.

In the cylindrical representation, cone-degenerate differential operators acquire the form

$$D = H\left(\overset{2}{e^{-t}}, -i\overset{1}{\frac{\partial}{\partial t}}\right) = f\left(\overset{2}{\omega}, \overset{2}{e^{-t}}, -i\overset{1}{\frac{\partial}{\partial \omega}}, -i\overset{1}{\frac{\partial}{\partial t}}\right)$$

with coefficients stabilizing as $t \to \infty$ to some smooth limit functions at an exponential rate. In this case, we speak of (exponentially) *stabilizing operators*.

### 1.2.5. Vector bundles

***General vector bundles.*** Bundles are an extremely important notion in the theory of elliptic differential equations (in particular, on manifolds with singularities). First, many natural differential operators (e.g., geometric operators) act in sections of bundles. Second, in the case of smooth manifolds the principal symbol of an operator is itself a function[3] on the total space of the cotangent bundle of the manifold. The same holds for manifolds with conical singularities except for the fact

[3]More precisely, for operators acting in sections of vector bundles, a section of a bundle.

that the symbolic structure on such manifolds is more complicated and the assertion pertains to the interior symbol (rather than the principal symbol on the whole).

We shall consider only finite-dimensional vector bundles here.

DEFINITION 1.31. A *vector bundle* $E$ over a manifold $\mathcal{M}$ with conical singularities is always understood as a vector bundle $E$ (denoted by the same letter) over the stretched manifold $M$ of the manifold $\mathcal{M}$.

Let $\mathcal{M}$ be a manifold with a conical singular point $\alpha$, and let $\Omega$ be the base of the corresponding cone. Using a trivialization $U \simeq \Omega \times [0,1)$ of a collar neighborhood of the boundary in the stretched manifold $M$, we obtain a well-defined projection

$$\pi : U \cong \Omega \times [0,1) \xrightarrow{\pi} \Omega$$

onto the first factor. Since the interval $[0,1)$ is contractible, it follows that for an arbitrary bundle $E$ there exists an isomorphism

$$E|_U \cong \pi^*(E|_{\Omega\times\{0\}}), \tag{1.32}$$

which we shall use in the description of sections of $E$ in a neighborhood of $\alpha$.

***The tangent and cotangent bundles.*** The cotangent bundle is important in the theory of differential and pseudodifferential operators. Let $\mathcal{M}$ be a compact manifold with conical singularities. The cotangent bundle $T^*\mathcal{M}$, as well as any other bundle over $\mathcal{M}$, is a bundle over the stretched manifold $M$ of $\mathcal{M}$. However, it is not just equal to $T^*M$. The definition of the cotangent bundle will be constructed in such a way that an operator $D$ of the form (1.28) will have the interior symbol $\sigma(D)$ on $T^*\mathcal{M}$ of the following form in canonical local coordinates:

$$\sigma(D) = f_k(r, \omega, 0, q, p), \tag{1.33}$$

where $f_k$ is the leading homogeneous part of $f$ with respect to the variables $(\theta, q, p)$. Should we set $T^*\mathcal{M} = T^*M$, the symbol in canonical local coordinates would have the form

$$\sigma(D) = f_k(r, \omega, 0, r^{-1}q, p)$$

and be necessarily singular on the boundary $\partial T^*M = \{r = 0\}$.

Thus the forthcoming definition is adapted to the study of cone-degenerate operators of the form (1.28). If the interior symbol (1.33) is nonzero for $|p| + |q| \neq 0$, then the operator (1.28) is said to be internally elliptic in our theory.

How can one ensure that the interior symbol of the operator (1.28) has the form (1.33)? Here the following considerations can be helpful.

In the usual (smooth) case, differential operators are arbitrary polynomials in vector fields with smooth coefficients. Vector fields are sections of the tangent bundle, and accordingly, the principal symbols of differential operators are functions on the total space of the dual (cotangent) bundle.

Let us transfer this construction to manifolds with conical singularities.

The space $V$ of cone-degenerate vector fields on $\mathcal{M}$ is a locally free module over the ring $C^\infty(M)$.

◀ The term "locally free module" means in this context that in a sufficiently small neighborhood of an arbitrary point there exist vector fields $\xi_1, \ldots \xi_n \in V$ such that an arbitrary field $\xi \in V$ can be represented in this neighborhood in the form

$$\xi = \sum a_j \xi_j, \tag{1.34}$$

where all $a_j$ are infinitely smooth, and conversely, every field of the form (1.34) belongs to $V$. ▶

By the Swan theorem (Atiyah 1989), $V$ is the module of sections of some vector bundle over $M$. This vector bundle is denoted by $T\mathcal{M}$ and called the *compressed tangent bundle* of $\mathcal{M}$. The dual bundle is denoted by $T^*\mathcal{M}$ and called the *stretched cotangent bundle* of $\mathcal{M}$. In what follows, we usually omit the words "compressed" and "stretched."

One can also give the following simple alternative description of $T^*\mathcal{M}$. The metric $ds^2$ can be interpreted as a (fiberwise) mapping of $TM$ into $T^*M$: to each vector $\xi \in T_x M$, this mapping assigns the form

$$\widetilde{\xi}(\eta) = \langle \xi, \eta \rangle.$$

(Recall that $\langle \cdot, \cdot \rangle$ is the inner product in the fibers of $TM$ corresponding to the metric $ds^2$.) This mapping takes $V$ to some isomorphic module $\widetilde{V} \subset \Lambda^1(M)$ of differential forms on $M$.

Since the mapping commutes with the action of $C^\infty(M)$, we see that $\widetilde{V}$ is also a locally free module. By applying the Swan theorem to it, we see that $\widetilde{V}$ is a module of sections of a vector bundle over $M$. Note that this vector bundle coincides with the stretched cotangent bundle $T^*\mathcal{M}$.

Canonical coordinates on $T^*\mathcal{M}$ in a neighborhood of a boundary point can be introduced as follows. Let $(\omega, r)$ be local coordinates on the base $M$ in a neighborhood of a boundary point, where $r \in [0, 1)$ and $\omega$ is a tuple of local coordinates on $\Omega$. Then the canonical local coordinates on $T^*\mathcal{M}$ are $(\omega, r, q, p)$, where the "dual" coordinates $(q, p)$ correspond to the coordinate representation

$$\alpha = p(\omega, r) dr + q(\omega, r) r \, d\omega$$

of elements of $\widetilde{V}$.

### 1.2.6. Symbols

Now we proceed to the adequate definition of the notion of symbol for cone-degenerate differential operators.

***Principal symbol.*** Recall that for the case of a smooth manifold the space of $m$th-order principal symbols, i.e., homogeneous polynomials of degree $m$ with respect to the momentum variables, is naturally identified with the quotient space of operators of order $m$ by operators of order $m-1$: the symbol is uniquely determined by the principal part of the operator and in turn uniquely determines the operator modulo lower-order terms. By analogy with this, we give the definition of the principal symbol in our case.

As we already mentioned, the sets $\mathcal{D}_k$ of operators of order $\leq k$ form an increasing filtration in the algebra $\mathcal{D}$, and from the abstract algebraic viewpoint it is natural to treat the (principal) symbol as an element of the associated graded algebra

$$\operatorname{gr}\mathcal{D} = \bigoplus_{j=0}^{\infty} \mathcal{D}_j/\mathcal{D}_{j-1} \equiv \bigoplus_{j=0}^{\infty} \operatorname{gr}_j \mathcal{D},$$

where by convention $\mathcal{D}_{-1} = \{0\}$. If $D \in \mathcal{D}_k$ is an operator of order $\leq k$, then its ($k$th-order) principal symbol is $\Sigma(D) \equiv \Sigma_k(D)$, where the $k$th principal symbol map $\Sigma_k$ is none other than the natural projection

$$\Sigma_k : \mathcal{D}_k \longrightarrow \operatorname{gr}_k \mathcal{D}.$$

How should the symbol given by this definition be computed in more constructive terms? We should of course expect that the answer is given by the usual, well-known principal symbol $\sigma(D)(x,\xi)$ if the manifold happens to have no singularities at all. However, there is a drawback here: in the smooth case, the symbol is a function on the new manifold $T^*M$, the cotangent bundle of $M$ (or, if one prefers, on the cosphere bundle $S^*M$), and the line of reasoning that leads from $M$ to $T^*M$ is by no means obvious or straightforward. Actually, one speaks of microlocalization (the transition from the configuration space to the phase space), but so far we *do not even know what the right notion of phase space is* for manifolds with conical singularities. (Looking ahead, let us mention that the stretched cotangent bundle, being a good candidate, reveals only half the truth.) To carry out the microlocalization process properly and avoid possible difficulties, we split it into two stages according to the following scheme:

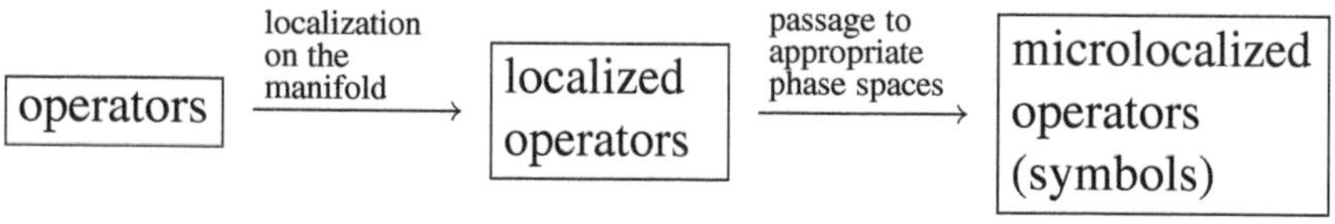


Let us carry out this scheme.

***Localization.*** By $C^\infty(\mathcal{M}) \subset C^\infty(M)$ we denote the subspace of functions constant on the boundary. This subspace is naturally interpreted as a subalgebra in $\mathcal{D}$ (recall that $C^\infty(M) \in \mathcal{D}_0$), and we have the commutation relation

$$[C^\infty(\mathcal{M}), \mathcal{D}_k] \subset \mathcal{D}_{k-1},$$

which shows that $C^\infty(\mathcal{M})$ naturally acts on the associated graded algebra $\operatorname{gr}\mathcal{D}$, which is a two-sided $C^\infty(\mathcal{M})$-module (the left and right actions coincide). In particular, each $\operatorname{gr}_k \mathcal{D}$ is a $C^\infty(\mathcal{M})$-module.

For each $x \in \mathcal{M}$, consider the maximal ideal $I_x \subset C^\infty(\mathcal{M})$ consisting of functions $f \in C^\infty(\mathcal{M})$ that vanish at $x$ and the corresponding submodule

$$I_x \operatorname{gr}_k \mathcal{D} \subset \operatorname{gr}_k \mathcal{D}.$$

We set

$$\Sigma_{kx} = \operatorname{gr}_k \mathcal{D}/I_x \operatorname{gr}_k \mathcal{D}.$$

This is the localization of the space of symbols of order $k$ to the point $x$.

LEMMA 1.32. *The mapping*

$$\pi : \operatorname{gr}_k \mathcal{D} \longrightarrow \prod_{x\in\mathcal{M}} \Sigma_{kx}, \tag{1.35}$$

$$\Sigma \longmapsto \{\pi_x(\Sigma)\}, \tag{1.36}$$

*where* $\pi_x : \operatorname{gr}_k \mathcal{D} \longrightarrow \Sigma_{kx}$ *is the natural projection onto the quotient space, is a monomorphism.*

This (obvious) fact can be verified by a straightforward computation. For an operator $D \in \mathcal{D}_k$, we denote the element $\pi_x(\Sigma(D))$ by[4] $\sigma_x(D)$ and call it the *local representative* of $D$ at a point $x \in \mathcal{M}$. Lemma 1.32 says that the set of all local representatives (where $x$ ranges over $\mathcal{M}$) completely determines the principal symbol of $D$ treated as the element of $\operatorname{gr}_k \mathcal{D}$.

Our immediate task is to obtain a convenient description of these local representatives. (By this we mean interpreting them as operators.) It turns out that this can be done by an application of appropriate *scaling procedure* to our operators. Let $x \in \mathcal{M}$ be an arbitrary point. Then the metric determines a local one-parameter group of scaling transformations $g_\lambda$, $\lambda \in \mathbb{R}_+$, in a sufficiently small neighborhood $U_x$ of $x$. These transformations are just dilations along the geodesics issuing from $x$. In a standard way, we embed $U_x$ in a space $K_x$ where the scaling transformations are globally defined.

◀ Indeed, if $x \in \mathcal{M}^\circ$ (a "smooth" point), then the exponential geodesic map gives an isomorphism of some neighborhood of zero in the tangent space $T_xM$ onto $U_x$, and then we can take $K_x = T_xM$ by identifying $U_x$ with that neighborhood (see Fig. 1.8). The scaling transformations $g_\lambda$ in this case are just multiplications by $\lambda$.

If $x = \alpha$ is the conical point, then we have to interpret geodesics as the shortest paths in $M$ joining interior points with the boundary $\partial M$. (The length of a path is taken in the metric (1.18).) For each interior point of the collar neighborhood

[4]Thus omitting the index $k$, which will be always clear from the context.

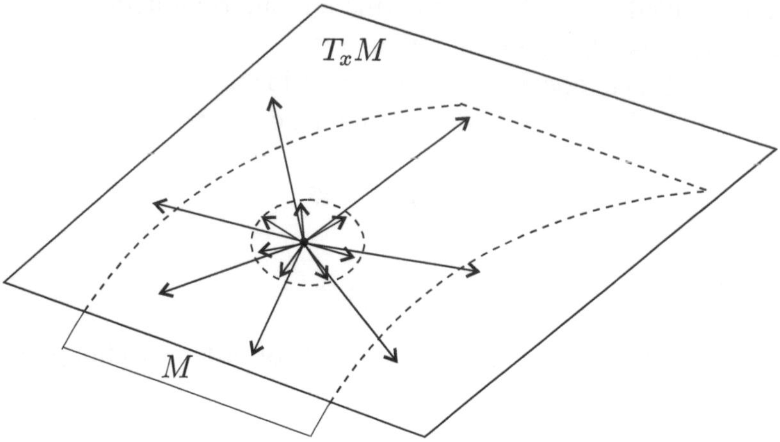


Figure 1.8. Geodesics in a neighborhood of an interior point.

$U_\alpha = U$, there obviously exists a unique shortest path; in the coordinates $(r, \omega)$ it is given by the formula $\omega = \text{const}$. We can take $K_x = K_\Omega$, the infinite model cone with base $\Omega$, and then the scaling transformations are given by the formula

$$g_\lambda(r, \omega) = (\lambda r, \omega), \qquad (r, \omega) \in K_\Omega^\circ, \quad \lambda \in \mathbb{R}_+.$$

(The vertex of the cone is a fixed point for all $g_\lambda$.) This situation is depicted in Fig. 1.9. ▶

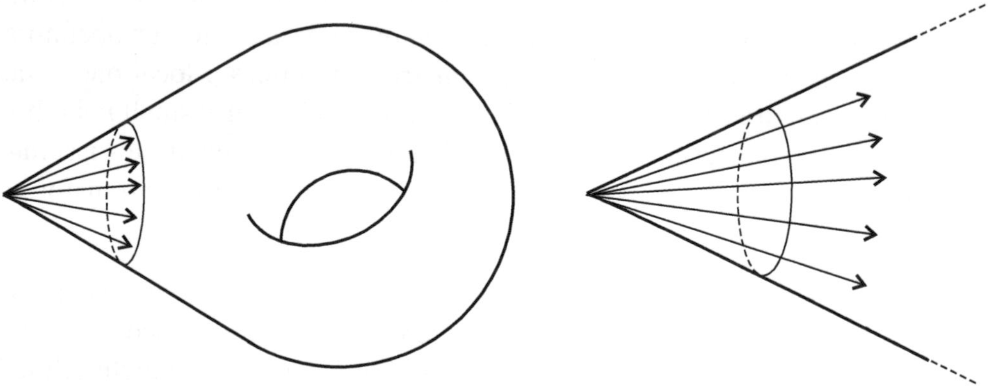

Figure 1.9. Geodesics in a neighborhood of a conical point.

Now, multiplying the operator $D$ by an appropriate cutoff function, we can assume that it is supported in $U_x$ and hence interpret it as an operator on $K_x$. We set

$$D_x u \overset{\text{def}}{=} \lim_{\lambda\to\infty} \lambda^{-m} \left(g_\lambda^*\right)^{-1} D g_\lambda^* u, \tag{1.37}$$

where $u \in C_0^\infty(K_x)$ (if $x$ is the conical point, then this means also that $\operatorname{supp} u$ does not contain $x$) and the limit is understood in the sense of pointwise convergence.

THEOREM 1.33. *The limit* (1.37) *exists for any* $D \in \mathcal{D}_m$ *and* $x \in U$. *The mapping* $D \longmapsto D_x$ *factors through* $\Sigma_{mx}$, *and the correspondence* $\sigma_x(D) \longmapsto D_x$ *is one-to-one and multiplicative.*

*Sketch of proof.* Of course, it is the expression for the limit which is most interesting. We shall consider two cases.

1) Suppose that $x$ is an interior point. Then the operator has the form

$$D = P\left(\overset{2}{y}, -i\overset{1}{\frac{\partial}{\partial y}}\right),$$

where $P(y,p)$ is a polynomial of degree $\le m$ in $p$ with coefficients smooth functions of $y$. To simplify the notation, we assume that the point $x$ has zero coordinates $y = 0$. Then

$$\left(g_\lambda^*\right)^{-1} D g_\lambda^* = P\left(\lambda^{-1}\overset{2}{y}, -i\lambda\overset{1}{\frac{\partial}{\partial y}}\right) = \lambda^m P_m\left(0, -i\overset{1}{\frac{\partial}{\partial y}}\right) + O(\lambda^{m-1}),$$

where $P_m$ is the component of homogeneity degree $m$ in $P$.

Dividing by $\lambda^m$ and passing to the limit as $\lambda \to \infty$, we obtain

$$D_x = P_m\left(0, -i\frac{\partial}{\partial y}\right). \tag{1.38}$$

2) Now suppose that $x$ is the conical point. We use the expression

$$D = P\left(\overset{3}{r}, \overset{3}{\omega}, \overset{3}{\frac{1}{r}}, -i\overset{2}{\frac{\partial}{\partial r}}, -i\frac{1}{r}\overset{1}{\frac{\partial}{\partial \omega}}\right)$$

for the operator $D$ in the coordinates $(r,\omega)$, where $P(r,\omega,\theta,p,q)$ is a polynomial of order $\le m$ in $(\theta,p,q)$ with smooth coefficients. Then

$$\left(g_\lambda^*\right)^{-1} D g_\lambda^* = \lambda^m P_m\left(\lambda^{-1}\overset{3}{r}, \overset{3}{\omega}, \overset{3}{\frac{1}{r}}, -i\overset{2}{\frac{\partial}{\partial r}}, -i\frac{1}{r}\overset{1}{\frac{\partial}{\partial \omega}}\right) + O(\lambda^{m-1}),$$

where $P_m$ is the leading $m$th-order homogeneous part of $P$, and after dividing by $\lambda^m$ and passing to the limit as $\lambda \to \infty$, we obtain

$$D_x = P_m\left(0, \overset{3}{\omega}, \overset{3}{\frac{1}{r}}, -i\overset{2}{\frac{\partial}{\partial r}}, -i\frac{1}{r}\overset{1}{\frac{\partial}{\partial \omega}}\right). \tag{1.39}$$

The remaining assertions of the theorem are trivial. □

Now we have performed localization and, for each $D \in \mathcal{D}_m$, obtained a family of operators $D_x$ on the spaces $K_x$ (where $K_x = T_x M$ for interior points and $K_x = K_\Omega$ for the conical point) representing the class of $D$ in $\operatorname{gr}_m \mathcal{D}$. The next step to be done is *microlocalization.*

***Microlocalization. The interior symbol.*** For an interior point $x$, the operator $D_x$ (1.38) is an operator with *constant coefficients* on the tangent space $T_x M$ for each $x \in M$. By passing to the Fourier transform with respect to the variables $y$, we arrive at the polynomial $P_m(0, \xi)$, that is, the usual principal symbol of the operator $D$ in the fiber of $T^*M$ over the point $x$. Using the natural identification of $T^*M^\circ$ with the interior of the stretched cotangent bundle $T^*\mathcal{M}$, we obtain a function on the interior of $T^*\mathcal{M}$, which will be denoted by $\sigma(D)$. A straightforward computation shows that $\sigma(D)$ extends by continuity to a $C^\infty$ function on the entire $T^*\mathcal{M}$.

DEFINITION 1.34. The function $\sigma(D)$ on $T^*\mathcal{M}$ is called the *interior symbol* of the operator $D$.

***The cone symbol.*** For the conical point $x$, the limit operator $D_x$ (1.37) has the form (1.39) and does not enjoy translation invariance in any variables. Hence there are no variables with respect to which microlocalization could be done, and we leave the limit operator "as is."

DEFINITION 1.35. The operator (1.39) is called the *cone symbol* of the operator $D$ and is denoted by $\sigma_c(D)$.

***Homogeneity properties of the symbols.*** It follows readily from the definitions that the interior and cone symbols possess the invariance properties stated in the following assertion.

LEMMA 1.36. *One has*

$$\sigma(D)(y, \lambda q) = \lambda^m \sigma(D)(y, q), \quad \lambda \in \mathbb{R}_+,$$

*and*

$$\left(g_\lambda^*\right)^{-1} \sigma_c(D) g_\lambda^* = \lambda^m \sigma_c(D).$$

◀ Both invariance properties follow from the definition of the symbols as limits of operators conjugated by dilations. To prove the first relation, one should also use the fact that the Fourier transform takes dilations to dilations (with the reciprocal ratio and with a numerical factor depending only on the dilation ratio and accounting for the Jacobian of the change of variables in the integral). ▶

### 1.2.7. Supplement: Manifolds with cusps

Here we only very briefly outline a similar construction that gives manifolds with cuspidal points.

***Cuspidal manifolds of order k.*** We again start from a compact manifold $M$ with boundary $\Omega$ and fix a trivialization

$$U = \Omega \times [0,1)$$

of a collar neighborhood of the boundary. We equip $M$ with a degenerating metric that has the form

$$ds^2 = dr^2 + r^{2k+2} d\omega^2 \tag{1.40}$$

near the boundary. Carrying out the same construction as in the conical case but with this metric, we obtain a space $V$ of vector fields on $M^\circ$ coinciding with its $C^\infty(M)$-dual $V'$ with respect to the pairing determined by the metric (1.40).

PROPOSITION 1.37. *The space $V$ is locally spanned over $C^\infty(M)$ by the vector fields*

$$\frac{\partial}{\partial r}, \quad \frac{1}{r^{k+1}} \frac{\partial}{\partial \omega}.$$

The proof is by straightforward verification.

DEFINITION 1.38. The pair $(M, \mathcal{D}(M))$, where $\mathcal{D}$ is the algebra of differential operators generated by $C^\infty(M)$ and $V$ is called a *manifold with cuspidal singularities*.

One can prove that in local coordinates near the cuspidal point each cusp-degenerate operator can be represented in the form (cf. (1.11))

$$D = \sum_{i+j+l \le m} b_{ijl}(r,\omega) \left(\frac{1}{r}\right)^{l(k+1)} \left(\frac{\partial}{\partial r}\right)^j \left(\frac{1}{r^{k+1}} \frac{\partial}{\partial \omega}\right)^i, \tag{1.41}$$

and can be an arbitrary differential operator with smooth coefficients away from the singular point.

The singular object $\mathcal{M}$ for a manifold with cusps can be constructed along the same lines and proves to be (topologically) isomorphic to the corresponding singular manifold with conical points. One also obtains a similar description of the cotangent bundle of a manifold with cusps.

The notion of *principal symbol* can be introduced and studied for the cuspidal case in the same way as in the conical case. The principal symbol is a pair (interior symbol, cuspidal symbol). The cuspidal symbol acts on functions on the infinite cone and, for the operator $D$ written above, has the form

$$\sigma_c(D) = \sum_{i+j+l=m} b_{ijl}(0,\omega) \left(\frac{1}{r}\right)^{l(k+1)} \left(\frac{\partial}{\partial r}\right)^j \left(\frac{1}{r^{k+1}} \frac{\partial}{\partial \omega}\right)^i,$$

Finally, note that the change of variables

$$r = (1+kt)^{-1/k}$$

in a neighborhood of the cusp takes a manifold with cusps to a manifold with cylindrical ends and cusp-degenerate differential operators to operators with coefficients (polynomially) stabilizing at infinity. In fact, it was noticed by Schulze, Sternin and Shatalov (1996, 1998*a*) that, by using an appropriate change of variables $r = \varphi(t)$ that takes the point $r = 0$ to infinity $t = \infty$, in the new variables one can describe the type of the isolated singular point by specifying the *stabilization rate* as $t \to \infty$ of the coefficients of differential operators forming the structure ring. On the other hand, the fact that both cone- and cusp-degenerate operators can be reduced to stabilizing operators in the cylindrical representation is very important. In particular, owing to this there is no separate index theory for cusp-degenerate operators: the results, whatever they are, are always the same as in the cone theory (Schulze, Sternin and Shatalov 1998*a*). That is why we do not consider the cuspidal case further in the book.

## 1.3. Manifolds with Edges

### 1.3.1. Definition

Just as in the definition of manifolds with conical singularities, we start from the definition of the *stretched object*, which plays the main role in all analytical constructions. In this case, the stretched object is again a manifold with boundary, but it is also equipped with some additional structure.

Namely, let $M$ be a smooth compact manifold with boundary $\partial M$. We suppose that the boundary is equipped with the structure

$$\pi : \partial M \xrightarrow{\Omega} X \tag{1.42}$$

of a locally trivial bundle with base $X$ and fiber $\Omega$, where $\Omega$ and $X$ are also smooth compact manifolds.

Let $U$ be a collar neighborhood of the boundary in $M$. Once and for all, we choose a direct product structure

$$U \simeq \partial M \times [0,1); \tag{1.43}$$

then the bundle (1.42) lifts to the bundle (denoted by the same letter)

$$\pi : U \longrightarrow X \times [0,1). \tag{1.44}$$

(The projection acts as the identity operator with respect to the second argument.)

We shall use *admissible* coordinates on $M$ consistent with this structure. Specifically, we consider coordinate neighborhoods of two types on $M$. First, these are coordinate neighborhoods of interior points, where arbitrary smooth coordinate systems will be called admissible. Second, these are neighborhoods of boundary points. The boundary $\partial M$ has the structure of a fiber bundle, and admissible coordinates on $M$ in these neighborhoods are coordinates compatible with this structure, namely, coordinates of the form $(x, \omega, r)$, where $r \in [0,1)$ is a defining function of the boundary and the variables $x$ are coordinates on the base $X$ of the bundle $\pi$ (i.e., they are constant in the fibers of $\pi$). Then for any given $x$ and $r$ the variables $\omega$ are automatically local coordinates in the fiber $\Omega_x$.

Now we introduce a Riemannian metric on $M$ associated with the bundle structure, namely, degenerating at the boundary in such a way that the distance between points lying in the same fiber is zero. The simplest metric of this sort has the form

$$d\rho^2 = dr^2 + dx^2 + r^2 d\omega^2 \tag{1.45}$$

in the local coordinates in $U$, where $r \in [0,1)$ is the second coordinate in the decomposition (1.43) (one can readily see that this coordinate is just the distance to the boundary in the metric (1.45)) and $dx^2$ and $d\omega^2$ are some smooth nondegenerate metrics on $X$ and $\Omega_x$, respectively. (The latter metric is naturally assumed to depend smoothly on $x \in X$.)

Now we can introduce the algebra $\mathcal{D}(M^\circ)$ of differential operators on $M^\circ$ which characterizes $M^\circ$ as a manifold with edges. According to the general construction, this algebra should be generated by $C^\infty(M)$ and the space $V = V'$ of vector fields self-dual with respect to the $C^\infty(M)$-valued pairing determined by the bilinear form associated with the metric (1.45). Just as in the conical case, one proves that such a space with the additional condition $V \supset \operatorname{Vect}(M)$ exists and is unique.

Let $(x, \omega, r)$ be an admissible coordinate system on $M$ in a neighborhood of a boundary point. The vector fields

$$A_j = -i\frac{\partial}{\partial x_j},\ j = 1, \dots, n, \quad B = -i\frac{\partial}{\partial r}, \quad C_j = -\frac{i}{r}\frac{\partial}{\partial \omega_j},\ j = 1, \dots, k, \tag{1.46}$$

form a local basis in $V$. As one might expect, they all have lengths of the order of unity (uniformly bounded above and below by positive constants) in the metric (1.45).

Interestingly, the space $V$ also admits the following alternative description. In the space $\Lambda^1(M)$ of smooth differential 1-forms on $M$, consider the subspace $\Lambda^1_\wedge(M) \subset \Lambda^1(M)$ of forms that vanish on the tangent vectors to the fibers $\Omega_x = \pi^{-1}(x)$ of $\pi$ on the boundary:

$$\Lambda^1_\wedge(M) = \{\alpha \in \Lambda^1(M) \mid \alpha|_{T\Omega_x} = 0 \quad \text{for all } x \in X\}.$$

In admissible local coordinates in a neighborhood of the boundary, such forms can be represented as

$$\alpha = \xi\, dx + qr\, d\omega + p\, dr, \tag{1.47}$$

where the functions $\xi$, $q$, and $p$ are smooth up to the boundary.

DEFINITION 1.39. By $\mathrm{Vect}_\wedge(M^\circ)$ we denote the space of vector fields $\xi$ on $M^\circ$ such that $\alpha(\xi) \in C^\infty(M)$ for every $\alpha \in \Lambda^1_\wedge(M)$. (More precisely, the function $\alpha(\xi)$ defined on $M^\circ$ extends by continuity to a smooth function on the entire $M$.)

PROPOSITION 1.40. *One has* $\mathrm{Vect}_\wedge(M^\circ) = V$.

Now we are in a position to introduce the main definition.

DEFINITION 1.41. The algebra $\mathcal{D} = \mathcal{D}(M)$ of edge-degenerate operators is generated by $C^\infty(M)$ and $V$. The pair $(M, \mathcal{D}(M))$ is called a *manifold with edges*.

Thus every operator $D \in \mathcal{D}$ is a finite linear combination of terms of the form $V_1 \cdots V_{s+l}$, where $s$ factors belong to $V$ and $l$ factors are smooth functions on $M$, with various nonnegative integers $s$ and $l$. The subspace of operators that can be represented by linear combinations in which $s \le m$ for each of the terms will be denoted by $\mathcal{D}_m = \mathcal{D}_m(M)$. The elements of $\mathcal{D}$ ($\mathcal{D}_m$) are called *edge-degenerate differential operators* on $\mathcal{M}$ (*of order* $\le m$). Clearly, the algebra $\mathcal{D}$ is filtered by the subspaces $\mathcal{D}_m$.

### 1.3.2. Geometric description of edges

Now we construct the singular object corresponding to a manifold $(M^\circ, \mathcal{D})$ with edges.

To obtain what geometric intuition says should be a manifold with edges, we identify all points in each fiber $\Omega_x = \pi^{-1}(x)$ of $\pi$ with one another, i.e., introduce the following equivalence relation on $M$:

> two points $a_1, a_2 \in M$ are equivalent ($a_1 \sim a_2$) if and only if either $a_1 = a_2$ or both points belong to $\partial M$ and $\pi(a_1) = \pi(a_2)$. (1.48)

DEFINITION 1.42. The quotient space $\mathcal{M} = M/\sim$ with respect to the equivalence relation (1.48) is called a (*geometric*) *manifold with edges* or, more precisely, a *manifold with edge* $X$.

The geometric situation is shown in Fig. 1.10.

Now let us describe the structure of a neighborhood of the edge. Passing to the quotient space, we see that the neighborhood $\mathcal{U} = U/\sim$ of the edge in $\mathcal{M}$ is also a bundle over $X$. The fiber of this bundle is a manifold with isolated singularity, namely, the neighborhood $\{r < 1\}$ of the vertex in the infinite cone

$$K_\Omega = \{\Omega \times \overline{\mathbb{R}}_+\} \, / \, \{\Omega \times \{0\}\} \tag{1.49}$$

with base $\Omega$.

Thus geometrically a manifold with edges is a space with singularities which looks like the direct product of a domain $V$ in Euclidean space $\mathbb{R}^n$ by the cone $K_\Omega$

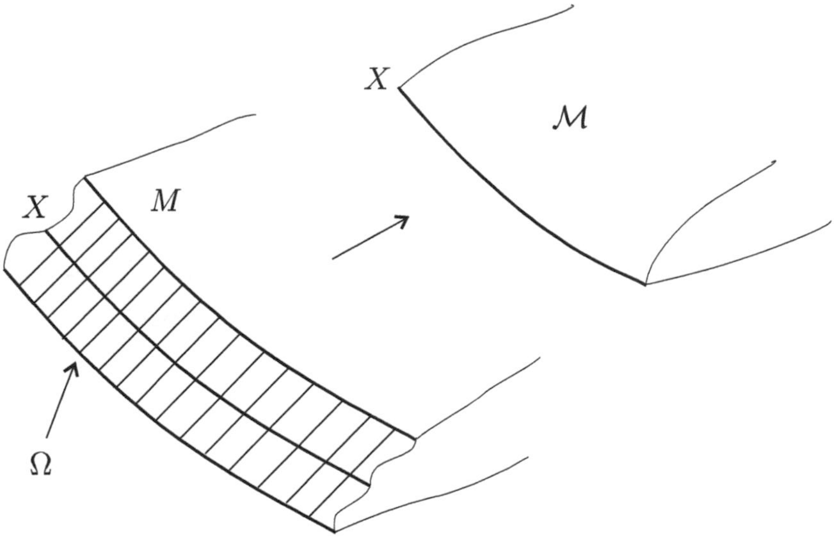


Figure 1.10. A manifold with fibered boundary and a manifold with edges.

with smooth compact base $\Omega$ in a neighborhood of any singular point. The *edge* $X$ (the set of singular points) itself is a smooth manifold and is locally represented by the product of $V$ by the cone vertex.

Next, there is a natural diffeomorphism

$$\mathcal{M} \setminus X \equiv \mathring{\mathcal{M}} \simeq \mathring{M} \equiv M \setminus \partial M$$

of the interiors.

If $\mathcal{M} = M/\sim$ is a compact manifold with edge $X$, then it has a naturally associated noncompact manifold $W$ with edge $X$ called an *infinite wedge* with edge $X$ (see Fig. 1.11). It is defined as the quotient space of the Cartesian product $\partial M \times \overline{\mathbb{R}}_+$ by an equivalence relation similar to (1.48).

Thus $W$ is a locally trivial bundle over $X$ with fiber the infinite cone $K_\Omega$. The projection of this bundle will also be denoted by $\pi$.

In $W$ as well as in $\mathcal{M}$, the neighborhood $\mathcal{U}$ of the edge is well defined. We sometimes identify it with the corresponding neighborhood in $\mathcal{M}$, which permits us to treat functions on $\mathcal{M}$ supported in $\mathcal{U}$ as functions on $W$ and, conversely, functions on $W$ supported in $\mathcal{U}$ with functions on $\mathcal{M}$.

Admissible coordinates on $M$ will also be called admissible coordinates on $\mathcal{M}$. Similar admissible coordinate systems $(x, \omega, r)$ are used on the infinite wedge $W$.

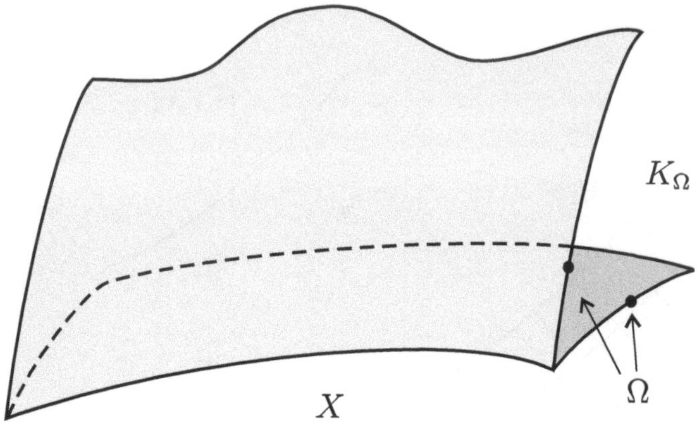


Figure 1.11. An infinite wedge.

### 1.3.3. Local expression of edge-degenerate operators and local Hamiltonians

Now we shall study the local form of edge-degenerate differential operators and write out the corresponding Hamiltonians. Just as with manifolds with conical singularities, manifolds with edges have two types of points, smooth (interior) points and singular (edge) points. The definition of a Hamiltonian in coordinate neighborhoods of smooth points is literally the same as for the case of smooth manifolds or manifolds with conical singularities, and we do not reproduce it here. Let us consider the case of an edge point.

First, we describe our operators in coordinates. Let $(x, \omega, r)$ be an admissible coordinate system on $M$ in a neighborhood of a boundary point. We would like to represent our differential operators as polynomials in the vector fields (1.46) with coefficients being smooth functions on $M$. Note that the operator tuple (1.46) is not closed with respect to commutators: for example, the commutator

$$[B, C_j] = -\frac{i}{r} C_j, \tag{1.50}$$

as well as the commutator

$$[C_j, \omega_j] = -\frac{i}{r}, \tag{1.51}$$

cannot be expressed as an ordered polynomial with smooth coefficients of the operators (1.46). Hence the set of *ordered* polynomials with smooth coefficients of the operators (1.46) is not closed with respect to composition: when moving the operator arguments in the product to the corresponding positions, we gain the singular factor $1/r$. To rectify the situation, we supplement our operator tuple with yet another operator

$$D = \frac{1}{r}. \tag{1.52}$$

PROPOSITION 1.43. *An edge-degenerate differential operator $P$ of order $\leq m$ on a manifold $\mathcal{M}$ with edge $X$ can always be represented in local coordinates $(x, r, \omega)$ in a neighborhood of the boundary as an ordered polynomial of degree $\leq m$ of the operators* (1.46) *and* (1.52) *with coefficients smooth up to $r = 0$:*

$$P = \sum_{|\alpha|+|\beta|+j+l \leq m} a_{\alpha\beta jl}(\overset{5}{x}, \overset{5}{r}, \overset{5}{\omega}) \overset{1}{A^{\alpha}} \overset{2}{B^{j}} \overset{3}{C^{\beta}} \overset{4}{D^{l}}. \tag{1.53}$$

*Sketch of proof.* Consider the product

$$S = V_1 \cdots V_s, \tag{1.54}$$

where each $V_k$ is a smooth function or one of the operators $A_j$, $B$, $C_j$, and $D$. If there are no differentiations at all in (1.54), then the factors are already in the order prescribed in (1.53); otherwise, we should rearrange them. We do so by using the commutation relations (1.50), (1.51), etc. Then we have the desired ordering of operators in the main term, and the commutators again have the form (1.54), but with fewer differentiations. So we can proceed by induction on the number of differentiations, which completes the proof. □

We point out that in this representation the order of an operator near the edge counts *not only differentiations but also the factors* $1/r$.

*Remark* 1.44. Needless to say, representability in the form (1.53) is independent of the choice of local coordinates near the boundary.

The representation obtained above is not a suitable candidate for the definition of a Hamiltonian near an edge point, since it is "too local": we restrict ourselves to a coordinate chart on $\Omega$ and hence do not cover an entire neighborhood of the edge point. However, from this representation we can derive the desired representation. Namely, a straightforward computation shows that the following assertion holds true.

PROPOSITION 1.45. *Every edge-degenerate differential operator $P$ can be represented in a neighborhood of a point of the edge in the form*

$$P = H\left(\overset{2}{x}, -i\overset{1}{\frac{\partial}{\partial x}}\right),$$

*where the operator-valued symbol $H(x, \xi)$ acts in function spaces on the cone $K_\Omega$ and can be represented in local coordinates on $\Omega$ in the form*

$$H(x, \xi) = \sum_{|\alpha|+|\beta|+j+l \leq m} a_{\alpha\beta jl}(x, \overset{4}{r}, \overset{4}{\omega}) \xi^{\alpha} \overset{1}{B^{j}} \overset{2}{C^{\beta}} \overset{4}{D^{l}}. \tag{1.55}$$

DEFINITION 1.46. The function $H(x, \xi)$ is called the *Hamiltonian* of the operator $P$ near the edge point.

### 1.3.4. The cotangent bundle of $\mathcal{M}$

Just as with the case of conical singularities, bundles over a manifold with edges are always understood as bundles over the corresponding stretched manifold.

The *cotangent bundle* of $\mathcal{M}$ plays an important role in the subsequent constructions. By definition, it is some special vector bundle over the stretched manifold $M$ with boundary. The construction is similar to that of the stretched cotangent bundle of a manifold with conical singularities.

***Definition.*** The space $\Lambda^1_\wedge(M)$ is a locally free $C^\infty(M)$-module and hence, by the Swan theorem, the module of sections of a vector bundle on $M$, which is called the *stretched cotangent bundle of the manifold $\mathcal{M}$ with edges* and denoted by $T^*\mathcal{M}$. The embedding $\Lambda^1_\wedge(M) \subset \Lambda^1(M)$ induces the natural mapping

$$\jmath : T^*\mathcal{M} \longrightarrow T^*M, \tag{1.56}$$

which is an isomorphism over the interior $M^\circ$ of $M$. The manifold $T^*\mathcal{M}$ is a manifold with boundary.

***Canonical coordinates.*** There are two types of canonical coordinates on $T^*\mathcal{M}$: in neighborhoods of interior points, these are standard canonical coordinates $(y, \theta)$ induced from the cotangent bundle $T^*M$, and in neighborhoods of boundary points, these are coordinates of the form $(x, \omega, r; \xi, q, p)$, where $(x, \omega, r)$ are admissible coordinates near the boundary on $M$ and $(\xi, q, p)$ are the coordinates corresponding to the representation (1.47) of differential forms $\alpha \in \Lambda^1_\wedge(M)$.

***The boundary of the cotangent bundle.*** Note that there is a natural diffeomorphism

$$\partial T^*\mathcal{M} \simeq \partial T^*W, \tag{1.57}$$

where the cotangent bundle of the infinite wedge $W$ is obtained by a similar construction over $\partial M \times \overline{\mathbb{R}}_+$. To obtain this diffeomorphism, it suffices to use the identification of the neighborhood $\mathcal{U}$ of the edge in $\mathcal{M}$ with the corresponding neighborhood of the edge in $W$. (This identification depends on the choice of the trivialization of a collar neighborhood of the boundary in $M$, but a straightforward verification shows that the diffeomorphism (1.57) is independent of this choice.)

In what follows, we need a special direct sum decomposition of the bundle $\partial T^*\mathcal{M}$ over $\partial M$. This decomposition is constructed as follows. We have already noted that the wedge $W$ is a bundle over $X$ with fiber $K_\Omega$. Let $T^*K_{\Omega_x}$ be the cotangent bundle of the cone $K_{\Omega_x}$ over $x \in X$ constructed by the same recipe as $T^*\mathcal{M}$. (The boundary bundle $\pi_x$ for $K_{\Omega_x}$ is just a mapping into the point $x$: $\pi_x : K_{\Omega_x} \longrightarrow \{x\}$.) This cotangent bundle is a vector bundle over the manifold

$\Omega_x \times \overline{\mathbb{R}}_+$. Taking the disjoint union of $T^*K_{\Omega_x}$ over all $x \in X$ and equipping this union with a natural topology, we obtain a bundle over $\partial M \times \overline{\mathbb{R}}_+$, which will be denoted by $T^*K \longrightarrow \partial M \times \overline{\mathbb{R}}_+$.

PROPOSITION 1.47. *There exists a natural decomposition*

$$\partial T^*\mathcal{M} = \pi^*(T^*X) \oplus \partial T^*K \tag{1.58}$$

*into a direct sum of vector bundles over $\partial M$. Here $\pi^*(T^*X)$ is the lift of the bundle $T^*X$ from $X$ to $\partial M$ via the projection $\pi$ and $\partial T^*K = T^*K\big|_{\partial M \times \{0\}}$.*

*Proof.* It suffices to construct complementary natural projections of the left-hand side *onto* each of the terms on the right-hand side. The projection

$$p_1 : \partial T^*\mathcal{M} \longrightarrow \pi^*(T^*X)$$

is defined as follows. Consider the restriction

$$\jmath : \partial T^*\mathcal{M} \longrightarrow \partial T^*M$$

of the mapping (1.56) to the boundary. The range of this mapping at each point consists of all (algebraic) forms vanishing on the tangent space to the fiber of $\pi$ through this point, i.e., exactly coincides with the corresponding fiber of $\pi^*(T^*X)$, and we can take $p_1 = \jmath$. To construct the projection

$$p_2 : \partial T^*\mathcal{M} \longrightarrow \partial T^*K,$$

we use the isomorphism (1.57). Let $\beta$ be an element of the fiber of $\partial T^*W$ over a point $v \in \partial M$. This means that $\beta = \alpha(v)$ for some section

$$\alpha \in \Gamma(T^*W) = \Lambda^1_\wedge(\partial M \times \overline{\mathbb{R}}_+) \subset \Lambda^1(\partial M \times \overline{\mathbb{R}}_+).$$

We interpret this section as a differential form on $\partial M \times \overline{\mathbb{R}}_+$ and restrict it to $\Omega_x \times \overline{\mathbb{R}}_+$, where $x = \pi(v)$ (i.e., $\Omega_x$ is the fiber of $\pi$ through $v$). The resulting restriction, which we denote by $\widetilde{\alpha}$, still vanishes on tangent vectors to $\Omega_x \times \{0\}$ and hence lies in $\Lambda^1_\wedge(\Omega_x \times \overline{\mathbb{R}}_+)$ and can be interpreted as a section $\widetilde{\alpha} \in \Gamma(T^*K_{\Omega_x})$. The restriction of this section to $v$ specifies an element of the bundle $\partial T^*K$, and we set

$$p_2(\beta) = \widetilde{\alpha}\big|_v.$$

Easy computations with the use of the coordinate representation (1.47) show that the mapping $p_2$ is well defined (i.e., is independent of the choice of the section $\alpha$) and is an epimorphism. The proof of the proposition is complete. □

### 1.3.5. Symbols of edge-degenerate operators

***Definition.*** We know that the algebra $\mathcal{D}$ of edge-degenerate differential operators is filtered by the order of the operator, and we define symbols as elements of the associated graded algebra

$$\operatorname{gr}\mathcal{D} = \bigoplus_{m=0}^{\infty} \mathcal{D}_m/\mathcal{D}_{m-1} \qquad (\mathcal{D}_{-1} \stackrel{\text{def}}{=} \mathbf{0}).$$

DEFINITION 1.48. The *principal symbol* (of order $m$) of an edge-degenerate differential operator $P \in \mathcal{D}_m$ of order $\le m$ is the image of $P$ in the quotient space $\mathcal{D}_m/\mathcal{D}_{m-1}$ under the natural projection:

$$\mathcal{D}_m \ni P \longmapsto \Sigma(P) \in \mathcal{D}_m/\mathcal{D}_{m-1}. \tag{1.59}$$

***The interior symbol and the edge symbol.*** To use this definition in practice, we however need to describe the structure of the principal symbol in more detail. It turns out that it can be represented by a pair consisting of the *interior symbol* and the *edge symbol*. Let us describe both components of the pair. Let $D \in \mathcal{D}_m$ be an edge-degenerate differential operator of order $m$. Its principal symbol $\sigma_{clas}(D)$ in the traditional sense is a function on $T^*M^\circ$ polynomial in the fibers and having singularities (that is, growing unboundedly) near $\partial T^*M$. We set

$$\sigma(D) = \jmath^* \sigma_{clas}(D), \tag{1.60}$$

where $\jmath : T^*\mathcal{M} \longrightarrow T^*M$ is the mapping introduced above. This is a function on the cotangent bundle $T^*\mathcal{M}$. Let us describe the behavior of this function near the boundary. Let

$$D = \sum_{\alpha+\beta+j+l\le m} a_{\alpha\beta jl}(x,\omega,r)\left(-i\frac{\partial}{\partial x}\right)^\alpha \left(-\frac{i}{r}\frac{\partial}{\partial \omega}\right)^\beta \left(-i\frac{\partial}{\partial r}\right)^j \left(\frac{1}{r}\right)^l \tag{1.61}$$

in admissible coordinates. Then one can readily compute that the function $\sigma(D)$ in the canonical local coordinates $(x,\omega,r;\xi,q,p)$ on $T^*\mathcal{M}$ has the form

$$\sigma(D) = \sum_{\alpha+\beta+j=m} a_{\alpha\beta j0}(x,\omega,r)\xi^\alpha q^\beta p^j. \tag{1.62}$$

In particular, it is smooth up to the boundary.

DEFINITION 1.49. The function $\sigma(D)$ on $T^*\mathcal{M}$ given by the formula (1.60) is called the *interior symbol* of the edge-degenerate differential operator $D \in \mathcal{D}_m$.

Let us now proceed to the definition of the edge symbol. Let $D$ again have the form (1.61).

DEFINITION 1.50. The operator family

$$\sigma_\wedge(D)(x,\xi) = \sum_{\alpha+\beta+j+l=m} a_{\alpha\beta jl}(x,\omega,0)\xi^\alpha \left(-\frac{i}{r}\frac{\partial}{\partial \omega}\right)^\beta \left(-i\frac{\partial}{\partial r}\right)^j \left(\frac{1}{r}\right)^l, \tag{1.63}$$

depending on the parameters $(x,\xi)$, is called the *edge symbol* of $D$.

We have defined the edge symbol in local coordinates. Needless to say, now we need to globalize our definition, i.e., find how this family is transformed under changes of coordinates. Consider the bundle $\widetilde{W}$ with fiber the cone $K_\Omega$ over $T^*X$ obtained by lifting the bundle

$$\pi : W \xrightarrow{K_\Omega} X$$

to $T^*X$ via the natural projection $p : T^*X \longrightarrow X$. Thus the local coordinates on $\widetilde{W}$ are $(x, \xi, \omega, r)$, where $(x, \xi)$ are canonical coordinates on $T^*X$ and $(x, \omega, r)$ are admissible coordinates on $W$.

PROPOSITION 1.51. *The local expression* (1.63) *specifies a well-defined operator family on the fibers $K_{\Omega_x}$ of the bundle $\widetilde{W} \longrightarrow T^*X$. This family is parametrized by points of the cotangent bundle $T^*X$.*

*The proof* is by a straightforward computation. □

The following proposition describes the relation of the interior symbol and the edge symbol to the above-introduced principal symbol as an element of the associated gradation.

PROPOSITION 1.52. (1) *The equations $\sigma(D) = 0$ and $\sigma_\wedge(D) = 0$ for the operator $D \in \mathcal{D}_m$ hold if and only if $D \in \mathcal{D}_{m-1}$. (Thus the pair $(\sigma(D), \sigma_\wedge(D))$ isomorphically represents the class $\Sigma(D) \in \mathcal{D}_m/\mathcal{D}_{m-1}$.)*

(2) *The correspondence $D \longmapsto (\sigma(D), \sigma_\wedge(D))$ is linear and multiplicative. This means that*

$$\sigma(D_1 D_2) = \sigma(D_1)\sigma(D_2), \quad \sigma_\wedge(D_1 D_2) = \sigma_\wedge(D_1)\sigma_\wedge(D_2).$$

◀ The proof can be carried out by a straightforward computation in local coordinates with regard to the fact that all assertions in the proposition are local. Indeed, the algebra of edge-degenerate differential operators, as well as the symbol algebra, is a module over $C^\infty(M)$. ▶

***The compatibility condition.*** The interior symbol and the edge symbol of an operator $D \in \mathcal{D}_m$ are not independent but satisfy some compatibility condition. To state this condition, let us view the family $\sigma_\wedge(D)(x, \xi)$ as a cone-degenerate *differential operator* on the cone $K_\Omega$ *with parameter $\xi \in T^*_x X$ in the sense of Agranovich–Vishik* (see Section 3.2.5). Needless to say, this operator also depends on the additional parameter $x \in X$. Let $D$ have the form (1.61). Then the interior symbol of the family $\sigma_\wedge(D)(x, \xi)$ viewed as an operator of order $m$ with parameter $\xi$ has the form

$$\sigma(\sigma_\wedge(D))(x, \omega, \xi, q, p) = \sum_{\alpha+\beta+j=m} a_{\alpha\beta j0}(x, \omega, 0)\xi^\alpha q^\beta p^j \tag{1.64}$$

(and is independent of $r$). By comparing this with the expression (1.62) for the interior symbol, we obtain the compatibility condition

$$\sigma(\sigma_\wedge(D)) = \sigma_\partial(D) \equiv \sigma(D)\,\big|_{\partial T^*\mathcal{M}}\,. \tag{1.65}$$

Condition (1.65) is invariant, which readily follows with regard to the decomposition (1.58).

This condition is obviously necessary and sufficient for the existence of a differential operator with given edge and interior symbols.

We can summarize this as follows:

> *The homogeneous component of degree $m$ of the graded algebra* gr $\mathcal{D}$ *associated with the algebra $\mathcal{D}$ of edge-degenerate differential operators on a manifold $M$ with edges consists of pairs* (*interior symbol, edge symbol*) *satisfying the compatibility condition* (1.65). *These pairs can naturally be viewed as symbols of such differential operators.*

***Homogeneity properties of the symbols.*** In the case of differential operators on smooth manifolds, the principal symbol $\sigma(D)$ of an operator of order $m$ is a homogeneous polynomial of degree $m$ and hence satisfies

$$\sigma(D)(y, \lambda\eta) = \lambda^m \sigma(D)(y, \eta), \quad \lambda > 0.$$

The same is true for the interior symbol of an edge-degenerate differential operator. Does the edge symbol also have some homogeneity property? The answer is "yes," but the property is more complicated. Consider again the dilation operator $g_\lambda^*$ multiplying the argument $r$ of a function by $\lambda$:

$$g_\lambda^* u(r) = u(\lambda r). \tag{1.66}$$

(The function $u$ may well have other arguments, which remain unaffected.) The homogeneity property of the edge symbol can be written in the form

$$\sigma_\wedge(D)(x, \lambda\xi) = \lambda^m g_\lambda^* \sigma(D)(x, \xi)(g_\lambda^*)^{-1}, \quad \lambda \in \mathbb{R}_+. \tag{1.67}$$

This property is referred to as *twisted homogeneity.*

# Chapter 2

# Elliptic Operators on Singular Manifolds

## 2.1. Operators on Manifolds with Conical Singularities

In this section, we shall study elliptic cone-degenerate differential operators. We describe spaces in which such operators naturally act, write out ellipticity conditions, and state the finiteness theorem. The proof of the finiteness theorem will be given in Chapter 3.

### 2.1.1. Function spaces

Let $\mathcal{M}$ be a compact manifold with a conical singular point. The aim of this subsection is to introduce function spaces on $\mathcal{M}$ where it is natural to study cone-degenerate elliptic differential operators. How can one do that? The following argument eventually suggests an answer.

1. Away from the conical point (that is, from $\partial M$), such operators are differential operators with smooth coefficients elliptic in the usual sense, and hence it is reasonable to require that the function spaces to be introduced should coincide there with spaces in which elliptic differential operators on smooth manifolds are usually studied. There is of course a variety of such spaces (one could, say, hesitate whether to take Hölder spaces), but the most common ones are undoubtedly the $L^2$-based Sobolev spaces $H^s$, and we make them our choice. In other words, we require that our spaces be subspaces of the respective Sobolev spaces $H^s_{loc}(M^\circ)$ and coincide with them outside any given neighborhood of the boundary.

2. Next, it is natural to require that the solutions of homogeneous equations with elliptic cone-degenerate operators belong to our spaces. The structure of such solutions is well known (e.g., see Schulze, Sternin and Shatalov 1998*a* and references therein). For our aims, it suffices to know the leading (i.e., most singular) term of the asymptotic expansion of a solution. It has the form

$$u_0(r, \omega) = r^{-\mu}(\ln r)^m \Phi(\omega), \tag{2.1}$$

where $\mu \in \mathbb{R}$ and $m \in \mathbb{Z}_+$ are some constants and $\Phi$ is a smooth function of $\omega$. While there are many ways to ensure that the function (2.1) belongs to our space, we take the simplest one based on invariance considerations. Up to lower-order terms, the function (2.1) is invariant under the representation

$$\lambda \longmapsto \lambda^\mu g_\lambda^* \tag{2.2}$$

of the group $\mathbb{R}_+$ in the space of functions of $(r, \omega)$, where $g_\lambda$ is the dilation group acting by the rule $g_\lambda(r, \omega) = (\lambda r, \omega)$.

◀ Indeed,

$$\begin{aligned}\lambda^\mu [g_\lambda^* u_0](r, \omega) &= \lambda^\mu (\lambda r)^{-\mu} (\ln(\lambda r))^m \Phi(\omega) \\ &= r^{-\mu} (\ln r)^m \Phi(\omega) \left(1 + \frac{\ln \lambda}{\ln r}\right)^m = u_0(r, \omega) \left(1 + O\left(\frac{1}{\ln r}\right)\right),\end{aligned} \tag{2.3}$$

as desired. ▶

Accordingly, it is natural to consider spaces in which the norm locally (near the conical point) possesses a similar invariance property:

$$\|\lambda^\gamma g_\lambda^* u\| = \|u\|, \tag{2.4}$$

where $\gamma \in \mathbb{R}$ is a given number.

THEOREM 2.1. *Let $s, \gamma \in \mathbb{R}$ be given numbers. Then, up to norm equivalence, there exists a unique Hilbert function space $H$ on $M^\circ$ satisfying the following conditions*:

(i) *The norm in $H$ is equivalent to the ordinary Sobolev norm of order $s$ on functions supported in the set $\{r \geq \varepsilon\} \subset M^\circ$, where $\varepsilon > 0$ is arbitrary.*

(ii) *The invariance condition* (2.4) *holds locally* (*on functions supported near the conical point*).

(iii) *The space $C_0^\infty(M^\circ)$ is dense in $H$.*

*Proof.* Indeed, let us choose and fix a Sobolev $H^s$-norm on functions supported in the domain $\{r \geq \varepsilon\}$. By applying the operator $\lambda^\gamma g_\lambda^*$ with an appropriately chosen $\lambda$ to an arbitrary function $u$ compactly supported near the conical point, we can move its support into the domain $\{r \geq \varepsilon\}$. The norm of $\lambda^\gamma g_\lambda^* u$ is hence known, and so is the norm of $u$, for the operator $\lambda^\gamma g_\lambda^*$ is supposed to be norm preserving. □

EXERCISE 2.2. What relation between $\mu$ and $\gamma$ should be observed for the function (2.1) to belong to the space thus defined?

DEFINITION 2.3. The space described in Theorem 2.1 is called a *weighted Sobolev space* on $\mathcal{M}$ and is denoted by $H^{s,\gamma}(\mathcal{M})$. The number $s$ is called the *order* and $\gamma$ is called the *weight exponent* of the Sobolev space $H^{s,\gamma}(\mathcal{M})$.

Let us now compute the norm in $H^{s,\gamma}(\mathcal{M})$ explicitly. To this end, it suffices to consider functions supported in the neighborhood $U$ of the conical point (outside this neighborhood, the norm is standard), and in this case we can assume that the function is defined on the infinite model cone $K = K_\Omega$ with base $\Omega$ rather than on $M^\circ$. The advantage is that the group $g_\lambda$ is now global rather than local. The corresponding function space on the model cone will be denoted by $H^{s,\gamma}(K)$.

PROPOSITION 2.4. *The space $H^{s,\gamma}(K)$ is the completion of the space $C_0^\infty(K^\circ)$ of functions $u(r,\omega)$ with compact support* $\operatorname{supp} u \subset (0,+\infty)\times\Omega$ *with respect to the norm*

$$\|u\|_{s,\gamma} = \left\{ \int_0^\infty \int_\Omega \left| \left(1 + \left(ir\frac{\partial}{\partial r}\right)^2 - \Delta_\Omega\right)^{s/2} (r^{-\gamma}u(r,\omega)) \right|^2 \frac{dr}{r} d\omega \right\}^{1/2}. \tag{2.5}$$

*Here $d\omega$ is a smooth measure on $\Omega$ and $\Delta_\Omega$ is the Beltrami–Laplace operator on $\Omega$ (generated by some Riemannian metric), self-adjoint in $L^2(\Omega, d\omega)$.*

*Proof.* Note that the operator

$$1 + \left(ir\frac{\partial}{\partial r}\right)^2 - \Delta_\Omega \tag{2.6}$$

with domain $C_0^\infty(K^\circ)$ is strongly positive definite and essentially self-adjoint in $L^2(K^\circ, dr\, d\omega/r)$.

◄ Indeed, the change of variables $r = e^{-t}$ transforms the space $L^2(K^\circ, dr\, d\omega/r)$ into $L^2(\Omega\times(-\infty,\infty), dt\, d\omega)$ and takes the operator (2.6) to the operator

$$1 - \frac{\partial^2}{\partial t^2} - \Delta_\Omega,$$

which is obviously self-adjoint in the latter space. ►

Hence the closure of the operator (2.6) is a strongly positive definite self-adjoint operator in $L^2(K^\circ, dr d\omega/r)$, and so the arbitrary real power of this operator occurring in (2.5) is well defined. A straightforward verification shows that the norm (2.5) satisfies the invariance condition (2.4) and is equivalent to the ordinary Sobolev norm on functions supported in any given compact set. □

*Remark* 2.5. It readily follows from Proposition 2.4 that the operator of multiplication by $r^a$, $a \in \mathbb{R}$, is an isomorphism

$$r^a : H^{s,\gamma}(K) \to H^{s,\gamma+a}(K)$$

of Sobolev spaces for any $s$ and $\gamma$.

Now the derivation of a global expression for the norm is standard. Consider an open cover

$$\mathcal{M} = U \cup U_0$$

of $\mathcal{M}$ such that

(i) the domain $U_0$ does not contain the singular point

(ii) the domain $U$ is the standard collar neighborhood of the conical point

Let $1 = e + e_0$ be a smooth partition of unity on $M^\circ$ subordinate to this cover. For functions $u \in C_0^\infty(\operatorname{supp} e_0)$, we introduce the Sobolev norm $\|u\|_s$ in the standard way.

PROPOSITION 2.6. *The weighted Sobolev space $H^{s,\gamma}(\mathcal{M})$ is the completion of $C_0^\infty(M^\circ)$ with respect to the norm*

$$\|u\|_{s,\gamma} = \left\{\|e_0 u\|_s^2 + \|eu\|_{s,\gamma}^2\right\}^{1/2}. \tag{2.7}$$

The following properties of weighted spaces are obvious.

PROPOSITION 2.7. (i) *For each given $\gamma$, the spaces $H^{s,\gamma}(\mathcal{M})$, $s \in \mathbb{R}$, form a scale of Hilbert spaces.*

(ii) *The scales $\{H^{s,\gamma}(\mathcal{M})\}_{s\in\mathbb{R}}$ with various $\gamma$ are isomorphic. Namely, the diagrams*

$$\begin{array}{ccc} H^{s,\gamma}(\mathcal{M}) & \longrightarrow & H^{s',\gamma}(\mathcal{M}) \\ \big\downarrow & & \big\downarrow \\ H^{s,\tilde{\gamma}}(\mathcal{M}) & \longrightarrow & H^{s',\tilde{\gamma}}(\mathcal{M}) \end{array} \tag{2.8}$$

*commute for any $s > s'$, where the horizontal arrows are the natural embeddings and the vertical arrows are the isomorphisms given by the operator of multiplication by a function $f(x)$, $x \in M$, with the following properties*:

(a) *$f(x)$ does not vanish in $M^\circ$*

(b) *$f(x) = r^{\tilde{\gamma}-\gamma}$ for $x \in U$*

The next theorem follows readily from the definitions.

THEOREM 2.8 (Continuity of cone-degenerate operators). *An $m$th-order cone-degenerate differential operator $D \in \mathcal{D}(M)$ on a manifold $\mathcal{M}$ with conical singularities is continuous in the spaces*

$$D : H^{s,\gamma}(\mathcal{M}) \longrightarrow H^{s-m,\gamma-m}(\mathcal{M})$$

*for all $s, \gamma \in \mathbb{R}$.*

◀ Clearly, it suffices to prove the theorem for operators on the infinite cone $K_\Omega$. For this case, the change of variables $r = e^{-t}$ reduces the desired assertion to the standard case of differential operators with coefficients stabilizing at infinity in Sobolev spaces with exponential weights $e^{-\gamma t}$ on the cylinder $\Omega \times \mathbb{R}_t$. ▶

### 2.1.2. Ellipticity and the finiteness theorem

Let $D \in \mathcal{D}(M)$ be a cone-degenerate differential operator of order $m$ on a manifold $\mathcal{M}$ with conical singularities.

DEFINITION 2.9. The operator $D$ is said to be *elliptic with respect to a weight* $\gamma \in \mathbb{R}$ if the following conditions hold:

1) (interior ellipticity): the interior symbol (of order $m$) $\sigma(D)$ of the operator $D$ is invertible on the stretched cotangent bundle $T^*\mathcal{M}$ outside the zero section

2) (conormal ellipticity) the cone symbol

$$\sigma_c(D) : H^{s,\gamma}(K_\Omega) \longrightarrow H^{s-m,\gamma-m}(K_\Omega) \tag{2.9}$$

is invertible for all $s \in \mathbb{R}$

The conormal ellipticity condition can be restated in a technically more convenient form as follows. A straightforward computation shows that the operator

$$r^m\sigma_c(D) : H^{s,\gamma}(K_\Omega) \longrightarrow H^{s-m,\gamma}(K_\Omega)$$

satisfies the invariance condition

$$\left(g_\lambda^*\right)^{-1} r^m\sigma_c(D)g_\lambda^* = r^m\sigma_c(D), \quad \lambda \in \mathbb{R}_+,$$

and hence can be represented in the form

$$r^m\sigma_c(D) = \mathfrak{M}_\gamma^{-1} \circ B(p) \circ \mathfrak{M}_\gamma,$$

where $\mathfrak{M}_\gamma$ is the Mellin transform with *weight line* $\{\operatorname{Im} p = \gamma\}$ and

$$B(p) : H^s(\Omega) \longrightarrow H^{s-m}(\Omega) \tag{2.10}$$

is some operator family. The specific form of $B(p)$ can be obtained in local coordinates near $\partial M$ as follows. We rewrite $D$ in the form (see Proposition 1.27)

$$D = r^{-m}H\left(r, ir\frac{\partial}{\partial r}\right),$$

and then

$$B(p) = H(0,p). \tag{2.11}$$

The following proposition is an immediate consequence of the properties of the Mellin transform.

PROPOSITION 2.10. *The invertibility of the cone symbol* $\sigma_c(D)$ *in the spaces* (2.9) *is equivalent to the invertibility of the family* (2.10) *for all* $p$ *on the weight line* $\{\operatorname{Im} p = \gamma\}$.

DEFINITION 2.11. The family (2.10), (2.11) is denoted by $\sigma_c(D)(p)$ and called the *conormal symbol* of the operator $D$.

Essentially, the conormal symbol is the cone symbol in the momentum representation: we pass from the "physical" variable $r$ to the dual "momentum" $p$. The variable $p$ is also known as the *conormal variable*, since it is a coordinate in the fibers of the conormal bundle $N^*\Omega$ of the embedding $\Omega = \partial M \hookrightarrow M$; hence the term "conormal symbol" and the notation. We often omit the argument $p$; this will not lead to a misunderstanding, since the meaning always will be clear from the context.

*Remark* 2.12. Suppose that the operator $D$ is interior elliptic. Then its conormal symbol $\sigma_c(D)(p)$ is a family of differential operators of order $m$ on $\Omega$ elliptic with parameter $p$ in the sense of Agranovich–Vishik (cf. Definition 3.76) in a sector of nonzero angle containing the real axis. By the Gokhberg–Krein theory (e.g., Egorov and Schulze 1997), this family is invertible for all $p \in \mathbb{C}$ but at most countably many points without finite accumulation points, whence it follows that the conormal ellipticity condition holds for all but countably many values of the parameter $\gamma$. Thus under the interior ellipticity condition one can always find a $\gamma$ for which the operator is elliptic. Note also that, by virtue of ellipticity, the invertibility of $\sigma_c(D)(p)$ is independent of the smoothness exponent $s$. Hence *the ellipticity conditions for $D$ are also independent of $s$* (but in general depend on $\gamma$).

Now we can state the main assertion.

THEOREM 2.13. *If an operator $D \in \mathcal{D}(M)$ of order $m$ is elliptic with respect to a weight $\gamma \in \mathbb{R}$, then it is Fredholm in the spaces*

$$D : H^{s,\gamma}(\mathcal{M}) \longrightarrow H^{s-m,\gamma-m}(\mathcal{M}) \tag{2.12}$$

*for each $s \in \mathbb{R}$. Moreover, the kernel and cokernel of the operator* (2.12) *are independent of $s$.*

*Proof.* A proof based on the calculus of pseudodifferential operators will be carried out in Chapter 3. □

## 2.2. Operators on Manifolds with Edges

In this section, we study elliptic edge-degenerate differential operators. We describe spaces in which such operators naturally act, write out ellipticity conditions, and state the finiteness theorem, the proof of which will be given in Chapter 3.

### 2.2.1. Function spaces

Let $\mathcal{M}$ be a compact manifold with edges. In this subsection, we introduce function spaces on $\mathcal{M}$ in which one can naturally study edge-degenerate elliptic operators. In doing so, we use the following scheme.

1. In the interior of $\mathcal{M}$ (more precisely, outside an arbitrary neighborhood $U$ of $\partial M$), the desired spaces should coincide with the ordinary Sobolev spaces $H^s_{loc}(M^\circ)$.

2. To understand the structure of the desired spaces near the edge, we study an edge-degenerate operator in $U$. This leaves some ambiguity in the definition of the spaces.

3. To eliminate the remaining ambiguity, we use the condition that the spaces outside a small neighborhood of the edge and in a larger neighborhood of the edge should agree with each other on the intersection.

Consider an edge-degenerate differential operator $D$ of order $m$ on a manifold $\mathcal{M}$ with edge $X$. We are interested in defining spaces where the operator could have the Fredholm property under appropriate conditions on the symbol.

One can study the Fredholm property of $D$ by trying to construct an almost inverse of $D$. This is usually done locally, and then local almost inverses are glued together with the help of a partition of unity. In a neighborhood of any interior point, interior ellipticity coincides with ordinary ellipticity, and one can construct an almost inverse by usual methods of elliptic theory in the form of a pseudodifferential operator whose symbol is the inverse of the interior symbol $\sigma(D)$ provided that $\sigma(D)$ is invertible.

***The frozen coefficients method.*** Now consider a neighborhood of the edge. How to construct an almost inverse there? We shall use the method of frozen coefficients. Locally, we can assume that

$$X = \mathbb{R}^n, \qquad \mathcal{M} = \mathbb{R}^n \times K_\Omega.$$

Let us freeze the coefficients of $D$ at some point of the edge and pass in the equation $Du = f$ from $u$ and $f$ to their Fourier transforms $\widetilde{u}$ and $\widetilde{f}$ with respect to the variables $x$ along the edge. Then we obtain the family of equations

$$\sigma_\wedge(D)(x,\xi)\widetilde{u} = \widetilde{f}. \tag{2.13}$$

To construct an almost inverse of $D$ in a neighborhood of the edge, we should solve Eq. (2.13) exactly for small $r$ for any $(x,\xi) \in T^*_0X$ (or at least for large $|\xi|$). The edge symbol possesses the homogeneity property (1.67), and hence it suffices to solve (2.13), say, only for $|\xi| = 1$ on the *entire infinite cone* $K_\Omega$.

Thus let us study Eq. (2.13) on the cone $K_\Omega$. The operator

$$\sigma_\wedge(D)(x,\xi) = F\left(\omega, \xi, \frac{1}{r}, \frac{\partial}{\partial r}, \frac{1}{r}\frac{\partial}{\partial \omega}\right)$$

is cone-degenerate at the vertex of the cone $K_\Omega$ and behaves for large $r$ as a usual elliptic operator with slowly varying coefficients[1] at infinity in $\mathbb{R}^{k+1}_y$ written in the coordinates $\omega, r$ related to the coordinates $y$ by the formula

$$y \equiv (y_1, \dots, y_{k+1}) = (r\omega, r).$$

***Function spaces on the cone.*** In what spaces should one study this operator? Let us cover the cone $K_\Omega$ with two overlapping neighborhoods, a neighborhood $U_1$ of the vertex $r = 0$ and a neighborhood $U_2$ of infinity. One can construct the almost inverse locally, in each of these neighborhoods. In $U_2$, our operator is a usual elliptic operator with coefficients slowly varying at infinity, and hence the almost inverse can be constructed in the usual Sobolev spaces $H^s(U_2)$. In $U_1$, our operator is cone-degenerate, and hence the almost inverse can be constructed in weighted Sobolev spaces $H^{s,\gamma}(U_1)$. On the intersection $U_1 \cap U_2$ these spaces coincide,

$$H^{s,\gamma}(U_1 \cap U_2) = H^s(U_1 \cap U_2), \tag{2.14}$$

and so we can give the following definition.

DEFINITION 2.14. The space $\mathcal{K}^{s,\gamma}(K_\Omega)$ is glued from $H^{s,\gamma-(k+1)/2}(K_\Omega)$ for small $r$ and $H^s(K_\Omega)$ for large $r$:

$$\mathcal{K}^{s,\gamma}(K_\Omega) = eH^{s,\gamma-(k+1)/2}(K_\Omega) + (1-e)H^s(K_\Omega),$$

where $\{e, 1-e\}$ is a partition of unity subordinate to the cover $\{U_1, U_2\}$.

*Remark* 2.15. The shift $-(k+1)/2$ in the weight exponent is chosen so as to ensure that

$$\mathcal{K}^{0,0}(K_\Omega) = L^2(K_\Omega) \equiv L^2(K_\Omega, r^k\, dr\, d\omega),$$

where

$$r^k\, dr\, d\omega, \quad k = \dim \Omega, \tag{2.15}$$

is the Riemannian volume form corresponding to the metric

$$dr^2 + r^2 d\omega^2 \tag{2.16}$$

on $K_\Omega$.

[1]An operator with coefficients slowly varying at infinity in $\mathbb{R}^{k+1}$ is an operator of the form

$$D = \sum a_\alpha(y)(-i\partial/\partial y)^\alpha,$$

where the coefficients $a_\alpha(y)$, $y \in \mathbb{R}^{k+1}$, satisfy the estimates

$$|a_\alpha^{(\beta)}(y)| \leq \text{const}(1 + |y|)^{-|\beta|}.$$

For the norm in $\mathcal{K}^{s,\gamma}(K_\Omega)$, one can readily write out a simple equivalent global expression that does not use partitions of unity. Namely, consider the operator

$$T = 1 + r^{-2} + \triangle_{K_\Omega} \tag{2.17}$$

in $L^2(K_\Omega)$, where $\triangle_{K_\Omega}$ is the Beltrami–Laplace operator on $K_\Omega$ associated with the cone-degenerate metric (2.16).

PROPOSITION 2.16. *The operator $T$ with domain $C_0^\infty(K_\Omega^\circ)$ is essentially self-adjoint in $L^2(K_\Omega)$.*

*Proof.* The proof of this fact follows the proof of Theorem X.11 in Reed and Simon (1975, p. 161) almost word for word. □

Consider also a smooth weight function $\rho(r)$ equal to $r$ in a neighborhood of $r = 0$ and tending to 1 as $r \to \infty$. The following assertion holds.

PROPOSITION 2.17. *An equivalent norm in $\mathcal{K}^{s,\gamma}(K_\Omega)$ can be given by the formula*

$$\|u\|_{s,\gamma} = \left\|T^{s/2}\rho^{s-\gamma}u\right\|_{L^2(K_\Omega)}. \tag{2.18}$$

*The proof* is by a straightforward computation. □

We see that it is natural to consider the edge symbol in the spaces $\mathcal{K}^{s,\gamma}(K_\Omega)$. One can readily show that the edge symbol of an edge-degenerate operator $D$ of order $m$ is continuous in the spaces

$$\sigma_\wedge(D)(x,\xi) : \mathcal{K}^{s,\gamma}(K_\Omega) \longrightarrow \mathcal{K}^{s-m,\gamma-m}(K_\Omega). \tag{2.19}$$

***Functions spaces on a manifold with edges.*** Note that so far we have only constructed the space in which the Fourier transform $\widetilde{u}$ of the solution should lie for every given $\xi$. To obtain function spaces in which $u$ itself should lie, we have to use the inverse Fourier transform. A reasonably general expression for the norm of $u$ via the norms of $\widetilde{u}(\xi)$ can be defined as

$$\|u\| = \left\{\int \|A(\xi)\widetilde{u}(\xi)\|_{s,\gamma}^2 \, d\xi\right\}^{1/2}, \tag{2.20}$$

where $A(\xi)$ is some family of bounded boundedly invertible operators in $\mathcal{K}^{s,\gamma}(K_\Omega)$. How should we choose this family? We shall use the following general construction (cf. also the definition of spaces of Hilbert-valued functions in Section 3.3).

DEFINITION 2.18 (Schulze 1991). Let $H$ be a Hilbert space, and let

$$\varkappa_\lambda : H \longrightarrow H, \qquad \lambda \in \mathbb{R}_+,$$

be a strongly continuous one-parameter group of bounded linear operators in $H$. By $\mathcal{W}^s(\mathbb{R}^n, H)$ we denote the completion of the space $C_0^\infty(\mathbb{R}^n, H)$ of compactly supported smooth functions ranging in $H$ with respect to the norm

$$\|u\|_{\mathcal{W}^s(\mathbb{R}^n,H)} = \left\{ \int_{\mathbb{R}^n_\xi} \langle\xi\rangle^{2s} \left\| \varkappa^{-1}_{\langle\xi\rangle} \widetilde{u}(\xi, \cdot) \right\|_H^2 d\xi \right\}^{1/2},$$

where

$$\langle\xi\rangle = \sqrt{1 + |\xi|^2}.$$

The space $\mathcal{W}^s(\mathbb{R}^n, H)$ is called the *abstract wedge space* of $H$-valued functions on $\mathbb{R}^n$.

Under natural conditions (see Section 3.3), which are valid, for example, for $H = \mathcal{K}^{s,\gamma}(K_\Omega)$, this definition can be globalized from $\mathbb{R}^n$ to an arbitrary smooth closed manifold $X$ with the help of a partition of unity, and we obtain the definition of the spaces $\mathcal{W}^s(X, H)$.

In our specific situation, this construction can be applied as follows. In the space $\mathcal{K}^{s,\gamma}(K_\Omega)$, consider the dilation group $\varkappa_\lambda$, $\lambda \in \mathbb{R}_+$, given by the formula

$$\varkappa_\lambda v(r, \omega) = \lambda^{(k+1)/2} v(\lambda r, \omega).$$

(The group $\varkappa_\lambda$ is unitary in $L^2(K_\Omega)$; this explains the choice of the normalizing factor.)

Let $W$ be the infinite wedge associated with $\mathcal{M}$.

DEFINITION 2.19. We set

$$\mathcal{W}^{s,\gamma}(W) = \mathcal{W}^s(X, \mathcal{K}^{s,\gamma}(K_\Omega)),$$
$$\mathcal{W}^{s,\gamma}(\mathcal{M}) = e_1 \mathcal{W}^{s,\gamma}(W) + e_2 H^s(M),$$

where

$$1 = e_1 + e_2$$

is a smooth partition of unity on $M$ subordinate to the cover $M = U \cup M^\circ$.

*Remark* 2.20. 1) Note that the formula for $\mathcal{W}^{s,\gamma}(\mathcal{M})$ makes sense, since functions on $W$ supported in $U$ can naturally be treated as functions on $M$ (once we have fixed the trivialization of the collar neighborhood $U$).

2) One can prove that the norms in the spaces $\mathcal{W}^{s,\gamma}(\mathcal{M})$ and $H^s(M)$ are equivalent on functions supported in an arbitrary given compact subset of $M^\circ$. This justifies the choice of the family $A(\xi)$ in Eq. (2.20) in the form

$$A(\xi) = \langle\xi\rangle^s \varkappa^{-1}_{\langle\xi\rangle}$$

and guarantees that up to equivalence the norm in $\mathcal{W}^{s,\gamma}(\mathcal{M})$ is independent of the choice of the partition of unity.

One can readily verify the following proposition.

PROPOSITION 2.21. *An edge-degenerate differential operator $D \in \mathcal{D}_m(M)$ of order $\leq m$ is continuous in the spaces*

$$D : \mathcal{W}^{s,\gamma}(\mathcal{M}) \longrightarrow \mathcal{W}^{s-m,\gamma-m}(\mathcal{M}). \tag{2.21}$$

### 2.2.2. Ellipticity and the finiteness theorem

Now we are in a position to state the finiteness theorem for edge-degenerate differential operators.

DEFINITION 2.22. We say that an edge-degenerate operator $D \in \mathcal{D}(M)$ on a manifold $\mathcal{M}$ with edges is *interior elliptic* if its interior symbol $\sigma(D)$ is everywhere invertible on the cotangent bundle $T^*\mathcal{M}$ outside the zero section.

We say that $D$ is *elliptic* (*with weight* $\gamma$) if it is interior elliptic and its edge symbol $\sigma_\wedge(D)(x,\xi)$ is everywhere invertible in the spaces (2.19) on the cotangent bundle $T^*X$ outside the zero section.

*Remark* 2.23. One can prove (see Chapter 3) that the ellipticity condition is independent of $s$.

THEOREM 2.24. *If $D \in \mathcal{D}_m(M)$ is elliptic with weight $\gamma$, then it is Fredholm in the spaces* (2.21). *The kernel, cokernel, and index of $D$ in these spaces are independent of $s$.*

The proof will be given in Chapter 3.

## 2.3. Examples of Elliptic Edge Operators

Here we give some examples of elliptic edge-degenerate operators.

### 2.3.1. The Laplacian

Suppose that the base $\Omega$ of the model cone consists of a single point, and so geometrically $\mathcal{M} = M$ is a manifold with boundary. On this manifold, we consider the Beltrami–Laplace operator $\triangle$. We consider the simplest case in which $\dim M = 2$ and the operator in local coordinates near the boundary has the form

$$\triangle = \frac{\partial^2 u}{\partial r^2} + \frac{\partial^2 u}{\partial x^2}.$$

The edge symbol of this operator is

$$\sigma_\wedge(\triangle)(\xi) = \frac{\partial^2}{\partial r^2} - \xi^2.$$

Let us study the invertibility of the edge symbol in the spaces

$$\sigma_\wedge(\triangle)(\xi) : \mathcal{K}^{s,\gamma}(\mathbb{R}_+) \longrightarrow \mathcal{K}^{s-2,\gamma-2}(\mathbb{R}_+).$$

We are interested in its kernel and cokernel, so we shall also consider the adjoint family

$$\sigma_\wedge(\triangle)(\xi)^* : \mathcal{K}^{2-s,2-\gamma}(\mathbb{R}_+) \longrightarrow \mathcal{K}^{-s,-\gamma}(\mathbb{R}_+),$$

which is given on $C_0^\infty(\mathbb{R}_+)$ by the same differential expression as $\sigma_\wedge(\triangle)(\xi)$. It suffices to study the invertibility of $\sigma_\wedge(\triangle)(\xi)$ for $|\xi| = 1$ and then extend the result to all $\xi \neq 0$ by homogeneity.

Let us compute the kernel and cokernel of the edge symbol. Formally, the kernel of $\sigma_\wedge(\triangle)(\xi)$ and $\sigma_\wedge(\triangle)(\xi)^*$ is given by the expression

$$u = C_1 e^{-r} + C_2 e^r.$$

However, the constant $C_2$ is always zero, which follows from the integrability at infinity.

Next, the presence of the weight factor $r^{-2\gamma}$ in the definition of the norm in $\mathcal{K}^{s,\gamma}(\mathbb{R}_+)$ results in the assertion that $e^{-r} \in \mathcal{K}^{s,\gamma}(\mathbb{R}_+)$ if and only if $\gamma < \frac{1}{2}$. We conclude that the edge symbol $\sigma_\wedge(\triangle)(\xi)$ is invertible for $\frac{1}{2} < \gamma < \frac{3}{2}$. Consequently, the following assertion holds.

PROPOSITION 2.25. *The operator*

$$\triangle : \mathcal{W}^{s,\gamma}(\mathcal{M}) \longrightarrow \mathcal{W}^{s-2,\gamma-2}(\mathcal{M})$$

*is Fredholm for* $\frac{1}{2} < \gamma < \frac{3}{2}$.

*Remark* 2.26. What happens for other $\gamma$ is a different story. We consider this in detail in Chapter 6.

### 2.3.2. Invertible edge symbols and the semiclassical approximation

In the preceding subsection, we have given an individual example of an elliptic edge-degenerate operator. Here we describe a method permitting one to construct series of such examples.

Let $D$ be an interior elliptic first-order edge-degenerate operator. The ellipticity condition for $D$ is the invertibility of the edge symbol $\sigma_\wedge(D)(x,\xi)$ in the weighted Sobolev spaces $\mathcal{K}^{s,\gamma}$ on the infinite cone $K_\Omega$. Let us compare this condition with the ellipticity condition in the case of manifolds with isolated singularity, where ellipticity is determined by the conormal symbol. There are two important differences:

- Ellipticity in the edge situation is reduced to the invertibility of a *family of operators* (rather than a single operator, as in the conical case).

- The edge symbol at an edge point is an operator on the infinite cone with *variable* rather than constant coefficients with respect to the variable $r$,[2] as in the case of isolated singularities.

The first difference is essential (for nonisolated singularities, the ellipticity of the operator at different edge points is independent, and hence the symbol is naturally a family), but the second difference seems to be rather artificial, and we shall show how one can sidestep it in certain cases.

***The semiclassical limit.*** We introduce a small parameter $h$ multiplying the $r$-derivative, i.e., make the change of operators

$$r\frac{\partial}{\partial r} \mapsto hr\frac{\partial}{\partial r}.$$

The idea of application of this *semiclassical limit* to edge symbols is to make the commutator of the operators $hr\partial/\partial r$ and $r$ small and obtain, in the limit as $h \to 0$, a composition formula for functions of these two operators in the form

$$D\left(hr\frac{\partial}{\partial r}, r\right) \circ D'\left(hr\frac{\partial}{\partial r}, r\right) = DD'\left(hr\frac{\partial}{\partial r}, r\right) + O(h). \tag{2.22}$$

Hence the composition of operators corresponds to the product of symbols.

Let us consider this in more detail. On a smooth closed manifold $\Omega$, consider the family of first-order differential operators

$$a(\xi, p) = a_0(\omega) + a_1(\omega)\xi + a_2(\omega)p + a_3(\omega)\left(-i\frac{\partial}{\partial \omega}\right), \qquad (\xi, p) \in \mathbb{R}^{n+1}, \tag{2.23}$$

linearly depending on the variable $\xi$.

THEOREM 2.27. *Suppose that the family* (2.23) *is Agranovich–Vishik elliptic on the set* $\mathbb{R}^n \times \Lambda \subset \mathbb{R}^n \times \mathbb{C}_p$, *where* $\Lambda$ *is a two-sided sector of a nonzero opening angle containing the real axis, and is invertible for* $(\xi, p) \in \mathbb{R}^{n+1}$. *Then the operator*

$$A_h = \overset{2}{r}{}^{-1} a\left(\overset{2}{r}\xi, ihr\overset{1}{\frac{\partial}{\partial r}} + ih\frac{k-1}{2}\right) : \mathcal{K}^{1,1}(K_\Omega) \longrightarrow \mathcal{K}^{0,0}(K_\Omega) \tag{2.24}$$

*is invertible for all sufficiently small* $h \in (0, 1]$. *Here* $k = \dim \Omega$.

This theorem permits one to construct invertible edge symbols at a point of the edge. The edge symbol can be constructed globally on the edge if the family (2.23) smoothly depends on the point $x$ of the edge.

The proof of the theorem will be given in Section 3.5.5.

[2] More precisely, an operator with constant coefficients is obtained from the conormal symbol by the change of variables $r = e^{-t}$.

***The geometric meaning of the semiclassical limit.*** Our semiclassical construction of an invertible edge symbol corresponds to the replacement of the set of vector fields

$$\frac{\partial}{\partial x}, \quad \frac{\partial}{\partial r}, \quad \frac{1}{r}\frac{\partial}{\partial \omega} \tag{2.25}$$

of unit length in the standard edge-degenerate metric $dx^2 + dr^2 + r^2 d\omega^2$ by the set

$$\frac{\partial}{\partial x}, \quad h\frac{\partial}{\partial r}, \quad \frac{1}{r}\frac{\partial}{\partial \omega}. \tag{2.26}$$

The latter fields are obviously of unit length with respect to the metric

$$dx^2 + \frac{1}{h^2}dr^2 + r^2 d\omega^2. \tag{2.27}$$

It is easy to visualize how the geometry of the manifold changes as $h \to 0$. The situation is depicted in Fig. 2.1. Namely, the metric on the edge does not vary, nor does the metric on the base of the cone (for $r = 1$), while the length of the cone (corresponding to the values $0 < r < 1$) is of the order of $1/h$.

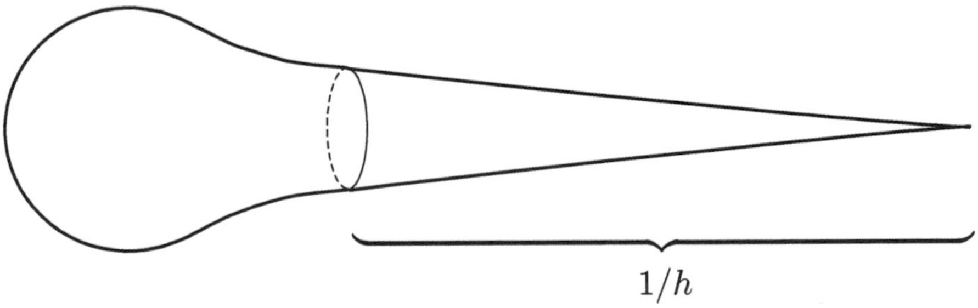


Figure 2.1. The adiabatic limit of the metric.

From a somewhat different viewpoint, the limit as $h \to 0$ means that the radii of the model cones (in a neighborhood of the singularity) tend to zero. This *adiabatic limit* has important applications in geometric analysis. For example, in the case of isolated conical singularities the adiabatic limit was used by Bismut and Cheeger (1990) to obtain index formulas for families of cone-degenerate operators. There are interesting relationships between the adiabatic limit and the nonmultiplicativity of the signature of smooth bundles (Dai 1991).

# Part II

# Analytical Tools

# Chapter 3

# Pseudodifferential Operators

## 3.1. Preliminary Remarks

Pseudodifferential operators ($\psi$DO) are one of the main analytic tools of elliptic theory on smooth as well as singular manifolds. Although it is the latter that are of main interest to us, we start from $\psi$DO on smooth manifolds. On the one hand, this permits us to avoid additional technical complications in describing the main ideas, which will help us motivate the constructions of $\psi$DO on singular manifolds. On the other hand, the calculus of $\psi$DO on smooth manifolds serves as building material for more complicated calculi; here the most important role is played by $\psi$DO acting in spaces of sections of *infinite-dimensional bundles*.

First, we would like to recall some basic facts about pseudodifferential operators and their importance in elliptic theory. This will be done in this section. The subsequent two sections deal with the calculus of classical $\psi$DO on smooth manifolds and $\psi$DO with operator-valued symbols ($\psi$DO in sections of Hilbert bundles). Then we consider the calculus of pseudodifferential operators on singular manifolds. In the last two sections, we apply these calculi to prove the finiteness theorem and the index theorems for $\psi$DO in sections of Hilbert bundles.

To avoid misunderstanding, note that $\psi$DO have a wide variety of applications and we do not even attempt to give comprehensive coverage of any aspect of the theory (even with the stipulation that we have focused our attention on the needs of elliptic theory). We refer the readers interested in deep subtleties of $\psi$DO theory to Hörmander's (1985*a*, 1985*b*) classical treatise.

### 3.1.1. Why are pseudodifferential operators needed?

Pseudodifferential operators arose at the dawn of elliptic theory as a natural class of operators containing (almost) inverses of elliptic differential operators. The fact that the solutions of elliptic differential equations with constant coefficients are given by the convolutions of the right-hand sides with appropriate *potentials*, i.e., functions smooth away from zero and having a singularity of a certain type at zero, had been known in partial differential equations for a long time. For equations with variable coefficients, it had also been known that the solutions are given by similar convolutions but with kernels that have an additional (smooth) dependence on the observation point; the corresponding operators were (and still are) called *singular integral operators*. These operators apparently had a nature very different from that of differential operators whose inverses they were, so that an explicit description of the algebra combining both types of operators seemed to be a nonawarding and complicated task. The creators of $\psi$DO theory observed that, being rewritten in terms of the Fourier transform, the definition of singular integral operators acquires a very simple form equally suitable for the representation of differential operators (Kohn and Nirenberg 1965, Hörmander 1965). Operators of this form were called *pseudodifferential*. Furthermore, the notion of symbol for $\psi$DO has a clear, transparent meaning (in contrast to the seemingly artificial definition of the symbol for singular integral operators), and the product of operators corresponds (modulo lower-order operators) to the product of symbols. If we note that negative-order operators on compact manifolds are compact, then it becomes clear that the introduction of $\psi$DO makes the finiteness theorem for elliptic operators almost tautological: if an operator is elliptic (i.e., its principal symbol is invertible), then any $\psi$DO whose principal symbol is the inverse of the principal symbol of the original operator is an almost inverse of the latter, whence the Fredholm property follows. Thus, the first fact determining the importance of $\psi$DO in elliptic theory is the following:

> *1. The composition law for pseudodifferential operators trivializes the finiteness theory for elliptic operators.*

However, the usefulness of $\psi$DO in elliptic theory would be still at doubt if this were the only reason. All in all, one has been able to prove the finiteness theorem (even though in a more complicated manner) without resorting to pseudodifferential operators. The genuine value of $\psi$DO reveals itself in topological issues of elliptic theory, in particular, in index theory. The proof of the Atiyah–Singer index theorem (e.g., Palais 1965) is based on the reduction of elliptic symbols, which are invertible functions on the cosphere bundle of the manifold in question, to the simplest form by (stable) homotopy, in other words, on the *homotopy classification* of elliptic symbols. Unfortunately, even starting from the symbol of a differential operator, one cannot carry out such a homotopy in the class of symbols of differential operators (i.e., restrictions of polynomial symbols to the cosphere bundle). To conclude that the indices of the operators with symbols corresponding to the beginning and end of the homotopy are the same, we should therefore be able to lift a homotopy of arbitrary elliptic symbols to a homotopy of Fredholm operators. The $\psi$DO calculus solves this problem: the symbol of a $\psi$DO can be an arbitrary smooth function on the cosphere bundle, and the invertibility of a symbol implies the Fredholm property of the corresponding operator by virtue of the finiteness theorem. Thus we see that

*2. Pseudodifferential operators naturally arise in homotopies used in the classification of elliptic operators.*

In connection with elliptic theory, note also that

*3. Pseudodifferential operators naturally arise in the reduction of elliptic boundary value problems to the boundary.*

Finally, $\psi$DO often prove useful from a purely technical point of view. For example, it is often more convenient to deal with bounded operators in $L^2$ than unbounded operators in the same space or even bounded operators in the Sobolev scale. One can reduce an unbounded operator to a bounded operator in $L^2$ by multiplying it by appropriate powers of the Laplace operator on both sides. The resulting operators are no longer differential; they are $\psi$DO of order zero. Thus

*4. Pseudodifferential operators provide a natural framework for the order reduction procedure.*

The list could readily be continued, but even now it should be clear that pseudodifferential operators do not make their entrance in elliptic theory by mere occasion.

Now let us proceed to studying the calculus of pseudodifferential operators.

## 3.2. Classical Theory

### 3.2.1. Pseudodifferential operators in $\mathbb{R}^n$

Now that we are sure that considering $\psi$DO is indeed useful, let us briefly recall their construction and main properties. For now, we carry out all constructions locally, in a given coordinate system $(x_1, \ldots, x_n)$ (or, if you wish, in the space $\mathbb{R}^n$).

Let $\widehat{H}$ be a differential operator of order $\leq m$ of the form

$$\widehat{H} = \sum_{|\alpha| \leq m} a_\alpha(x) \left( -i \frac{\partial}{\partial x} \right)^\alpha, \tag{3.1}$$

where $\alpha = (\alpha_1, \ldots, \alpha_n)$ is a multi-index with integer nonnegative components, the coefficients $a_\alpha(x) = a_\alpha(x_1, \ldots, x_n)$ are smooth functions, and

$$\left( -i \frac{\partial}{\partial x} \right)^\alpha = \left( -i \frac{\partial}{\partial x_1} \right)^{\alpha_1} \cdots \left( -i \frac{\partial}{\partial x_n} \right)^{\alpha_n}.$$

The operator (3.1) can be treated as a *function*

$$\widehat{H} = H\left(\overset{2}{x}, -i\overset{1}{\frac{\partial}{\partial x}}\right) \tag{3.2}$$

of the operators $-i\partial/\partial x$ of differentiation and $x$ of multiplication by the independent variables with symbol

$$H(x,\xi) = \sum_{|\alpha|\leq m} a_\alpha(x)\xi^\alpha \tag{3.3}$$

being a polynomial of degree $\leq m$ in the variables $\xi$.

*Remark* 3.1. The numbers over operators in (3.2) denote, according to Maslov (1973), the order in which the operators act: first the function to be acted upon by $\widehat{H}$ is differentiated, and then the results are multiplied by the coefficients and summed. In what follows, we usually adopt this standard ordering of operator arguments and omit the numbers over operators.

However, it is not immediately apparent from (3.2), (3.3) how to generalize this definition to nonpolynomial symbols $H(x,\xi)$. There are at least two ways to do this which, however, lead to equivalent results. First, one can use *noncommutative analysis*, the general theory of functions of noncommuting operators (Maslov 1973); see also a popular exposition in Nazaikinskii, Sternin and Shatalov (1995). In this framework, one represents the symbol $H(x,\xi)$ in the form

$$H(x,\xi) = \left(\frac{-i}{2\pi}\right)^{n/2} \int \widetilde{H}(x,y)e^{-iy\xi}\,dy, \tag{3.4}$$

where $\widetilde{H}(x,y)$ is the inverse Fourier transform (in the sense of distributions) of the function $H(x,\xi)$ with respect to the variable $\xi$, and then replaces $\xi$ on the right-hand side in (3.4) by the vector operator $-i\partial/\partial x$ (with regard to the adopted ordering of operators). Thus

$$H\left(x, -i\frac{\partial}{\partial x}\right) \overset{\text{def}}{=} \left(\frac{-i}{2\pi}\right)^{n/2} \int \widetilde{H}(x,y)e^{-y\frac{\partial}{\partial x}}\,dy. \tag{3.5}$$

The expression on the right-hand side in (3.5) is meaningful, since the operator exponential is well defined as the translation operator

$$e^{-y\frac{\partial}{\partial x}}u(x) = u(x-y).$$

The integrand in (3.5) is strongly continuous in each Sobolev space, and the integral is treated in the sense of distributions. (For the case of an operator-valued integrand, this is known as the *Pettis integral.*)

◀ Note that the formulas (3.1) and (3.5) give the same result for polynomial symbols (3.3). Indeed, the Fourier transform of a polynomial is a linear combination of the delta function at zero and its derivatives, and so the integral amounts to a sum of derivatives of the operator exponential at zero, which are just the differentiation operators. ▶

Second, one can recall the relationship between the Fourier transforms of a function and its derivatives and rewrite (3.1) in the equivalent form

$$\left[H\left(x,-i\frac{\partial}{\partial x}\right)u\right](x)=\left(\frac{i}{2\pi}\right)^{n/2}\int H(x,\xi)e^{ix\xi}\,\widetilde{u}(\xi)\,d\xi, \tag{3.6}$$

where $\widetilde{u}(\xi)$ is the Fourier transform of $u$. The right-hand side of (3.6) is well defined without the assumption that $H(x,\xi)$ is a polynomial. Specifically, it is readily seen that the formulas (3.5) and (3.6) give the same result if the function $H(x,\xi)$ decays sufficiently rapidly as $|\xi|\to\infty$ together with sufficiently many derivatives (so that all integrals in question converge absolutely). However, in fact, both formulas are well defined and give the same result for a much wider symbol class, in particular, for *tempered symbols*, which grow at infinity together with all derivatives no faster than some power of $|\xi|$, depending on the symbol (see Maslov 1973).

THEOREM 3.2. *Suppose that for some $m\in\mathbb{R}$ the symbol $H(x,\xi)$ satisfies the estimates*

$$\left|\frac{\partial^{|\alpha|+|\beta|}H(x,\xi)}{\partial x^{\alpha}\partial\xi^{\beta}}\right|\le C_{\alpha\beta}(1+|\xi|)^{m-|\beta|},\quad (x,\xi)\in\mathbb{R}^{2n}, \tag{3.7}$$

*for arbitrary multi-indices $\alpha$ and $\beta$ with constants $C_{\alpha,\beta}$ that depend only on $\alpha$ and $\beta$. Then the operator $H(x,-i\partial/\partial x)$ is well defined by either of the formulas* (3.5) *and* (3.6) *and is continuous in the Sobolev spaces*

$$H\left(x,-i\frac{\partial}{\partial x}\right):H^{s}(\mathbb{R}^{n})\longrightarrow H^{s-m}(\mathbb{R}^{n}),\quad s\in\mathbb{R}. \tag{3.8}$$

◀ This well-known theorem can be proved in several ways. One, classical method is based on the so-called *Calderón lemma* (Hörmander 1985*a*, Theorem 18.1.13). Another proof involves noncommutative analysis. Namely, Theorem 3.2 is a very special case of Theorem IV.6 in Nazaikinskii, Sternin and Shatalov (1995). ▶

We denote the class of symbols satisfying the estimates (3.7) by $S^m(\mathbb{R}^{2n})$.

DEFINITION 3.3. Let $H(x,\xi)\in S^m(\mathbb{R}^{2n})$. Then the operator $H(x,-i\partial/\partial x)$ given by the formula (3.5) or (3.6) is called the *pseudodifferential operator with (complete) symbol $H(x,\xi)$*.

Thus Theorem 3.2 establishes the boundedness of the classical pseudodifferential operators in Sobolev spaces.

One of the main results of $\psi$DO theory is the *composition formula*, which states that the product of two $\psi$DO is again a $\psi$DO and permits one to compute the symbol of the product.

THEOREM 3.4. *Let $H_1(x,\xi) \in S^m(\mathbb{R}^{2n})$ and $H_2(x,\xi) \in S^l(\mathbb{R}^{2n})$. The product*

$$A = H_1\left(x, -i\frac{\partial}{\partial x}\right) \circ H_2\left(x, -i\frac{\partial}{\partial x}\right) \tag{3.9}$$

*of pseudodifferential operators is again a pseudodifferential operator, and its symbol $A(x,\xi)$ has the following asymptotic expansion as $|\xi| \to \infty$:*

$$A(x,\xi) \simeq \sum_{k=0}^{\infty}(-i)^k \sum_{|\alpha|=k} \frac{1}{\alpha!}\frac{\partial^\alpha H_1(x,\xi)}{\partial \xi^\alpha}\frac{\partial^\alpha H_2(x,\xi)}{\partial x^\alpha}. \tag{3.10}$$

*More precisely,*

$$A(x,\xi) - \Sigma_N(x,\xi) \in S^{m+l-N-1}(\mathbb{R}^{2n}),$$

*where $\Sigma_N$ is the $N$th partial sum of the outer series on the right-hand side in* (3.10).

Thus the symbol of a product of $\psi$DO is just the product of the symbols of the factors in the leading term ($k = 0$).

◀ For the case of compactly supported symbols (or, more generally, symbols stabilizing as $|x| \to \infty$), this is essentially the composition theorem in (Kohn and Nirenberg 1965). For general symbols of the class $S^m(\mathbb{R}^{2n})$, this theorem can also be found in numerous places (e.g., Hörmander 1985*a*, Taylor 1981). In noncommutative analysis, the starting point of the proof is the composition formula

$$A(x,\xi) = H_1\left(x, \xi - i\frac{\partial}{\partial x}\right)\left(H_2(x,\xi)\right). \tag{3.11}$$

From this, the formula (3.10) can be obtained by expanding the first factor in a Taylor series in powers of $-i\partial/\partial x$ with rather routine estimates of the remainder. Note that the series terminates without any remainder for the case in which $H_1$ is a differential operator, so that in this case the theorem is trivial. ▶

Hörmander's class $S^m(\mathbb{R}^{2n}) \equiv S^m_{1,0}(\mathbb{R}^{2n})$ of symbols satisfying the estimates (3.7) is unnecessarily wide for most problems in elliptic theory, and one usually exploits only the subset $S^m_{cl}(\mathbb{R}^{2n})$ of *classical* symbols possessing an asymptotic expansion in homogeneous functions as $|\xi| \to \infty$:

$$H(x,\xi) \simeq \sum_{j=0}^{\infty} H_{m-j}(x,\xi), \tag{3.12}$$

where $H_{m-j}(x,\xi)$ is homogeneous of degree $m-j$ in $\xi$:

$$H_{m-j}(x,\lambda\xi) = \lambda^{m-j} H_{m-j}(x,\xi), \quad \lambda > 0, \quad |\xi| \neq 0.$$

The expansion (3.12) means that

$$H(x,\xi) - \sum_{j=0}^{N} H_{m-j}(x,\xi) \in S^{m-N-1}(\mathbb{R}^{2n})$$

for large $|\xi|$.

In particular, almost inverses of elliptic differential operators are *classical* $\psi$*DO*, i.e., $\psi$DO with classical symbols.

DEFINITION 3.5. The function $H_m(x,\xi)$ is called the *principal symbol* of the operator

$$\widehat{H} = H\left(x, -i\frac{\partial}{\partial x}\right)$$

and is denoted by $\sigma(\widehat{H})(x,\xi)$.

*Remark* 3.6. In the following, by writing $\sigma(x, -i\partial/\partial x)$, where $\sigma$ is a function smooth for $\xi \neq 0$ and homogeneous of degree $m$ in $\xi$, we mean an (arbitrary) operator $H(x, -i\partial/\partial x)$, where $H(x,\xi)$ is some symbol in $S^m_{cl}(\mathbb{R}^{2n})$ whose principal part is equal to $\sigma$. One can define such a symbol, e.g., by the formula $H(x,\xi) = \psi(\xi)\sigma(x,\xi)$, where $\psi(\xi)$ is an excision function vanishing near zero and equal to unity at infinity.

We shall see in the next subsection how one can transfer the $\psi$DO theory from Euclidean space to manifolds.

### 3.2.2. Definition of pseudodifferential operators on a manifold

The construction of $\psi$DO in $\mathbb{R}^n$ was briefly described in Section 3.2.1. Now assume that we wish to consider $\psi$DO on a smooth compact manifold $M$ rather than in $\mathbb{R}^n$. How we define them in this case? Let us proceed as follows. We cover the manifold $M$ by coordinate neighborhoods and define $\psi$DO by formulas like (3.6) in each neighborhood and then try to paste the local definitions together. To this end, it is vital to study the behavior of $\psi$DO under changes of coordinates. It turns out (e.g. Hörmander 1965) that changes of variables take pseudodifferential operators to pseudodifferential operators modulo operators with smooth kernel, and moreover, for classical $\psi$DO the principal symbol behaves as an invariantly defined function on the cotangent bundle $T_0^*M$ of $M$ without the zero section. More precisely, the following assertion holds.

PROPOSITION 3.7. *Let $U, V \subset \mathbb{R}^n$ be given domains, and let $\psi : U \longrightarrow V$ be a diffeomorphism. Next, let $H(x, \xi) \in S^m(\mathbb{R}^{2n})$ be a symbol such that*[1] $\operatorname{supp}_x H \Subset U$. *Set*

$$\widehat{H} = H\left(x, -i\frac{\partial}{\partial x}\right).$$

*Take an arbitrary function $\chi(x) \in C_0^\infty(U)$ such that $\chi(x)H(x,\xi) = H(x,\xi)$. Then the operator*

$$\widehat{G} = (\psi^*)^{-1} \circ \widehat{H} \circ \chi(x) \circ \psi^* \tag{3.13}$$

*is also a pseudodifferential operator of order $\leq m$, i.e.,*

$$\widehat{G} = G\left(x, -i\frac{\partial}{\partial x}\right), \quad G \in S^m(\mathbb{R}^{2n}), \tag{3.14}$$

*and moreover, the symbol $G(x,\xi)$ satisfies the relation*

$$G\left(\psi(x), \left({}^t\frac{\partial \psi}{\partial x}\right)^{-1} \xi\right) - H(x,\xi) \in S^{m-1}(\mathbb{R}^{2n}). \tag{3.15}$$

In particular, the formula (3.15) acquires an especially simple form for classical symbols admitting the expansion (3.12) into homogeneous functions:

$$\sigma(\widehat{G})\left(\psi(x), \left({}^t\frac{\partial \psi}{\partial x}\right)^{-1} \xi\right) = \sigma(\widehat{H})(x,\xi). \tag{3.16}$$

We see that the principal symbols indeed are transformed under changes of variables as functions on the cotangent bundle.

Now we can give the following definition of a pseudodifferential operator on $M$.

DEFINITION 3.8. A *classical pseudodifferential operator* with principal symbol $D(x,\xi)$, $(x,\xi) \in T_0^*M$, on a smooth compact manifold $M$ is an operator $\widehat{D}$ in the scale of Sobolev spaces on $M$ such that

1) $\varphi \widehat{D} \psi$ is an integral operator with smooth kernel on $M$ whenever $\varphi$ and $\psi$ are smooth functions on $M$ with disjoint supports

2) if $\varphi$ and $\psi$ are smooth functions on $M$ supported in the same coordinate neighborhood $U$, then $\varphi \widehat{D} \psi$ is a classical pseudodifferential operator in $\mathbb{R}^n$ (here we identify $U$ with a subset in $\mathbb{R}^n$ using the coordinate mapping) with principal symbol $\varphi(x)\psi(x)D(x,\xi)$

An operator with a given principal symbol $D(x,\xi)$ (which is a homogeneous function of order $m$ on $T_0^*M$) can be constructed, e.g., as follows. Cover $M$ with

[1]Here $\operatorname{supp}_x$ is the projection of the support onto the $x$-space.

coordinate neighborhoods $U_j$, $j = 1, \dots, N$, and consider a smooth partition of unity

$$1 = \sum_{j=1}^{N} \chi_j(x)^2$$

subordinate to this cover. We define a pseudodifferential operator with symbol $D(x,\xi)$ by the formula

$$D\left(x, -i\frac{\partial}{\partial x}\right) \stackrel{\text{def}}{=} \sum_{j=1}^{N} (\chi_j D)\left(x, -i\frac{\partial}{\partial x}\right) \circ \chi_j(x),$$

where the $j$th summand is determined in the local coordinates of the chart $U_j$ as the composition of the operator of multiplication by $\chi_j(x)$ (which localizes the function to be acted upon by the $\psi$DO into the chart $U_j$) and a pseudodifferential operator with symbol $\chi_j(x)D(x,\xi)$ in $\mathbb{R}^n$; cf. Remark 3.6. Here the symbol is expressed via the canonical coordinates in $T^*U_j$, which we denote by the same letters $(x,\xi)$.

◀ It follows by standard technical reasoning from theorems on the composition of $\psi$DO and their behavior under changes of coordinates that the operator thus constructed satisfies the conditions in Definition 3.8. ▶

*Remark* 3.9. The operator is not uniquely determined by its principal symbol. For example, the operator given by the above-mentioned construction depends also on the choice of the cover and the partition of unity. However, all operators with the same symbol differ by lower-order operators.

Pseudodifferential operators on $M$ obey the following composition law similar to Theorem 3.4.

THEOREM 3.10. *The principal symbol of the product of pseudodifferential operators is equal to the product of their principal symbols.*

*The proof* is trivial. □

### 3.2.3. Hörmander's definition of pseudodifferential operators

Along with the above approach to the definition of $\psi$DO, which is due to Kohn and Nirenberg (1965), there is another approach devised by Hörmander (1965). In this approach, classical pseudodifferential operators are defined (from the very beginning, on a manifold $M$!) in an invariant way as continuous linear operators

$$P : C^\infty(M) \longrightarrow C^\infty(M)$$

such that if $f$ and $g$ are smooth functions, $f$ is compactly supported, $g$ is real-valued, and $dg \neq 0$ on $\operatorname{supp} f$, then as $\lambda \to \infty$ one has an asymptotic expansion

$$e^{-i\lambda g} P\big(fe^{i\lambda g}\big) \sim \sum_{j=0}^{\infty} P_j(f,g)\lambda^{s_j} \tag{3.17}$$

locally uniformly in $g$ for some monotone decreasing sequence of real numbers $s_j \to -\infty$, where the coefficients $P_j(f,g)$ are smooth functions on $M$.

PROPOSITION 3.11 (Hörmander 1965). *One has*

$$P_0(f,g) = H(dg)f,$$

*where $H(\xi)$ is a homogeneous function of degree $s_0$ on the cotangent bundle of $M$ without the zero section.*

In this approach, the function $H(\xi)$ is called the *principal symbol* of the $\psi$DO in question.

Hörmander proved that in local coordinates both definitions (for $s_j = m - j$) result in the same class of classical $\psi$DO of order $m$, and moreover, the principal symbol (of order $m$) of a $\psi$DO coincides in local coordinates with the function $H_m(x,\xi)$ in the asymptotic expansion (3.12).

Hörmander's definition possesses a number of interesting properties, e.g.,

1) pseudodifferential operators are defined intrinsically (i.e., as operators possessing certain properties) rather than constructively (i.e., by some recipe allowing one to construct them)

2) the definition uses no coordinate systems, so that the invariant nature of the principal symbol is immediate

### 3.2.4. Basic properties of pseudodifferential operators

Now let us summarize the main properties of the theory of elliptic $\psi$DO on smooth compact manifolds.

Let $M$ be a smooth compact manifold. Classical $\psi$DO of integer order on $M$ form a *filtered algebra*

$$L^\infty(M) = \bigcup_{k=-\infty}^{\infty} L^k(M),$$

where $L^k(M)$ is the space of $\psi$DO of order $\leq k$. For an operator $D \in L^k(M)$, there is a well-defined *principal symbol* (or, more precisely, principal symbol of order $k$) $\sigma(D)$, which is a homogeneous function of degree $k$ on the cotangent bundle $T_0^*M$ of $M$ without the zero section. Next, for each $k \in \mathbb{Z}$ there is an exact sequence

$$0 \longrightarrow L^{k-1}(M) \longrightarrow L^k(M) \xrightarrow{\ \sigma\ } \mathcal{O}^k(T_0^*M) \longrightarrow 0,$$

where $\mathcal{O}^k(T_0^*M)$ is the space of homogeneous functions of degree $k$ on $T_0^*M$.

Furthermore,

$$L^0(M) \cap \mathcal{K}_s = L^{-1}(M)$$

for all $s$, where $\mathcal{K}_s$ is the space of compact operators in the Sobolev space $H^s(M)$. Thus the space

$$\mathcal{O}^0(T_0^*M) \cong L^0(M)/L^{-1}(M) = L^0(M)/L^0(M) \cap \mathcal{K}_s$$

of principal symbols is isomorphic to the Calkin algebra corresponding to the algebra $L^0(M)$.

We see that the almost invertibility of a $\psi$DO (invertibility modulo compact operators) is equivalent to the invertibility of the principal symbol in the Calkin algebra, whence the main role of the principal symbol in elliptic theory on smooth compact manifolds.

*Remark* 3.12. The notion of principal symbol can be introduced for nonclassical $\psi$DO as well. However, the principal symbols are no longer functions on $T_0^*M$ in this case; instead, they are elements of the quotient space $S^m(T^*M)/S^{m-1}(T^*M)$, where $S^m(T^*M)$ is the space of smooth functions on $T^*M$ satisfying the estimates (3.7) in local coordinates.

*Remark* 3.13. So far we have considered operators with scalar symbols, i.e., operators acting in function spaces on the manifold $M$. In a similar way, one constructs pseudodifferential operators acting in spaces of sections of finite-dimensional vector bundles over $M$. In trivializing neighborhoods of these bundles, the symbol of a $\psi$DO is represented by a $k \times l$ matrix function, where $k$ and $l$ are the dimensions of the bundles in whose sections the operator acts.

Globally, the principal symbol of a $\psi$DO

$$A : C^\infty(M, E) \longrightarrow C^\infty(M, F)$$

acting in sections of vector bundles $E$ and $F$ over a manifold $M$ is naturally interpreted as a homomorphism

$$\sigma(A) : \pi^*E \longrightarrow \pi^*F$$

of the lifts of these bundles to the cotangent bundle of $M$ without the zero section. Here

$$\pi : T_0^*M \longrightarrow M$$

is the natural projection.

### 3.2.5. Parameter-dependent operators

In the classical theory, one also considers *parameter-dependent* $\psi$DO, where on the symbolic level the parameter (or parameters) occurs as yet another covariable. Let us briefly explain this notion. The reader can find a detailed exposition, say, in Shubin (1985). Here we are concise on the subject.

We start from differential operators. A family of differential *operators of order* $m$ *with parameter* $p \in \mathbb{C}$ is a family $A(p)$ obtained from a translation-invariant $m$th-order differential operator $\widehat{A}$ on the cylinder $X \times \mathbb{R}$ by the Fourier transform with respect to the variable $t \in \mathbb{R}$. (Here $t$ and $p$ are dual variables.) Thus if the operator on the cylinder is

$$\widehat{A} = P\left(x, -i\frac{\partial}{\partial x}, -i\frac{\partial}{\partial t}\right)$$

(note that the coefficients of the operator are independent of $t$ by virtue of translation invariance), then the corresponding family is

$$A(p) = P(x, -i\partial/\partial x, p).$$

In other words, we replace the differentiation operator $-i\partial/\partial t$ by the variable $p$. Conversely,

$$\widehat{A} = A\left(-i\frac{\partial}{\partial t}\right);$$

that is, the family $A(p)$ serves as an operator-valued symbol of the operator $\widehat{A}$ with respect to the quantization

$$p \longmapsto -i\frac{\partial}{\partial t}.$$

(We shall learn more about operator-valued symbols in the next section.) We have already occasionally dealt with such families in the preceding chapters; e.g., cf. Remark 2.12.

The *principal symbol* of the parameter-dependent family $A(p)$ is by definition just the principal symbol $\sigma(\widehat{A})$ of the original operator.

Now we proceed to parameter-dependent pseudodifferential operators. We do not restrict ourselves to the case of a one-dimensional parameter $p$ and assume that $p \in V$, where $V$ is a finite-dimensional vector space (or, more generally, a cone in a finite-dimensional vector space). A parameter-dependent operator on a manifold $X$ is said to be *smoothing* if, viewed as a function of the parameter $p$, it belongs to the Schwartz space of smooth functions ranging in the space of smoothing operators on $X$ and rapidly decaying at infinity. A *parameter-dependent* $\psi$*DO* $A$ of order $m$ is locally representable in the form

$$A \equiv A(p) = f\left(\overset{2}{x}, -i\overset{1}{\frac{\partial}{\partial x}}, p\right)$$

modulo smoothing parameter-dependent operators, where $f(x,\xi,p)$, *a parameter-dependent symbol of order* $m$, satisfies the estimates

$$\left|\partial_x^\alpha \partial_{\xi,p}^\beta f(x,\xi,p)\right| \le C_{\alpha,\beta,\gamma}(1+|\xi|^2+p^2)^{m-|\beta|}$$

for arbitrary multi-indices $\alpha, \beta \in \mathbb{Z}_+^n$.

By analogy with the case of a single operator, one defines the notion of a *classical* parameter-dependent pseudodifferential operator. The complete symbol $f(x,\xi,p)$ of such an operator admits an asymptotic expansion

$$f(x,\xi,p) \sim \sum_{j\ge 0} f_{m-j}(x,\xi,p)$$

with terms $f_{m-j}$ homogeneous in the covariables $(\xi,p)$ of degree $m-j$, where $j \in \mathbb{Z}_+$. The function $f_m(x,\xi,p)$, which is invariantly defined on

$$(T^*X \times V) \setminus \{\xi = 0, p = 0\}$$

is called the *principal symbol* of the parameter-dependent pseudodifferential operator and is denoted by $\sigma(A)$.

Note that a parameter-dependent pseudodifferential operator $A(p)$ is also a pseudodifferential operator in the usual sense for each given value of the parameter $p$. In the case of classical pseudodifferential operators, for the principal symbol one has

$$\sigma(A(p_0)) = \sigma(A)(x,\xi,0)$$

for each $p_0 \in V$.

A parameter-dependent $\psi$DO is often also called a *$\psi$DO with parameter in the sense of Agranovich–Vishik.*

*Remark* 3.14. In what follows, we often speak, by abuse of language, of parameter-dependent pseudodifferential operators $A(p)$ with parameter $p \in \mathcal{L}_\gamma$, where $\mathcal{L}_\gamma \subset \mathbb{C}$ is the weight line

$$\mathcal{L}_\gamma = \{p \in \mathbb{C} \mid \operatorname{Im} p = \gamma\}.$$

What is really meant is that $A(p) = A(\operatorname{Re} p + i\gamma)$ is a parameter-dependent pseudodifferential operator with parameter $\operatorname{Re} p \in \mathbb{R}$.

For a closed manifold $X$, by $\Psi_p(X)$ we denote the set of classical parameter-dependent pseudodifferential operators with parameter $p$. (The set $V$ where the parameter $p$ varies should, of course, be specified in each particular case.) The set $\Psi_p(X)$ is an algebra with respect to pointwise addition and multiplication of operator families. This algebra is filtered by the subspaces $\Psi_p^m(X)$ of operators of order $\le m$.

## 3.3. Operators in Sections of Hilbert Bundles

We shall consider pseudodifferential operators acting in sections of Hilbert bundles over a smooth compact manifold $X$ without boundary.

DEFINITION 3.15. A *Hilbert bundle* over a smooth compact manifold $X$ is a vector bundle $E \longrightarrow X$ whose fiber is a Hilbert space $H$ and whose transition functions are operator norm smooth mappings defined on intersections of trivializing neighborhoods and ranging in the group $\mathbb{U}(H)$ of unitary operators in $H$.

In the following, we consider only the case in which the space $H$ is infinite-dimensional (and separable). Then the group $\mathcal{U}(H)$ is contractible by Kuiper's theorem (see Atiyah 1989), and consequently, any Hilbert bundle with fiber $H$ is trivial. Hence we assume without loss of generality that the bundles in question have the form $X \times H \longrightarrow X$.

The main examples arise in applications if one describes Hilbert function spaces on the total space $M$ of a locally trivial bundle

$$\pi : M \longrightarrow X \tag{3.18}$$

with compact base $X$ as spaces of Hilbert-valued functions (or distributions) on $X$. It is assumed that these spaces behave "along $X$" very similarly to usual Sobolev spaces. (There are no distinguished or degenerate directions.) It is convenient to define such spaces in the Fourier transform as *direct integrals of Hilbert spaces*. In this section, we define pseudodifferential operators in such spaces and establish boundedness and composition theorems for such operators.

### 3.3.1. Specific features of the infinite-dimensional case

At first glance, the problem of constructing a calculus of $\psi$DO acting in sections of Hilbert bundles seems to be pretty trivial. Indeed, one should only assume in the general definition (3.6) that the symbol $H(x, \xi)$ is not a scalar function but ranges in the algebra $\mathcal{B}(H)$ of bounded operators in $H$ (an *operator-valued symbol*). If this function is smooth and satisfies, say, the estimates (3.7), where the operator norm should be used instead of the absolute value on the left-hand side, then the corresponding operator will be well defined as a bounded operator in Sobolev spaces:

$$H\left(x, -i\frac{\partial}{\partial x}\right) : H^s(\mathbb{R}^n, H) \longrightarrow H^{s-m}(\mathbb{R}^n, H), \quad s \in \mathbb{R}.$$

Under changes of coordinates, the symbol still behaves "well," i.e., is transformed modulo lower-order symbols as a function on the cotangent bundle. Thus one can define $\psi$DO globally on the manifold, and a pseudodifferential operator is uniquely determined by its principal symbol modulo lower-order operators. However, there are two facts showing that this straightforward generalization of the definitions to the operator-valued case is not completely adequate.

1. In contrast to the case of finite-dimensional bundles, the embeddings

$$H^s(X, H) \subset H^t(X, H), \quad s > t,$$

of Sobolev spaces *are not compact.* Consequently, neither are negative-order operators. Thus a pseudodifferential operator in this setting *is not uniquely determined modulo compact operators by its principal symbol.* Accordingly, one cannot verify whether an operator is Fredholm if only the principal symbol is known. It follows that such a calculus is useless in elliptic theory.

◀ Indeed, for example, an operator of order $-1$ with principal symbol $|\xi|^{-1}I$, where $I$ is the identity operator in $H$, is not compact in $L^2(X, H)$. ▶

2. Even the simplest examples show that it is of no interest to consider classical symbols (i.e., symbols asymptotically homogeneous in the momentum variables $\xi$). Indeed, an example of this sort already arises if we consider conventional $\psi$DO on the product of two manifolds, which is a special case of the bundle (3.18). Suppose that $M = X \times Y$, where $X$ and $Y$ are smooth compact manifolds without boundary. Then pseudodifferential operators on $M$ can be treated as $\psi$DO on $X$ whose symbols are, in turn, pseudodifferential operators on $Y$. In local coordinates, we can write

$$H\left(x, y, -i\frac{\partial}{\partial x}, -i\frac{\partial}{\partial y}\right) = G\left(x, -i\frac{\partial}{\partial x}\right),$$

where

$$G(x, \xi) = H\left(x, y, \xi, -i\frac{\partial}{\partial y}\right).$$

Even if the symbol $H(x, y, \xi, \eta)$ is asymptotically homogeneous with respect to the *pair* $(\xi, \eta)$, the symbol $G(x, \xi)$ cannot be asymptotically homogeneous in $\xi$!

Thus the "naive" approach described above should be revised. We should abandon the homogeneity requirement for the symbols. We should also invent adequate spaces of sections of Hilbert bundles. On the other hand, we need to subject the symbols to some additional conditions guaranteeing that an operator with symbol of "negative" order is compact. Such additional conditions indeed exist; they are known as the *compact fiber variation* conditions.

Starting from the next subsection, we present the theory of $\psi$DO with symbols of compact fiber variation. A version of such a theory for operators acting in spaces of $L^2$ sections of Hilbert bundles was developed by Luke (1972).

### 3.3.2. Spaces of sections of Hilbert bundles

First, we consider Hilbert-valued functions on $\mathbb{R}^n$.

***Families of norms of tempered growth.*** Throughout the following, we use the notation $\langle\xi\rangle = (1 + |\xi|^2)^{1/2}$.

Let $H$ be a Hilbert space with norm $\|\cdot\|_H$ and inner product $(\cdot, \cdot)_H$. Suppose that it is equipped with a family $\{\|\cdot\|_\xi\}$ of Hilbert norms depending on the parameter $\xi \in \mathbb{R}^n$, equivalent to the original norm $\|\cdot\|_H$, and satisfying the following conditions:

- $\|u\|_\xi$ is a measurable function of $\xi$ for each $u \in H$
- there exist constants $C$ and $N$ such that

$$\|u\|_\eta \leq C\left(\frac{\langle\eta\rangle}{\langle\xi\rangle} + \frac{\langle\xi\rangle}{\langle\eta\rangle}\right)^N \|u\|_\xi \quad \text{for any } \xi, \eta \in \mathbb{R}^n \text{ and for any } u \in H \tag{3.19}$$

DEFINITION 3.16. A family of norms satisfying the above-mentioned conditions will be called a *tempered family of norms*.

***Direct integral of Hilbert spaces.*** We denote the space $H$ equipped with the norm $\|\cdot\|_\xi$ by $H_\xi$ and define the *direct integral* of the family

$$\mathfrak{H} = \{H_\xi\} \tag{3.20}$$

of Hilbert spaces by the formula

$$\int H_\xi \, d\xi \stackrel{\text{def}}{=} \{u : \mathbb{R}^n \longrightarrow H \mid u(\xi) \text{ is measurable and } \|u(\xi)\|_\xi \text{ is square integrable}\}. \tag{3.21}$$

*Remark* 3.17. (1) Strong and weak measurability are the same in a separable Hilbert space, so we do not specify the kind of measurability in the definition.

(2) If $u(\xi)$ is measurable, then so is $\|u(\xi)\|_\xi$, and hence (3.21) is well defined.

The space $\int H_\xi \, d\xi$ is equipped with the natural norm

$$\|u\|_\int = \left\{\int \|u(\xi)\|_\xi^2 \, d\xi\right\}^{1/2}, \tag{3.22}$$

which makes it a Hilbert space.

LEMMA 3.18. *The space $\int H_\xi \, d\xi$ is the closure of the Schwartz space $\mathcal{S}(\mathbb{R}^n, H)$ in the norm* (3.22).

*The proof* is standard. □

We are actually interested in the space

$$\mathcal{H} = \mathcal{F}^{-1} \int H_\xi \, d\xi$$

obtained from the space $\int H_\xi \, d\xi$ by the inverse Fourier transform. It follows from the preceding lemma that $\mathcal{H}$ is the completion of the Schwartz space $\mathcal{S}(\mathbb{R}^n, H)$ with respect to the norm

$$\|u\| = \|\widetilde{u}\|_\int ,$$

where $\widetilde{u}$ is the Fourier transform of $u$.

EXAMPLE 3.19. (i) If $H = \mathbb{C}$ and $\|u\|_\xi = \langle \xi \rangle \, |u|$, then $\mathcal{H} = H^s(\mathbb{R}^n)$ is the ordinary Sobolev space in $\mathbb{R}^n$.

(ii) If $H = H^s(M)$, where $M$ is a smooth compact manifold, and

$$\|u\|_\xi = \left\| \left( \frac{1 + |\xi|^2 - \triangle}{1 - \triangle} \right)^{s/2} u \right\|_{H^s(M)} ,$$

where $\triangle$ is the Beltrami–Laplace operator on $M$, then $\mathcal{H} = H^s(\mathbb{R}^n \times M)$ is the Sobolev space on $\mathbb{R}^n \times M$. (The tempered growth condition for this family of norms can be checked easily.)

***Equivalent tempered families of norms.*** A tempered family of norms in $H$ can always be given by the formula

$$\|u\|_\xi = \|A(\xi)u\|_H , \tag{3.23}$$

where $A(\xi)$ is a strongly measurable family of bounded operators in $H$.

The operator $A(\xi)$ is uniquely determined if we require it to be positive and self-adjoint. However, these requirements are not necessary in applications; moreover, one can proceed to equivalent norms. The following assertion is obvious.

LEMMA 3.20. *Two tempered families of norms determined by operator families $A(\xi)$ and $B(\xi)$ in accordance with* (3.23) *determine the same* (*up to norm equivalence*) *space $\mathcal{H}$ if and only if the operator families $A(\xi)B^{-1}(\xi)$ and $B(\xi)A^{-1}(\xi)$ are bounded uniformly with respect to $\xi$.*

This lemma permits one to give quite remarkable equivalent expressions for the norm. For example, let $\varkappa_\lambda$ be the group of dilations in $H^s(\mathbb{R}^k)$ acting by the formula

$$\varkappa_\lambda u(x) = \lambda^{k/2} u(\lambda x), \quad \lambda \in \mathbb{R}_+.$$

The operator families

$$A(\xi) = \left(\frac{1+|\xi|^2-\triangle}{1-\triangle}\right)^{s/2}, \quad B(\xi) = \langle\xi\rangle^s \varkappa_{\langle\xi\rangle}^{-1}$$

satisfy the assumptions of the lemma, and hence the norms associated with these families are equivalent. Moreover, $B(\xi)$ defines the space $\mathcal{W}^s(\mathbb{R}^n, H^s(\mathbb{R}^k))$. With regard to Example 3.19, we obtain the well-known identity

$$H^s(\mathbb{R}^{n+k}) = \mathcal{W}^s(\mathbb{R}^n, H^s(\mathbb{R}^k)).$$

(Recall that the meaning of the right-hand side is given by Definition 2.18.)

***Globalization.*** So far, we have defined spaces of Hilbert-valued functions on $\mathbb{R}^n$. To proceed to a manifold, we should study how the norms are affected by changes of variables. Let $\|\cdot\|_\xi$ be a given tempered family of norms in a Hilbert space $H$, and let $\|\cdot\|$ be the norm in the space $\mathcal{H}$ corresponding to the family $\mathfrak{H} = \{H_\xi\}$.

PROPOSITION 3.21. *Let $U \subset \mathbb{R}^n$ be a bounded domain, and let $f : U \longrightarrow V \subset \mathbb{R}^n$ be a diffeomorphism extendible into some neighborhood of the closure $\overline{U}$ of the domain $U$. Then there exist positive constants $C_1$ and $C_2$ such that the inequalities*

$$C_1 \|u\| \le \|f^*u\| \le C_2 \|u\|$$

*hold for each smooth $H$-valued function $u(x)$ supported in $V$.*

*Proof.* The assumptions of the proposition remain valid if we interchange $U$ and $V$, and so it suffices to prove the right inequality. Let $u$ be an $H$-valued function supported in $V$, let $v(\xi)$ be its Fourier transform, and let $w(\eta)$ be the Fourier transform of $f^*u$. Next, let $\rho(y)$ be a smooth compactly supported function whose support is contained in the domain of the extended diffeomorphism $f$ and which is equal to unity in $U$. Then

$$w(\eta) = \left(\frac{1}{2\pi}\right)^n \iint e^{i(f(y)\xi - y\eta)} \rho(y) v(\xi)\, d\xi dy. \tag{3.24}$$

Using the operator

$$L = \left(1+|f'(y)\xi-\eta|^2\right)^{-1}\left(1 - i(f'(y)\xi-\eta)\frac{\partial}{\partial y}\right),$$

which satisfies the relation

$$L e^{i(f(y)\xi - y\eta)} = e^{i(f(y)\xi - y\eta)},$$

and integrating by parts $M$ times in (3.24), we obtain

$$w(\eta) = \left(\frac{1}{2\pi}\right)^n \iint e^{i(f(y)\xi - y\eta)} [({}^tL)^M \rho(y)] v(\xi)\, d\xi dy, \tag{3.25}$$

where ${}^tL$ is the transpose of $L$. Moreover, the estimate

$$|({}^tL)^M \rho(y)| \le \text{const}\left[1 + |f'(y)\xi - \eta|^2\right]^{-M/2}$$

is obviously valid. In the integral (3.25), we make the change of variables

$$\xi = \xi(\eta, t) \equiv f'(y)^{-1}(t + \eta);$$

then it becomes

$$w(\eta) = \left(\frac{1}{2\pi}\right)^n \iint e^{i(f(y)\xi(\eta,t) - y\eta)} \frac{({}^tL)^M \rho(y)}{\det f'(y)} v(f'(y)^{-1}(t + \eta))\, dtdy, \quad (3.26)$$

or

$$w(\,\cdot\,) = \int U(\,\cdot\,, t, y)\, dtdy,$$

where the integrand $U$ satisfies the estimate

$$\|U(\eta, t, y)\|_\eta \le \text{const}\, \langle t \rangle^{(N-M)/2} \left\| v(f'(y)^{-1}(t + \eta)) \right\|_{f'(y)^{-1}(t+\eta)}$$

and is compactly supported with respect to the variable $y$. (The extra factor $\langle t \rangle^N$ arises in the passage from $\|\,\cdot\,\|_\eta$ to $\|\,\cdot\,\|_{f'(y)^{-1}(t+\eta)}$ with regard to the estimates in the definition of a tempered family of norms.) Now we have

$$\begin{aligned}
\|U(\,\cdot\,, t, y)\|_{\int} &= \left\{\int \|U(\eta, t, y)\|_\eta^2\, d\eta\right\}^{1/2} \\
&\le \text{const}\, \langle t \rangle^{(N-M)/2} \left\{\int \left\| v(f'(y)^{-1}(t + \eta)) \right\|^2_{f'(y)^{-1}(t+\eta)} d\eta\right\}^{1/2} \\
&\le \text{const}\, \langle t \rangle^{(N-M)/2} \|v(\,\cdot\,)\|_{\int}\,.
\end{aligned}$$

Taking $M$ sufficiently large and integrating this with respect to $t$ and $y$, we obtain

$$\|w(\,\cdot\,)\|_{\int} \le \text{const}\, \|v(\,\cdot\,)\|_{\int}\,,$$

and the passage to the inverse Fourier transform completes the proof. □

PROPOSITION 3.22. *The operator of multiplication by a smooth compactly supported function on $X$ is continuous in the space $\mathcal{H}$.*

*Proof.* This proposition is obvious. However, we should also mention that it is a very special case of Theorem 3.28 below. □

Propositions 3.21 and 3.22 permit one to use a partition of unity subordinate to a coordinate cover on the manifold $X$ and introduce a well-defined space $\mathcal{H}(X)$ of $H$-valued functions on $X$ whose local model is the space $\mathcal{H}$.

DEFINITION 3.23. The space $\mathcal{H}(X)$ is the completion of $C_0^\infty(X, H)$ with respect to the norm

$$\|u\| = \left(\sum \|e_j u\|^2\right)^{1/2},$$

where $\{e_j\}$ is a partition of unity subordinate to a coordinate cover on $X$ and the norms on the right-hand side are computed in local coordinates.

Needless to say, the norm is independent of the choice of a partition of unity up to equivalence.

### 3.3.3. Symbols of compact fiber variation

The canonical coordinates on the cotangent bundle $T^*X$ will be denoted by $(x, \xi)$. For convenience, we fix some smooth norm $|\xi|$ in the fibers of $T^*X$.

DEFINITION 3.24. Let $\|\cdot\|_\xi^1$ and $\|\cdot\|_\xi^2$ be two tempered families of norms in Hilbert spaces $H_1$ and $H_2$, respectively, and let $\mathfrak{H}_1$ and $\mathfrak{H}_2$ be the corresponding families of spaces (3.20). By

$$S^0_{CV}(\mathbb{R}^{2n}) \equiv S^0_{CV}(\mathbb{R}^{2n}, \mathfrak{H}_1, \mathfrak{H}_2)$$

we denote the space of symbols $a(x, \xi)$ ranging in the space $\mathcal{B}(H_1, H_2)$ of bounded operators acting from $H_1$ to $H_2$, satisfying the estimates

$$\left\| \frac{\partial^{\alpha+\beta} a(x,\xi)}{\partial x^\alpha \partial \xi^\beta} : H_{1\xi} \longrightarrow H_{2\xi} \right\| \le C_{\alpha\beta} \langle \xi \rangle^{-|\beta|}, \quad |\alpha| + |\beta| = 0, 1, 2 \dots, \tag{3.27}$$

and having the property of *compact fiber variation*

$$a(x,\xi) - a(x,\widetilde{\xi}) \in \mathcal{K}(H_1, H_2) \quad \text{for} \quad \xi, \widetilde{\xi} \in T_x^* X, \tag{3.28}$$

where $\mathcal{K}(H_1, H_2) \subset \mathcal{B}(H_1, H_2)$ is the subspace of compact operators. The space

$$S^0_{CV}(T^*X) \equiv S^0_{CV}(T^*X, \mathfrak{H}_1, \mathfrak{H}_2)$$

is defined as the space of operator-valued symbols whose coordinate representatives satisfy the estimates (3.27).

In applications, one often deals with symbols with similar properties defined only outside the zero section of the cotangent bundle. Hence we introduce a space of such symbols.

Namely, by

$$S^0_{CV}(T_0^*X) \equiv S^0_{CV}(T_0^*X, \mathfrak{H}_1, \mathfrak{H}_2)$$

we denote the space of functions

$$a : T_0^*X \longrightarrow \mathcal{B}(H_1, H_2)$$

that satisfy the estimates (3.27) for $|\xi| > \varepsilon$ for each $\varepsilon > 0$ with constants $C_{\alpha\beta}$ depending on $\varepsilon$ and have the property (3.28) for $\xi, \widetilde{\xi} \neq 0$.

Elements of the spaces $S^0_{CV}(T^*X, \mathfrak{H}_1, \mathfrak{H}_2)$ and $S^0_{CV}(T^*_0X, \mathfrak{H}_1, \mathfrak{H}_2)$ are called *symbols of compact fiber variation* on $T^*X$ (respectively, on $T^*_0X$).

*Remark* 3.25. Condition (3.28) is equivalent to the condition

$$\frac{\partial a(x,\xi)}{\partial \xi} \in \mathcal{K}(H_1, H_2). \tag{3.29}$$

(The equivalence can be proved by integration from $\xi$ to $\widetilde{\xi}$.)

Symbols in $S^0_{CV}(T^*X)$ and $S^0_{CV}(T^*_0X)$ can be multiplied pointwise if the corresponding Hilbert spaces and families of norms fit.

*Remark* 3.26. The space $S^0_{CV}(T^*X)$ contains the subspace $J^{-1}_K(T^*X)$ of compact-valued symbols $a(x,\xi)$ satisfying the estimates

$$\left\| \frac{\partial^{\alpha+\beta} a(x,\xi)}{\partial x^\alpha \partial \xi^\beta} : H_{1\xi} \longrightarrow H_{2\xi} \right\| \leq C_{\alpha\beta} \langle \xi \rangle^{-1-|\beta|}, \quad |\alpha|+|\beta| = 0, 1, 2, \dots, \tag{3.30}$$

in any canonical coordinate system $(x, \xi)$.

The following lemma shows that a symbol of compact fiber variation defined outside the zero section of $T^*X$ can always be filled up to an everywhere defined symbol of compact fiber variation and that the fill-up is essentially unique.

LEMMA 3.27. *For any $f \in S^0_{CV}(T^*_0X)$, there exists a symbol $\widetilde{f} \in S^0_{CV}(T^*X)$ such that $f(x,\xi) = \widetilde{f}(x,\xi)$ for sufficiently large $|\xi|$. If $\widetilde{f}' \in S^0_{CV}(T^*X)$ is another symbol with the same property, then the difference $\widetilde{f} - \widetilde{f}'$ is compact-valued and has a compact support.*

*Proof.* The proof of this lemma, although technical, is from our point of view typical in dealing with symbols of this class.

1. The existence of $\widetilde{f}$. On the unit spheres

$$S^*_x = \{|\xi| = 1\} \subset T^*_xX,$$

we take an arbitrary smooth measure $d\mu(\xi)$ smoothly depending on $x$ with the property

$$\int_{S^*_x} d\mu(\xi) = 1$$

and define the desired symbol by averaging:

$$\widetilde{f}(x,\xi) = \chi(|\xi|) f(x,\xi) + (1 - \chi(|\xi|)) \int_{S^*_x} f(x,\xi')\, d\mu(\xi'), \tag{3.31}$$

where $\chi(t) \in C^\infty(\mathbb{R}_+)$ is an excision function that vanishes near zero and is equal to unity in a neighborhood of infinity. Then, obviously, $f(x,\xi) = \widetilde{f}(x,\xi)$ for sufficiently large $|\xi|$. Moreover, $f(x,\xi)$ is infinitely differentiable, satisfies the estimates (3.27), and enjoys the compact fiber variation property.

2. The uniqueness of $\widetilde{f}$ modulo compactly supported compact remainders is obvious. □

The symbol $\widetilde{f}$ constructed in Lemma 3.27 will be called a *fill-up* of $f$. Symbols from the space $S^0_{CV}(T^*_0X)$ play the main role for us, since such symbols (undefined for $\xi = 0$) arise in applications.

### 3.3.4. Pseudodifferential operators

***Local definition.*** For operator-valued symbols of compact fiber variation, pseudodifferential operators are defined locally in a standard way. The construction is literally the same as in the finite-dimensional case (cf. Section 3.2.2). Namely, one defines the $\psi$DO with symbol $D(x,\xi) \in S^0_{CV}(\mathbb{R}^{2n}, \mathfrak{H}_1, \mathfrak{H}_2)$ by the formula

$$D\left(\overset{2}{x}, -i\overset{1}{\frac{\partial}{\partial x}}\right)u(x) = \left(\frac{1}{2\pi}\right)^{n/2} \int e^{ix\xi} D(x,\xi)\widetilde{u}(\xi)\, d\xi, \quad u \in C_0^\infty(\mathbb{R}^n), \quad (3.32)$$

where $\widetilde{u}(\xi)$ is the Fourier transform of $u(x)$.

Let us study the continuity from $\mathcal{H}_1$ to $\mathcal{H}_2$ of pseudodifferential operators with symbols $D(x,\xi) \in S^0_{CV}(\mathbb{R}^{2n}, \mathfrak{H}_1, \mathfrak{H}_2)$. We are interested only in symbols compactly supported with respect to $x$, since it suffices to consider such symbols when dealing with operators on compact manifolds. The estimates (3.27) for a symbol $D$ compactly supported with respect to $x$ imply the weaker estimates

$$\left\| \frac{\partial^\alpha D}{\partial x^\alpha}(x,\xi) : H_{1\xi} \longrightarrow H_{2\xi} \right\| \le C_{\alpha l} \langle x \rangle^{-l}, \quad l, |\alpha| = 0, 1, 2, \dots . \quad (3.33)$$

It turns out that these estimates suffice for the boundedness.

THEOREM 3.28. *If the estimates* (3.33) *hold, then the operator*

$$D\left(\overset{2}{x}, -i\overset{1}{\frac{\partial}{\partial x}}\right) : \mathcal{H}_1 \longrightarrow \mathcal{H}_2$$

*is continuous.*

*Proof.* We shall prove an equivalent assertion, namely, the continuity of the operator

$$\widehat{D} = D\left(i\overset{2}{\frac{\partial}{\partial \xi}}, \overset{1}{\xi}\right) : \int H_{1\xi} d\xi \longrightarrow \int H_{2\xi} d\xi.$$

The operator acts by the formula

$$\left[\widehat{D}u\right](\xi) = \int \widetilde{D}(\xi - \eta, \eta)u(\eta)\,d\eta = \int \widetilde{D}(z, \xi - z)u(\xi - z)\,dz, \tag{3.34}$$

where $\widetilde{D}$ is the Fourier transform of the symbol $D$ with respect to the first argument. By virtue of the estimates (3.33), the Fourier transform is continuous and satisfies the estimates

$$\left\|\widetilde{D}(z,\eta) : H_\eta \longrightarrow G_\eta\right\| \le C_l \langle z\rangle^{-l}, \quad l = 0, 1, 2, \dots .$$

We can rewrite (3.34) in the form

$$\widehat{D}u = \int U(\,\boldsymbol{\cdot}\,, z)\,dz, \tag{3.35}$$

where

$$U(\xi, z) = \widetilde{D}(z, \xi - z)u(\xi - z).$$

By virtue of the properties of the norm and the estimates imposed on $\widetilde{D}$, we have

$$\|U(\xi, z)\|_\xi \le \text{const}\,\langle z\rangle^N \|U(\xi, z)\|_{\xi - z} \le \text{const}\,\langle z\rangle^{-M} \|u(\xi - z)\|_{\xi - z}\,,$$

where $M$ is arbitrarily large. Hence

$$\begin{aligned}\|U(\,\boldsymbol{\cdot}\,, z)\|_{\mathcal{J}} &= \left\{\int \|U(\xi, z)\|_\xi^2\,d\xi\right\}^{1/2} \\ &\le \text{const}\,\langle z\rangle^{-M}\left\{\int \|u(\xi - z)\|_{\xi - z}^2\,dz\right\}^{1/2} = \text{const}\,\langle z\rangle^{-M}\|u\|_{\mathcal{J}}\,;\end{aligned}$$

by substituting this estimate into (3.35) and by integrating over $z$, we arrive at the desired result. □

The proof of the following lemma is now standard.

LEMMA 3.29 (cf. Luke 1972, Proposition 2.1). *If a symbol*

$$g(x, \xi) \in J_K^{-1}(\mathbb{R}^{2n}, \mathfrak{H}_1, \mathfrak{H}_2)$$

*is compactly supported in $x$, then the operator*

$$g\left(\overset{2}{x}, -i\overset{1}{\frac{\partial}{\partial x}}\right) : \mathcal{H}_1 \longrightarrow \mathcal{H}_2$$

*is compact.*

***Global definition.*** Let

$$f(x,\xi) \in S^0_{CV}(T^*X, \mathfrak{H}_1, \mathfrak{H}_2)$$

be an operator-valued symbol on $X$. We cover $X$ by coordinate neighborhoods $U_j$, $j = 1, \dots, N$, and consider a smooth partition of unity

$$1 = \sum_{j=1}^{N} \chi_j(x)^2$$

subordinate to the cover.

A pseudodifferential operator with symbol $f(x,\xi)$ can be defined by the formula

$$f\left(x, -i\frac{\partial}{\partial x}\right) \stackrel{\text{def}}{=} \sum_{j=1}^{N} (\chi_j f)\left(\overset{2}{x}, -i\overset{1}{\frac{\partial}{\partial x}}\right) \circ \chi_j(x), \tag{3.36}$$

where the $j$th term in the sum is defined in the local coordinates of the chart $U_j$ as the composition of the operator of multiplication by the function $\chi_j(x)$ (which localizes the function into the chart $U_j$) and a pseudodifferential operator with symbol $\chi_j(x)f(x,\xi)$ in $\mathbb{R}^n$. Here the symbol is expressed via the canonical coordinates in $T^*U_j$, which we denote by the same letters $(x,\xi)$.

PROPOSITION 3.30. *Let* $f \in S^0_{CV}(T^*X, \mathfrak{H}_1, \mathfrak{H}_2)$. *The formulas* (3.36) *and* (3.32) *define a continuous operator*

$$f\left(x, -i\frac{\partial}{\partial x}\right) : \mathcal{H}_1 \longrightarrow \mathcal{H}_2, \tag{3.37}$$

*which is independent of the choice of the atlas* $U_j$ *and the subordinate partition of unity modulo compact operators.*

◀ The proof follows the standard scheme used to prove the invariance of the definition of $\psi$DO (Hörmander 1965, Kohn and Nirenberg 1965). It is based on the composition and change of variable formulas for $\psi$DO in $\mathbb{R}^n$, which can be transferred word for word from the finite-dimensional case. In addition, we should only prove the compactness of remainders in these formulas in the infinite-dimensional case. To this end, we prove that the symbols of the remainders belong to the space $J^{-1}_K(\mathbb{R}^{2n}, \mathfrak{H}_1, \mathfrak{H}_2)$ and then use Lemma 3.29. ▶

Proposition 3.30 permits us to give the following definition.

DEFINITION 3.31. Let $f \in S^0_{CV}(T^*X, \mathfrak{H}_1, \mathfrak{H}_2)$. A *pseudodifferential operator with symbol* $f$ is the operator (3.37) given modulo compact operators by the formula (3.36).

Next, let $f \in S^0_{CV}(T^*_0X, \mathfrak{H}_1, \mathfrak{H}_2)$. (That is, $f$ is defined only outside the zero section.) A pseudodifferential operator with symbol $f$ is the operator

$$\widetilde{f}\left(x, -i\frac{\partial}{\partial x}\right) : \mathcal{H}_1 \longrightarrow \mathcal{H}_2,$$

where $\widetilde{f}(x,\xi) \in S^0_{CV}(T^*X, \mathfrak{H}_1, \mathfrak{H}_2)$ is an arbitrary fill-up of $f(x,\xi)$.

By Lemma 3.27, the symbol $\widetilde{f}$ exists and is uniquely determined modulo compact-valued compactly supported symbols, which correspond to compact $\psi$DO by Lemma 3.29. Thus a $\psi$DO with symbol $f \in S^0_{CV}(T^*_0X, \mathfrak{H}_1, \mathfrak{H}_2)$ is well defined modulo compact operators. By abuse of notation, we denote such an operator by

$$f\left(x, -i\frac{\partial}{\partial x}\right) : \mathcal{H}_1 \longrightarrow \mathcal{H}_2. \tag{3.38}$$

The theorem below readily follows from the preceding.

THEOREM 3.32. *If $f \in S^0_{CV}(T^*_0X, \mathfrak{H}_1, \mathfrak{H}_2)$, then the operator* (3.38) *is bounded and independent, modulo compact operators in these spaces, of the choice of local coordinates and the partition of unity.*

***The composition theorem.*** The operators with operator-valued symbols thus defined form an algebra. More precisely, the following assertion holds.

PROPOSITION 3.33. *Let $p \in S^0_{CV}(T^*_0X, \mathfrak{H}_2, \mathfrak{H}_3)$, let $q \in S^0_{CV}(T^*_0X, \mathfrak{H}_1, \mathfrak{H}_2)$, and let $P$ and $Q$ be $\psi$DO with symbols $p$ and $q$, respectively. Then $PQ$ is a $\psi$DO with symbol $pq$.*

This theorem essentially deals with the principal symbol (since we have defined pseudodifferential operators modulo compact operators). But by fixing the arbitrary elements in the construction of $\psi$DO (the cover of $X$ by charts, the partition of unity, the fill-up, etc.) one can obtain more precise composition formulas, which have the standard form in local coordinates (e.g., Kohn and Nirenberg 1965). We shall use such formulas below in the proof of the index theorem in the operator-valued case.

## 3.4. Operators on Singular Manifolds

Pseudodifferential calculus on manifolds with conical or edge singularities is well known (e.g., see Schulze 1991, Egorov and Schulze 1997, Schulze, Sternin and Shatalov 1998*a*, and references therein), and so we give only a brief account of this theory in a form suitable to us. The proofs are omitted altogether for the conical case and sometimes outlined in the edge case. The interested reader is referred to Egorov and Schulze (1997) and Schulze, Sternin and Shatalov (1998*a*) for a detailed exposition with complete proofs. In fact, for our aims, the mere *existence* of pseudodifferential operators with given symbols and certain properties, rather than expressing these operators in any computable form, is sufficient. That is why we sacrifice any finer structures irrelevant to our applications and describe a "minimal" calculus suitable for establishing finiteness theorems and for index theory. This is the main novelty in our exposition. We do not impose any analyticity conditions on conormal symbols, wherever they occur. It suffices to have them defined only on the weight line corresponding to the problem. Accordingly, the almost inverses of elliptic elements in the calculus do not improve the weight exponents; the remainders are only compact and smoothing.

### 3.4.1. Operators on manifolds with isolated singularities

Throughout this subsection, $\mathcal{M}$ is a manifold with conical singularities; for simplicity, we assume that there is only one conical point with a base $\Omega$ of the cone. Thus the stretched manifold $M$ is a manifold with boundary $\partial M = \Omega$.

***Symbols.*** The principal symbol of a cone-degenerate pseudodifferential operator $D$ on $\mathcal{M}$ is the pair $\mathfrak{s}(D) = (\sigma(D), \sigma_c(D))$ consisting of the interior symbol $\sigma(D)$ and the cone symbol $\sigma_c(D)$.

DEFINITION 3.34. An *interior symbol of order* $m$ is a smooth function $\sigma = \sigma(y, \eta)$ on the stretched cotangent bundle $T_0^*\mathcal{M}$ without the zero section, homogeneous of order $m$ in the fibers:

$$\sigma(y, \lambda\eta) = \lambda^m \sigma(y, \eta), \quad (y, \eta) \in T_0^*\mathcal{M}, \quad \lambda \in \mathbb{R}_+.$$

The space of interior symbols of order $m$ will be denoted by $\mathcal{O}^m(T_0^*\mathcal{M})$.

Next, we define cone symbols. This is conveniently done in terms of conormal symbols (cf. Definition 2.11).

DEFINITION 3.35. A *conormal symbol of order* $m$ *and weight* $\gamma$ is an $m$th-order pseudodifferential operator

$$\sigma_c(p) : H^s(\Omega) \longrightarrow H^{s-m}(\Omega), \quad s \in \mathbb{R},$$

with parameter $p \in \mathcal{L}_\gamma$ in the sense of Agranovich and Vishik (1964) on the manifold $\Omega$ (cf. Section 3.2.5).

The operator $\sigma_c(p)$ has a well-defined principal symbol $\sigma(\sigma_c(p))$ (in the sense of Agranovich–Vishik), which is a function defined on $T^*\Omega \times \mathbb{R}_p$ outside the zero section.

If $\sigma_c(p)$ is a conormal symbol of order $m$ and weight $\gamma$, then the operator

$$\sigma_c = r^{-m}\sigma_c\left(ir\frac{\partial}{\partial r}\right) : H^{s,\gamma}(K_\Omega) \longrightarrow H^{s-m,\gamma-m}(K_\Omega) \tag{3.39}$$

in the weighted Sobolev spaces on the infinite cone $K_\Omega$ is well defined and continuous for all $s \in \mathbb{R}$.

DEFINITION 3.36. The operator (3.39) is called the *cone symbol of order* $m$ *and weight* $\gamma$ associated with the conormal symbol $\sigma_c(p)$. The function $\sigma(\sigma_c(p))$ is called the *principal symbol of the cone symbol* and is also denoted by $\sigma(\sigma_c)$. The space of all cone symbols of order $m$ and weight $\gamma$ will be denoted by $\mathrm{Cone}_\gamma^m(K_\Omega)$.

The boundary $\partial T^*\mathcal{M}$ is naturally isomorphic to $T^*\Omega \times \mathbb{R}_p$. Hence the following definition makes sense.

DEFINITION 3.37. Symbols $\sigma \in \mathcal{O}^m(T_0^*\mathcal{M})$ and $\sigma_c \in \mathrm{Cone}_\gamma^m(K_\Omega)$ are said to be *compatible* if

$$\sigma(\sigma_c) = \sigma \,\big|_{\partial T_0^*\mathcal{M}} . \tag{3.40}$$

By $\Sigma_\gamma^m = \Sigma_\gamma^m(M)$ we denote the set of compatible pairs

$$\mathfrak{s} = (\sigma, \sigma_c) \in \mathcal{O}^m(T_0^*\mathcal{M}) \times \mathrm{Cone}_\gamma^m(K_\Omega).$$

Such pairs will be referred to as *principal symbols*.

The common value of both sides in the compatibility condition (3.40) will also be called the *boundary symbol* and denoted by $\sigma_\partial$.

***Quantization and pseudodifferential operators.*** Now our task is to assign an operator in weighted Sobolev spaces on $\mathcal{M}$ to each principal symbol $\mathfrak{s} \in \Sigma_\gamma^m(M)$. The result is not unique, for the symbol does not (and should not) determine lower-order terms in the operator. Hence we start by describing the set of such lower-order terms, referred to as "negligible" operators.

DEFINITION 3.38. A *negligible operator of order* $m$ *and weight* $\gamma$ in weighted Sobolev spaces on $M$ is an operator $D$ on $\mathcal{M}$ compact in the spaces

$$D : H^{s,\gamma}(\mathcal{M}) \longrightarrow H^{s-m,\gamma-m}(\mathcal{M})$$

and continuous in the spaces

$$D : H^{s,\gamma}(\mathcal{M}) \longrightarrow H^{s-m+1,\gamma-m}(\mathcal{M})$$

for all $s \in \mathbb{R}$. The space of negligible operators of order $m$ and weight $\gamma$ will be denoted by $I_\gamma^m$.

Now we are in a position to construct pseudodifferential operators. Let

$$\mathfrak{s} = (\sigma, \sigma_c) \in \Sigma_\gamma^m(M)$$

be a given principal symbol. The construction will be done locally. Let $\sigma_c(p)$ be the conormal symbol corresponding to the cone symbol $\sigma_c$. We take a fixed trivialization $U \simeq \Omega \times [0,1)$ of the collar neighborhood of the boundary, so that the isomorphism of $\partial T^*\mathcal{M}$ and $T^*\Omega \times \mathbb{R}$ is extended (noncanonically, i.e., depending on the trivialization) to an isomorphism

$$T^*\mathcal{M}\,\big|_U \simeq T^*\Omega \times [0,1) \times \mathbb{R}.$$

By virtue of the compatibility condition (3.40), there exists a family $\widetilde{\sigma}_c(r,p)$ of conormal symbols smoothly depending on the parameter $r \in [0,1)$ such that

$$\widetilde{\sigma}_c(0,p) = \sigma_c(p), \quad \sigma(\widetilde{\sigma}_c(r_0,p)) = \sigma\,\big|_{r=r_0}$$

for each $r_0 \in [0,1)$. Consider a smooth partition of unity

$$1 = e_0 + e_1$$

subordinate to the cover

$$M = U \cup M^\circ$$

of the manifold $M$ and excision functions $\chi_0$ and $\chi_1$ supported in the corresponding elements of the cover and satisfying

$$\chi_0 e_0 = e_0, \quad \chi_1 e_1 = e_1.$$

We define an operator $D$ on $M$ by setting (cf. (1.29) for differential operators)

$$D = r^{-m}\chi_0\sigma_c\left(\overset{2}{r}, ir\frac{\overset{1}{\partial}}{\partial r}\right)e_0 + \chi_1\widehat{\sigma}e_1, \tag{3.41}$$

where $\widehat{\sigma}$ is an arbitrary classical pseudodifferential operator on $M^\circ$ with principal symbol $\sigma$.

PROPOSITION 3.39. *The operator* (3.41) *is well defined and continuous in the spaces*

$$D : H^{s,\gamma}(\mathcal{M}) \longrightarrow H^{s-m,\gamma-m}(\mathcal{M}) \tag{3.42}$$

*for all $s \in \mathbb{R}$ and is independent of the ambiguity in the construction* (*the choice of the trivialization of $U$, partition of unity, and excision functions*) *modulo negligible operators* (*elements of $I_\gamma^m$*).

The standard method of proof is to pass to the cylindrical coordinates $t = -\ln r$ in a neighborhood of the conical point and then use standard theorems for pseudodifferential operators in Section 3.2.

This proposition permits one to give the following definition.

DEFINITION 3.40. A *pseudodifferential operator of order $m$ and weight $\gamma$* with principal symbol $\mathfrak{s} \in \Sigma_\gamma^m(M)$ is an operator of the form $D+R$, where $D$ is given by the formula (3.41) and $R \in I_\gamma^m$ is arbitrary. The set of pseudodifferential operators of order $m$ and weight $\gamma$ on $M$ will be denoted by $\Psi_\gamma^m(M)$.

The following assertion shows that the notion of principal symbol is well defined for operators $D \in \Psi_\gamma^m(M)$.

PROPOSITION 3.41. *The operator* (3.41) *belongs to $I_\gamma^m$ if and only if $\mathfrak{s} = 0$. Thus the principal symbol $\mathfrak{s}(D)$ is uniquely determined for each $D \in \Psi_\gamma^m(M)$, and the mapping $\mathfrak{s} : D \longmapsto \mathfrak{s}(D)$ gives rise to an isomorphism*

$$\mathfrak{s} : \Psi_\gamma^m(M)/I_\gamma^m \longrightarrow \Sigma_\gamma^m(M).$$

*Moreover, a pseudodifferential operator $D \in \Psi_\gamma^m(M)$ is compact in the spaces* (3.42) *if and only if $D \in I_\gamma^m$.*

Thus compactness in the set of pseudodifferential operators is equivalent to a stronger property that also involves smoothing.

*Remark* 3.42. For cone-degenerate differential operators, the notions of symbols introduced here coincide with those used earlier in the book.

***Calculus.*** Now we can state theorems concerning the behavior of pseudodifferential operators and their symbols under products and the passage to the adjoint operator.

We introduce multiplication of symbols as follows. Let $\mathfrak{s} \in \Sigma_{\gamma-l}^m$ and $\widetilde{\mathfrak{s}} \in \Sigma_\gamma^l$. Then we define the product

$$\mathfrak{s}\widetilde{\mathfrak{s}} \in \Sigma_\gamma^{m+l}$$

by componentwise multiplication:

$$\mathfrak{s}\widetilde{\mathfrak{s}} = (\sigma\widetilde{\sigma}, \sigma_c\widetilde{\sigma}_c).$$

Note that this is well-defined, that is, the products exist and the compatibility condition is preserved.

*Remark* 3.43. In terms of conormal symbols, the product of cone symbols can be expressed as follows:

$$(\sigma_c\widetilde{\sigma}_c)(p) = \sigma_c(p - il)\widetilde{\sigma}_c(p), \quad p \in \mathcal{L}_\gamma.$$

Both factors on the right-hand side are defined on the weight line $\mathcal{L}_\gamma$.

THEOREM 3.44 (composition theorem). *Let*

$$D_1 \in \Psi^m_{\gamma-l}(M), \quad D_2 \in \Psi^l_\gamma(M).$$

*Then the product $D_1 D_2$ is also a pseudodifferential operator,*

$$D_1 D_2 \in \Psi^{m+l}_\gamma(M),$$

*and*

$$\mathfrak{s}(D_1 D_2) = \mathfrak{s}(D_1)\mathfrak{s}(D_2).$$

In a similar way, one can describe the passage to adjoint operators. In doing so, we take adjoint operators with respect to the inner product in $H^{0,0}(\mathcal{M})$.

THEOREM 3.45. *Let $D \in \Psi^m_\gamma(M)$. Then $D^* \in \Psi^{-m}_{m-\gamma}(M)$,*

$$\sigma(D^*) = \sigma(D)^*$$

*(the ordinary Hermite conjugation), and*

$$\sigma_c(D^*)(p) = \sigma_c(D)^*(\bar{p} + im),$$

*where the bar over p indicates complex conjugation.*

### 3.4.2. Operator families on infinite cones

Now we shall proceed to the study of edge-degenerate pseudodifferential operators. From now on, $\mathcal{M}$ will be a compact manifold with edge $X$ and model cone base $\Omega$. First of all, we should define symbols of edge-degenerate operators. The principal symbols are pairs $(\sigma, \sigma_\wedge)$ where $\sigma$ is the interior symbol and $\sigma_\wedge$ is the edge symbol. The definition of the former is much similar to the one used in the conical case; it is just a homogeneous function on the stretched cotangent bundle without the zero section. On the contrary, the edge symbol is considerably more complicated than the cone symbol in conical theory, and in this subsection we shall study edge symbols.

For an edge-degenerate differential operator, the edge symbol is a family of differential operators parametrized by the cotangent space of the edge without the zero section (see Chapter 1). At each point $x$ of the edge, this family is a differential operator with parameter $\xi$ on the infinite cone $K_\Omega$. Accordingly, the edge symbols of edge-degenerate *pseudodifferential* operators should obviously be families of pseudodifferential operators on the same cone. Our aim in this subsection is to construct a calculus of such families.

These operators will depend on the parameters $(x, \xi) \in T_0^*X$ and act in the spaces $\mathcal{K}^{s,\gamma}(K_\Omega)$ on the cone $K_\Omega$ (see Definition 2.14). Our considerations are mainly local with respect to $x$, and so without loss of generality we can work in local coordinates and assume that the parameter $\xi$ ranges over $\mathbb{R}^n \setminus \{0\}$, where $n = \dim X$. Next, our operator classes are "bound" to a specific value of the weight exponent $\gamma$, whereas the smoothness parameter $s$ runs over the entire real line. By $K_\Omega^\circ$ we denote the open half-cylinder $(0, +\infty) \times \Omega$.

In what follows, we often use a specific class of cutoff functions depending on the variable $r$. Thus it will be reasonable to give such functions a name.

DEFINITION 3.46. A smooth function $\psi(r)$ defined for $r \geq 0$, equal to unity for sufficiently small $r$, and vanishing for sufficiently large $r$ is called an *R-function.*

***Continuity properties.*** Prior to defining the set of edge symbols constructively, we shall describe the mapping properties that we wish to hold for these symbols. In a sense, we require as little as possible so as still to be able to prove the finiteness theorem and the smoothness of solutions.

Our edge symbols will be families of order $m$ and weight $\gamma$ in the sense of the following definition.

DEFINITION 3.47. A *family of order $m$ and weight $\gamma$* is a smooth operator family

$$D(x,\xi) : C_0^\infty(K_\Omega^\circ) \longrightarrow \mathcal{D}'(K_\Omega^\circ)$$

parametrized by points $(x, \xi) \in T_0^*X$ and possessing the following properties:

1) (**twisted homogeneity**)

$$D(x, \lambda\xi) = \lambda^m \varkappa_\lambda D(x,\xi) \varkappa_\lambda^{-1}, \quad \lambda > 0$$

(recall that $\varkappa_\lambda u(r,\omega) = \lambda^{(k+1)/2} u(\lambda r, \omega)$, $k = \dim \Omega$)

2) (**continuity**) the family $D(x, \xi)$ and all of its derivatives extend by closure to smooth families of continuous operators in the spaces

$$D^{(\alpha,0)}(x,\xi) : \mathcal{K}^{s,\gamma}(K_\Omega) \longrightarrow \mathcal{K}^{s-m,\gamma-m}(K_\Omega), \qquad |\alpha| = 0,1,2,\dots, \tag{3.43}$$
$$D^{(\alpha,\beta)}(x,\xi) : \mathcal{K}^{s,\gamma}(K_\Omega) \longrightarrow \mathcal{K}^{s-m+1,\gamma-m}(K_\Omega), \quad |\alpha| = 0,1,2,\dots, \; |\beta| = 1,2,\dots, \tag{3.44}$$

for a given weight $\gamma \in \mathbb{R}$ and for each $s \in \mathbb{R}$; here

$$D^{(\alpha,\beta)}(x,\xi) = \frac{\partial^{\alpha+\beta} D}{\partial x^\alpha \partial \xi^\beta}(x,\xi)$$

3) (**almost compact fiber variation**) for an arbitrary $R$-function $\varphi(r)$, the operator families

$$\varphi(r) D(x,\xi),\; D(x,\xi)\varphi(r) : \mathcal{K}^{s,\gamma}(K_\Omega) \longrightarrow \mathcal{K}^{s-m,\gamma-m}(K_\Omega) \tag{3.45}$$

have compact fiber variation (i.e., the operator $\varphi(r)\partial D/\partial\xi(x,\xi)$ and the operator $\partial D/\partial \xi(x,\xi)\varphi(r)$ are compact in this pair of spaces)

4) conditions 1)–3) remain valid for the family $(1+r|\xi|)^l D(x,\xi)(1+r|\xi|)^{-l}$ for arbitrary $l \in \mathbb{R}$

*Remark* 3.48. 1. The smoothness of our families is assumed in the strong operator topology and hence[2] in the uniform operator topology.

2. In what follows, we do not usually mention extension by closure explicitly and speak merely of the continuity of the operator in the corresponding spaces; the closure is denoted by the same letter as the original operator.

PROPOSITION 3.49. *If $D \in \mathcal{D}_m(M)$ is an edge-degenerate differential operator of order $m$, then its edge symbol is a family of order $m$ for each $\gamma$.*

*The proof* is by a straightforward verification. □

*Remark* 3.50. Note that the estimate (3.44) remains valid for edge symbols of differential operators even if we replace the right-hand side by $\mathcal{K}^{s-m+|\beta|,\gamma-m+|\beta|}(K_\Omega)$. We do not require this stronger property in the general case.

***Compact smoothing edge symbols.*** Instead of defining the entire set of edge symbols in one step, it is more convenient to describe the subset of compact smoothing edge symbols first.

DEFINITION 3.51. A family $D(x,\xi)$ of order $m$ and weight $\gamma$ is said to be a *compact smoothing edge symbol* if it additionally satisfies the following properties:

5) the operators $D(x,\xi)$ are compact in the spaces

$$D(x,\xi) : \mathcal{K}^{s,\gamma}(K_\Omega) \longrightarrow \mathcal{K}^{s-m,\gamma-m}(K_\Omega)$$

6) the operators (3.44) are continuous also for $\beta = 0$

7) properties 5) and 6) remain valid for the families $rD(x,\xi)$ and $D(x,\xi)r$

The set of compact smoothing edge symbols of order $m$ and weight $\gamma$ will be denoted by $I^m_\gamma$.

PROPOSITION 3.52. 1. *The multiplication of operators induces a bilinear mapping*

$$I^l_{\gamma-m} \times I^m_\gamma \longrightarrow I^{m+l}_\gamma$$

*for any $\gamma, m, l \in \mathbb{R}$.*

2. *The passage to the adjoint operator with respect to the inner product in $\mathcal{K}^{0,0}(K_\Omega)$ induces an antilinear mapping*

$$I^m_\gamma \longrightarrow I^m_{m-\gamma}$$

*for any $\gamma, m \in \mathbb{R}$.*

[2]The resonance theorem (Yosida 1968) guarantees the uniform boundedness of all derivatives.

*Proof.* Properties 1), 2), and 6) are respected by multiplication as well as by the passage to the adjoint operator. Property 3) is not respected in general, but for compact smoothing edge symbols it is covered by the stronger property 5), which is already preserved under these operations. The conjugation with the operator $(1 + r|\xi|)^l$, which is self-adjoint in $\mathcal{K}^{0,0}(K_\Omega)$, is a homomorphism and hence preserved under multiplication; when passing to the adjoint operator, we should just replace $l$ by $-l$. This proves that property 4) is preserved. Property 7) obviously remains valid for the adjoint operators; as to the products, one should only note that property 7) is equivalent to the same property with the operators $(1+r|\xi|)D(x,\xi)$ and $D(x,\xi)(1+r|\xi|)$ instead of $rD(x,\xi)$ and $D(x,\xi)r$ and then use property 4). □

***Definition of general edge symbols.*** Now we are in a position to describe general edge symbols. These symbols are families of pseudodifferential operators on the infinite cones $K_\Omega$ and, as such, they have interior symbols and conormal symbols.

Let $H_\partial = H_\partial(x,\omega,\xi,p,q) \in \mathcal{O}^m(\partial T_0^*\mathcal{M})$ be a smooth homogeneous function of order $m$ on the boundary $\partial T_0^*\mathcal{M}$ of the cotangent bundle of $\mathcal{M}$. Next, let

$$h(x) = r^{-m}h\left(x, ir\frac{\partial}{\partial r}\right) \in \mathrm{Cone}^m_{\gamma-(k+1)/2}(K_\Omega), \quad x \in X, \tag{3.46}$$

be an $m$th-order cone symbol with weight line[3]

$$p \in \mathcal{L}_\gamma = \left\{\operatorname{Im} p = \gamma - \frac{k+1}{2}\right\}. \tag{3.47}$$

We assume that $h(x)$ depends on the parameter $x \in X$ but is *independent of the variable* $\xi$ *in the fibers of* $T_0^*X$. Next, we assume that the boundary symbol $H_\partial$ and the cone symbol $h$ satisfy an analog of the compatibility condition (3.40):

$$\sigma(h) = H_\partial\big|_{\xi=0}.$$

Our aim is to define an edge symbol

$$D(x,\xi) : \mathcal{K}^{s,\gamma}(K_\Omega) \longrightarrow \mathcal{K}^{s-m,\gamma-m}(K_\Omega) \tag{3.48}$$

with the symbol pair $(H_\partial, h)$. The construction is carried out separately in a neighborhood of the cone vertex and outside the neighborhood; the results are then glued together with the use of a partition of unity constructed from $R$-functions.

We proceed with the definition for $|\xi| = 1$. Later on we extend the definition to all $\xi$ by twisted homogeneity.

[3]Note the different normalization in the definition of the weight line as compared with the conical case.

The spaces $\mathcal{K}^{s,\gamma}(K_\Omega)$ coincide with their counterparts $H^{s,\gamma-(k+1)/2}(K_\Omega)$ near the vertex, and we define $D(x,\xi)$ there as the family of cone-degenerate operators

$$D(x,\xi) = h(x) : H^{s,\gamma-(k+1)/2}(K_\Omega) \longrightarrow H^{s-m,\gamma-m-(k+1)/2}(K_\Omega), \quad x \in X.$$

Note that this operator is actually independent of $\xi$, but this is no surprise, because it is near infinity on the cone that the dependence on $\xi$ becomes essential.

Away from a neighborhood of the vertex, the cone symbol plays no role at all, and all we need to define our operator family there is the boundary symbol $H_\partial(x,\omega,\xi,p,q)$. Informally speaking, the operator, $\widehat{H}_\partial$, should be obtained from the symbol $H_\partial$ by the substitution

$$p \mapsto -i\frac{\partial}{\partial r}, \quad q \mapsto -\frac{i}{r}\frac{\partial}{\partial\omega}.$$

To make this precise, we first note that since we work away from the vertex, we can multiply the symbol by a smooth cutoff function $\rho(r)$ equal to zero for small $r$ and unity for large $r$. Next, consider a partition $\sum e_j = 1$ of unity subordinate to some finite coordinate cover on $\Omega$ and excision functions $f_j$ supported in the same coordinate neighborhoods as the respective $e_j$ and satisfying the condition $e_j f_j = e_j$. We set

$$F_j(x,r,\omega,\xi,p,q) = \rho(r)e_j(\omega)H_\partial(x,\omega,\xi,p,q).$$

As is customary in the theory of pseudodifferential operators, we shall define $\widehat{H}_\partial$ by the formula

$$\widehat{H}_\partial = \sum_j \widehat{F}_j \circ f_j, \tag{3.49}$$

and so the problem is to define the local representatives $\widehat{F}_j$ corresponding to coordinate neighborhoods on $\Omega$. Consider the change of variables

$$\alpha : [\varepsilon,\infty) \times \mathbb{R}^k_\omega \longrightarrow \mathbb{R}^{k+1}_y$$

given by the formulas

$$y_0 = r, \quad y' \equiv (y_1,\dots,y_k) = r\omega.$$

In the new variables $y$, we define $\widehat{F}$ to be the pseudodifferential operator

$$\widehat{F}_j = P\left(x,\overset{2}{y},\xi,-i\overset{1}{\frac{\partial}{\partial y}}\right) \tag{3.50}$$

in $\mathbb{R}^{k+1}_y$ with symbol

$$P(y,\eta) = F_j(x,y_0,y/y_0,\xi,\eta).$$

(Here $\eta$ is the variable dual to $y$.)

The symbol $P(y,\eta)$ satisfies the estimates (recall that $|\xi|=1$)

$$|P^{(\alpha,\beta)}(y,\eta)| \le C_{\alpha\beta}(1+|\eta|)^{m-|\beta|}(1+|y|)^{-\alpha} \tag{3.51}$$

and hence by Theorem 3.2 the operator $\widehat{F}_j$ in the new variables $y$ is continuous in the Sobolev spaces

$$\widehat{F}_j : H^s(\mathbb{R}^{k+1}_y) \longrightarrow H^{s-m}(\mathbb{R}^{k+1}_y).$$

Returning to the original variables and recalling the definition of $\mathcal{K}^{s,\gamma}$, we see that the operator $\widehat{F}_j$ is continuous in the spaces

$$\widehat{F}_j : \mathcal{K}^{s,\gamma}(K_\Omega) \longrightarrow \mathcal{K}^{s-m,\gamma-m}(K_\Omega).$$

Now we patch together our constructions (3.48) and (3.49) and extend them to arbitrary $|\xi|$ by twisted homogeneity, thus arriving at the following definition.

DEFINITION 3.53. An *edge symbol of order* $m$ *and weight* $\gamma$ *with interior symbol* $H_\partial$ *and cone symbol* $h$ is a family of order $m$ representable modulo compact smoothing edge symbols (elements of the space $I^m_\gamma$) in the form

$$D(x,\xi) = \chi_1(r|\xi|)h(x)\psi(r|\xi|) + (1-\chi_2(r|\xi|))\widehat{H}_\partial(1-\psi(r|\xi|)), \tag{3.52}$$

where $\chi_1$, $\chi_2$, and $\psi$ are $R$-functions such that

$$\chi_1\psi = \psi, \quad (1-\chi_2)(1-\psi) = 1-\psi \tag{3.53}$$

and the operator $\widehat{H}_\partial$ is defined by the formula (3.49). The set of edge symbols of order $m$ and weight $\gamma$ will be denoted by $\mathrm{Edge}^m_\gamma \equiv \mathrm{Edge}^m_\gamma(T^*_0X)$. We write $H_\partial = \sigma(D(x,\xi))$ and $h = \sigma_c(D(x,\xi))$.

***Properties of edge symbols.*** The following theorem differs from the one already stated for cone-degenerate operators in that now the operators act on *infinite* cones.

THEOREM 3.54. 1) *The family* (3.52) *is indeed a family of order* $m$ *in the sense of Definition* 3.47 *and is independent of the ambiguity in the construction modulo compact smoothing operators. In other words, if* $D \in \mathrm{Edge}^m_\gamma$*, then the interior and cone symbols of* $D$ *are well defined, and* $D \in I^m_\gamma$ *if and only if* $\sigma(D)=0$ *and* $\sigma_c(D)=0$.

2) *The product of operators induces a bilinear mapping*

$$\mathrm{Edge}^l_{\gamma-m} \times \mathrm{Edge}^m_\gamma \longrightarrow \mathrm{Edge}^{m+l}_\gamma$$

*with the composition law*

$$\sigma(D_1D_2) = \sigma(D_1)\sigma(D_2), \quad \sigma_c(D_1D_2) = \sigma_c(D_1)\sigma_c(D_2). \tag{3.54}$$

3) *The passage to the adjoint operator (with respect to the inner product in* $\mathcal{K}^{0,0}$) *induces an antilinear mapping*

$$\mathrm{Edge}_{\gamma}^{m} \longrightarrow \mathrm{Edge}_{m-\gamma}^{m},$$

*and*

$$\sigma(D^{*})=\sigma(D)^{*}, \sigma_{c}(D^{*})=\sigma_{c}(D)^{*}. \tag{3.55}$$

4) *If* $D \in \mathrm{Edge}_{\gamma}^{m}$ *and* $\varphi(r)$ *is a smooth function bounded together with all derivatives and equal to zero for large* $r$, *then the operator families* $\varphi(r)D$ *and* $D\varphi(r)$ *have compact fiber variation in the spaces*

$$\mathcal{K}^{s,\gamma}(K_{\Omega}) \longrightarrow \mathcal{K}^{s-m,\gamma-m}(K_{\Omega}).$$

*Proof.* 1) Let us verify conditions 1) and 2) in the definition of a family of order $m$. Condition 1) (twisted homogeneity) follows by a straightforward computation from (3.52). Let us verify condition 2), i.e., the continuity of the family $D(x,\xi)$ and its derivatives in the spaces (3.43) and (3.44). The continuity of the operators (3.43) is clear from the construction for both terms in (3.52) separately. To prove the continuity of the operators (3.44), which involve $\xi$-derivatives, note that the only troublesome terms are those arising from the differentiation of $\psi$ and $1-\psi$ in (3.52). (Indeed, the differentiation of $\widehat{H}_{\partial}$ gives an operator with the desired properties, and so does the differentiation of $\chi_1$ and $\chi_2$, since the derivatives of these functions are zero on the supports of $\psi$ and $1-\psi$, respectively.) Now note that $\chi_1=\chi_2=1$ on the support and that, owing to the compatibility condition, $h$ and $\widehat{H}_{\partial}$ are operators with the same interior symbol on the (compact) support of $\psi'$; it follows that the troublesome derivatives cancel each other out modulo terms satisfying the desired estimates.

The family $D(x,\xi)$ is independent modulo smoothing operators of the ambiguity in the construction (the choice of the $R$-functions $\chi_1$, $\chi_2$, and $\varphi$ as well as partitions of unity and excision functions on $\Omega$). This follows from the composition formulas and formulas for the change of variables in pseudodifferential operators in conjunction with the estimate (3.51). The key point is as follows: this estimate implies that the remainder terms in these formulas contain the factor $(1+r)^{-1}$, which guarantees the validity of condition 7) in Definition 3.51. Finally, the validity of condition 4) in the definition of a family of order $m$ follows from the fact that the set of pseudodifferential operators with symbols satisfying the estimate (3.51) is invariant with respect to conjugation by $(1+y^2)^{l/2}$ for any $l$.

2) This property follows from the composition formula for pseudodifferential operators; the desired estimates guaranteeing that the remainders belong to $I_{\gamma}^{m}$ again follow from (3.51).

3) To prove this property, one passes to the adjoint operator in the formula (3.52) and uses the fact that the remainder arising from the change in the order of action

of operator arguments in the pseudodifferential operator belongs to the space $I^m_\gamma$ by virtue of the estimates (3.51).

4) It suffices to prove that the fiber variation of the product of the symbol (3.52) by $\varphi(r)$ on the right or on the left is compact. (Compact smoothing symbols have compact fiber variation even without this factor.) We represent the symbol (3.52) in the form

$$D(x,\xi) = B(x,\xi,\xi),$$

where the first argument $\xi$ of $B$ corresponds to the argument $\xi$ of the cutoff functions $\chi_1$, $\chi_2$, and $\psi$, and the second, to the argument $\xi$ of the symbol $\widetilde{H}$ itself. Then we have

$$D(x,\xi) - D(x,\xi') = \bigl[B(x,\xi,\xi) - B(x,\xi',\xi)\bigr] + \bigl[B(x,\xi',\xi) - B(x,\xi',\xi')\bigr].$$

The first bracketed expression is a pseudodifferential operator of order $m-1$ with compactly supported Schwartz kernel and hence is compact. Hence it suffices to prove that the second expression multiplied by $\varphi(r)$ on the right or on the left is compact. But this assertion is obvious, since the symbol $\partial H_\partial/\partial\xi$ is homogeneous of degree $m-1$ in the momentum variables and hence the operator $\partial\widehat{H}_\partial/\partial\xi$ without the factor $\varphi(r)$ is continuous in the spaces $\mathcal{K}^{s,\gamma}(K_\Omega) \longrightarrow \mathcal{K}^{s-m+1,\gamma-m}(K_\Omega)$. The multiplication by $\varphi(r)$ with regard to the fact that the support of the kernel is separated from $r=0$ gives the desired compactness result.

The proof of the theorem is complete. □

PROPOSITION 3.55. *If $D \in \mathcal{D}_m(M)$ is an edge-degenerate differential operator of order $m$, then $\sigma_\wedge(D) \in \mathrm{Edge}^m_\gamma$ for each $\gamma$.*

*The proof* is by a straightforward verification. □

***Smoothing edge symbols of order*** $m$**.** In what follows, we also use the ideal of smoothing edge symbols.

DEFINITION 3.56. An edge symbol $D(x,\xi)$ of order $m$ and weight $\gamma$ is said to be *smoothing* if its interior symbol is zero. The set of smoothing edge symbols of order $m$ and weight $\gamma$ will be denoted by $J^m_\gamma$.

Smoothing edge symbols possess the following properties.

PROPOSITION 3.57. *Let $D \in J^m_\gamma$. Then*

1) *(gain in smoothness) the operator $D(x,\xi)$ and all of its derivatives are continuous in the spaces*

$$D^{(\alpha,\beta)}(x,\xi) : \mathcal{K}^{s,\gamma}(K_\Omega) \longrightarrow \mathcal{K}^{s-m+1,\gamma-m}(K_\Omega), \quad s \in \mathbb{R}$$

2) *the operators $rD(x,\xi)$ and $D(x,\xi)r$ possess the same property*

3) *the family*

$$D(x,\xi) : \mathcal{K}^{s,\gamma}(K_\Omega) \longrightarrow \mathcal{K}^{s-m,\gamma-m}(K_\Omega)$$

*has a compact fiber variation (i.e., the differences* $D(x,\xi) - D(x,\xi')$ *are compact in these spaces)*

4) *conditions* 1)–3) *remain valid for the family* $(1+r^2)^l D(x,\xi)(1+r^2)^{-l}$ *for arbitrary* $l \in \mathbb{R}$

5) *if* $\varphi(r)$ *is a smooth function such that* $\varphi(r) = 0$ *in a neighborhood of zero and* $\varphi(r) = 1$ *for large* $r$*, then the operators*

$$\varphi(r)D(x,\xi),\ D(x,\xi)\varphi(r) : \mathcal{K}^{s,\gamma}(K_\Omega) \longrightarrow \mathcal{K}^{s-m,\gamma-m}(K_\Omega)$$

*are compact*

*Proof.* Properties 1)–4) are an immediate consequence of the definitions. Let us prove 5). The operators $\varphi(r)D(x,\xi)$ and $D(x,\xi)\varphi(r)$ in these spaces can be represented as the compositions

$$\mathcal{K}^{s,\gamma}(K_\Omega) \xrightarrow{rD(x,\xi)} \mathcal{K}^{s-m+1,\gamma-m}(K_\Omega) \xrightarrow{\varphi(r)/r} \mathcal{K}^{s-m,\gamma-m}(K_\Omega)$$

and

$$\mathcal{K}^{s,\gamma}(K_\Omega) \xrightarrow{\varphi(r)/r} \mathcal{K}^{s-1,\gamma}(K_\Omega) \xrightarrow{D(x,\xi)r} \mathcal{K}^{s-m,\gamma-m}(K_\Omega),$$

respectively, of continuous operators. Moreover, the operator $\varphi(r)/r$ is compact in both cases. (Indeed, the smoothness exponent in both cases is diminished by one, the weight exponent remains unchanged, and the function $\varphi(r)/r$ is equal to zero in a neighborhood of zero and tends to zero at infinity.) □

### 3.4.3. Operators on manifolds with edges

Now we shall define pseudodifferential operators. This shall be done in several steps. First, we describe their general continuity properties. Then in the set of all operators satisfying these continuity properties we single out pseudodifferential operators as operators that possess edge and interior symbols. Finally, we define quantization, i.e., a mapping that assigns pseudodifferential operators to compatible pairs (interior symbol, edge symbol) and show that this mapping possesses all necessary properties.

***Operators of order*** $m$***.*** We assume the weight exponent $\gamma \in \mathbb{R}$ to be fixed.

DEFINITION 3.58. An operator

$$D : C_0^\infty(\mathcal{M}^\circ) \longrightarrow \mathcal{D}'(\mathcal{M}^\circ)$$

is called an *operator of order* $m$ (*with weight exponent* $\gamma$) if it can be extended to a continuous operator in the spaces

$$D : \mathcal{W}^{s,\gamma}(\mathcal{M}) \longrightarrow \mathcal{W}^{s-m,\gamma-m}(\mathcal{M}) \tag{3.56}$$

for every $s \in \mathbb{R}$. An operator of order $m$ is said to be *negligible* if it is compact in the spaces (3.56) and continuous in the spaces

$$D : \mathcal{W}^{s,\gamma}(\mathcal{M}) \longrightarrow \mathcal{W}^{s-m+1,\gamma-m}(\mathcal{M}) \tag{3.57}$$

for every $s \in \mathbb{R}$.

The space of operators of order $m$ will be denoted by $\mathrm{Op}_\gamma^m = \mathrm{Op}_\gamma^m(\mathcal{M})$, and the subspace of negligible operators by $J\,\mathrm{Op}_\gamma^m(\mathcal{M}) \subset \mathrm{Op}_\gamma^m(\mathcal{M})$.

Obviously, the operator product induces bilinear mappings

$$\mathrm{Op}_{\gamma-m}^l \times \mathrm{Op}_\gamma^m \longrightarrow \mathrm{Op}_\gamma^{l+m},$$
$$J\,\mathrm{Op}_{\gamma-m}^l \times \mathrm{Op}_\gamma^m \longrightarrow J\,\mathrm{Op}_\gamma^{l+m}, \quad \mathrm{Op}_{\gamma-m}^l \times J\,\mathrm{Op}_\gamma^m \longrightarrow J\,\mathrm{Op}_\gamma^{l+m}$$

for any $l$ and $m$.

***The compatibility condition.*** Our pseudodifferential operators will have interior symbol $\sigma$ and edge symbol $\sigma_\wedge$.

What pairs $(\sigma, \sigma_\wedge)$ should be quantized? Note that the interior symbol and the edge symbol of a *differential* operator are related by the compatibility condition (1.65).

We impose the same compatibility condition on the symbols of pseudodifferential operators.

***Definition of pseudodifferential operators.*** We introduce the notion of smooth functions on $\mathcal{M}$. (For manifolds with conical singularities, this was done in Section 1.2.6.)

DEFINITION 3.59. A function $\varphi$ on a manifold $\mathcal{M}$ with edges is said to be *smooth* if the lift of $\varphi$ to the corresponding manifold $M$ with boundary via the natural projection $M \longrightarrow \mathcal{M}$ is a smooth function. In other words, $\varphi \in C^\infty(\mathcal{M})$ if and only if $\varphi \in C^\infty(M)$ and the restriction $\varphi|_{\partial M}$ is constant along the fibers of $\pi$.

Now we are in a position to give the definition of pseudodifferential operators.

DEFINITION 3.60. An operator $P \in \mathrm{Op}_\gamma^m$ is called a *pseudodifferential operator of order* $m$ *and weight* $\gamma$ on the manifold $\mathcal{M}$ if the following conditions are satisfied:

1) the inclusion $[\varphi, P] \in J\,\mathrm{Op}_\gamma^m$ holds for each function $\varphi \in C^\infty(\mathcal{M})$

Next, there exists a smooth interior symbol $\sigma$ of order $m$ on $T_0^*\mathcal{M}$ and an edge symbol $\sigma_\wedge \in \mathrm{Edge}_\gamma^m$ satisfying the compatibility condition (1.65) and such that

2) the operator $P$ is pseudodifferential with principal symbol $\sigma$ on the open manifold $M^\circ$

3) the operator $P$ can be represented modulo $J\,\mathrm{Op}_\gamma^m$ in the form

$$P = (\varphi(r)\sigma_\wedge)\left(x, -i\frac{\partial}{\partial x}\right) + rQ, \tag{3.58}$$

where $\varphi(r)$ is an $R$-function and $Q \in \mathrm{Op}_\gamma^m$ is a pseudodifferential operator on the open manifold $M^\circ$ such that $Q$ can be represented in a neighborhood of $X$ (that is, after the multiplication by $R$-functions on the right and on the left) as a pseudodifferential operator on $X$ with operator-valued symbol

$$q(x,\xi) \in S^0_{CV}(X; \mathcal{K}^{s,\gamma}(K_\Omega), \mathcal{K}^{s-m,\gamma-m}(K_\Omega))$$

of compact variation for each $s \in \mathbb{R}$

The set of pseudodifferential operators of order $m$ and weight $\gamma$ will be denoted by $\mathrm{PSD}_\gamma^m \equiv \mathrm{PSD}_\gamma^m(\mathcal{M})$.

*Remark* 3.61. The interior symbol of the operator $Q$ in (3.58) is of course

$$\sigma(Q) = r^{-1}\big(\sigma - \sigma|_{\partial T^*\mathcal{M}}\varphi(r)\big).$$

PROPOSITION 3.62. *The interior symbol and the edge symbol of a pseudodifferential operator $P$ are uniquely determined.*

◀ This follows from the fact that either of these symbols can be computed from the operator by some limit procedure. To compute the interior symbol of $P$, it suffices to multiply the operator on both sides by cutoff functions supported in $M^\circ$ and then apply, say Hörmander's procedure (see Section 3.2.3). To compute the edge symbol of $P$ at a point $(x_0, \xi_0) \in T_0^*X$, one multiplies $P$ on the right and left by a function $\psi \in C^\infty(\mathcal{M})$ supported in a small neighborhood of $x_0$. Then the operator $P' = \psi P \psi$ can be treated as an operator that acts on functions of local coordinates $(x_1, \dots, x_n)$ ranging in the space of functions on $K_\Omega$ and, by virtue of condition (3.58), represented as

$$P' = P'\left(x, -i\frac{\partial}{\partial x}\right),$$

where the complete symbol

$$P'(x,\xi) = \psi\varphi(r)\sigma_\wedge(x,\xi)\psi + r\psi q(x,\xi)\psi$$

is uniquely determined by $P'$. Now, by the homogeneity properties of the edge symbol,

$$\begin{aligned}\lambda^{-m}\varkappa_\lambda^{-1}P'(x,\lambda\xi)\varkappa_\lambda = \varkappa_\lambda^{-1}\psi\varphi(r)\varkappa_\lambda \circ \sigma_\wedge(x,\xi) \circ \varkappa_\lambda^{-1}\psi\varkappa_\lambda \\ + \lambda^{-1}r \circ \lambda^{-m}\varkappa_\lambda^{-1}\psi q(x,\lambda\xi)\psi\varkappa_\lambda,\end{aligned}$$

and, by virtue of the estimates imposed on $q(x,\xi)$, the right-hand side strongly converges as $\lambda \to \infty$ to $\sigma_\wedge(x,\xi)$. Thus we have reconstructed the edge symbol from $P$, which proves its uniqueness. ▶

***Quantization.*** The computation of symbols is embedded in the very definition of pseudodifferential operators. Now our task is to construct an inverse mapping, *quantization*. Thus pseudodifferential operators will be obtained by quantization of *pairs* $(\sigma,\sigma_\wedge)$, where $\sigma$ is the interior symbol (a homogeneous function on $T_0^*\mathcal{M}$) and $\sigma_\wedge = \sigma_\wedge(x,\xi)$ is the edge symbol.

Let $\mathcal{A}_\gamma^m$ be the set of pairs $(\sigma,\sigma_\wedge)$ satisfying the compatibility condition (1.65), where $\sigma \in \mathcal{O}^m(T_0^*\mathcal{M})$ is an interior symbol of order $m$ and $\sigma_\wedge \in \operatorname{Edge}_\gamma^m$. This is obviously a linear space, and we have the natural embedding

$$\begin{aligned} J_\gamma^m &\hookrightarrow \mathcal{A}_\gamma^m \\ \sigma_\wedge &\longmapsto (0,\sigma_\wedge). \end{aligned}$$

Having this embedding in mind, we sometimes denote the corresponding element of $\mathcal{A}_\gamma^m$ merely by $\sigma_\wedge$ instead of $(0,\sigma_\wedge)$.

PROPOSITION 3.63. *The componentwise multiplication induces a bilinear mapping*

$$\mathcal{A}_{\gamma-m}^l \times \mathcal{A}_\gamma^m \longrightarrow \mathcal{A}_\gamma^{m+l},$$

*and the subspace of smoothing edge symbols is an "ideal" in the sense that*

$$J_{\gamma-m}^l \times \mathcal{A}_\gamma^m \longrightarrow J_\gamma^{m+l}, \quad \mathcal{A}_{\gamma-m}^l \times J_\gamma^m \longrightarrow J_\gamma^{m+l}.$$

*Proof.* Both assertions follow from item 2) of Theorem 3.54, since the compatibility condition is linear and multiplicative. □

Note that the above-mentioned embedding $J_\gamma^m \hookrightarrow \mathcal{A}_\gamma^m$ gives rise to the short exact sequence

$$0 \longrightarrow J_\gamma^m \longrightarrow \mathcal{A}_\gamma^m \longrightarrow \mathcal{O}^m \longrightarrow 0,$$

where $\mathcal{O}^m \equiv \mathcal{O}^m(T_0^*\mathcal{M})$ is the space of interior symbols of order $m$. We use this exact sequence to construct quantization of $\mathcal{A}_\gamma^m$ modulo negligible operators. Namely, first we quantize the elements of the ideal $J_\gamma^m$ and then extend the quantization to the entire algebra $\mathcal{A}_\gamma^m$.

Thus let us quantize the ideal $J_\gamma^m$. Let $a \in J_\gamma^m$, i.e., $a = (0,\sigma_\wedge)$, where $\sigma_\wedge \in J_\gamma^m$. Let $\varphi(r)$ be an $R$-function, and let

$$\widehat{a} = \varphi(r)\sigma_\wedge\left(x, -i\frac{\partial}{\partial x}\right)\varphi(r), \tag{3.59}$$

where the $\psi$DO with operator-valued symbol $\varphi(r)\sigma_\wedge(x,\xi)\varphi(r)$ is defined in the usual way with the help of a partition of unity subordinate to a cover of $X$ by local charts (see Section 3.3). Owing to the presence of the factors $\varphi(r)$ on the right and on the left, we can interpret the operator (3.59) not only as an operator on the infinite wedge $W$ with base $X$ and fiber the infinite cone $K_\Omega$ but also as an operator on $\mathcal{M}$. (The support of its kernel is contained in the Cartesian product $U \times U$ of the collar neighborhood of the boundary $\partial M$ by itself.)

THEOREM 3.64. *The following assertions hold.*

1) *The interior symbol of the operator* $\widehat{a}$ *is zero.*

2) *The operator* $\widehat{a}$ *is modulo the subspace* $J\,\mathrm{Op}_\gamma^m$ *independent of the choice of a cutoff function* $\varphi$ *and other ambiguous elements of the construction.*

*Proof.* If $a \in J_\gamma^m$, then, by property 1 in Proposition 3.57, the products $(1-\psi(r))\widehat{a}$ and $\widehat{a}(1-\psi(r))$ raise the smoothness by one and are compact in the corresponding spaces. From this, we readily obtain both assertions of the theorem. □

Let $a = (\sigma, \sigma_\wedge) \in \mathcal{A}_\gamma^m$ be a given symbol. We wish to assign a pseudodifferential operator to it. This can be done as follows.

1) First, we construct some pseudodifferential operator $P_0 \in \mathrm{PSD}_\gamma^m$ with interior symbol $\sigma$.

2) Then the edge symbols $\sigma_\wedge$ and $\sigma_\wedge(P_0)$ are compatible with the same interior symbol $\sigma$, so that their difference is compatible with the zero interior symbol. This means that
$$\sigma_\wedge - \sigma_\wedge(P_0) \in J_\gamma^m.$$

3) Thus, using the construction from the preceding item, we can construct a pseudodifferential operator $P_1$ with interior symbol $\sigma(P_1) = 0$ and edge symbol
$$\sigma_\wedge(P_1) = \sigma_\wedge - \sigma_\wedge(P_0).$$

4) It remains to set
$$\widehat{a} = P_0 + P_1;$$
now we have
$$\sigma(\widehat{a}) = \sigma, \quad \sigma_\wedge(\widehat{a}) = \sigma_\wedge$$
by construction.

All steps except for the first are obvious, and it remains to explain the first step. It suffices to consider the case in which the interior symbol $\sigma$ is supported in a neighborhood of the edge. (The general case then follows with the use of a partition of unity.) We set
$$P_0 = e(r) D\left(x, -i\frac{\partial}{\partial x}\right) e(r), \tag{3.60}$$

where $e(r)$ is an $R$-function equal to unity in a wider neighborhood of the edge and the symbol $D(x,\xi)$ is given by the formula

$$D(x,\xi) = \chi_1(r|\xi|)h\psi(r|\xi|) + (1-\chi_2(r|\xi|))\widehat{\sigma}(1-\psi(r|\xi|)). \tag{3.61}$$

Here $\chi_1$, $\chi_2$, and $\psi$ are $R$-functions such that

$$\chi_1\psi = \psi, \quad (1-\chi_2)(1-\psi) = 1-\psi, \tag{3.62}$$

$h$ is an arbitrary cone symbol compatible with $\sigma$, and the operator $\widehat{\sigma}$ is constructed in the same way as in the definition of edge symbols. (The only difference is that the symbol $\sigma$ has an additional dependence on the variable $r$. This does not cause any complications.)

PROPOSITION 3.65. *The operator* (3.60) *satisfies* $P_0 \in \mathrm{PSD}^m_\gamma$ *and*

$$\sigma(P_0) = \sigma.$$

*Proof.* The fact that the operator (3.60) is a pseudodifferential operator on the open manifold $M^\circ$ and the relation $\sigma(P_0) = \sigma$ can be verified by an easy computation involving the representation of pseudodifferential operators via symbols with the help of the Fourier transform. To prove that $P_0 \in \mathrm{PSD}^m_\gamma$, it suffices to show that $P_0$ has an edge symbol. This is clear directly from (3.60). Indeed, one can show that the operator $P_0$ can be represented in the form (3.58) with the edge symbol obtained from (3.61) by setting $r = 0$ in the coefficients of the operator $h$ and in the symbol $\sigma$. □

***Calculus.*** The quantization mapping constructed above has the following properties.

THEOREM 3.66. *The quantization mapping* $a \longmapsto \widehat{a}$ *is well defined and unique modulo negligible operators. This mapping is the right inverse and the left almost inverse of the symbol mapping*:

$$\begin{aligned} &\textit{if} \quad a = (\sigma, \sigma_\wedge) \in \mathcal{A}^m_\gamma, \quad \textit{then} \quad \sigma(\widehat{a}) = \sigma, \ \ \sigma_\wedge(\widehat{a}) = \sigma_\wedge; \\ &\textit{if} \quad P \in \mathrm{PSD}^m_\gamma, \quad \textit{then} \quad (\sigma(P), \sigma_\wedge(P))^\wedge - P \in J\,\mathrm{Op}^m_\gamma. \end{aligned}$$

*The proof* readily follows from the construction. □

Moreover, the following assertion holds.

COROLLARY 3.67. *An operator* $P \in \mathrm{PSD}^m_\gamma$ *is compact in the spaces*

$$P : \mathcal{W}^{s,\gamma}(\mathcal{M}) \longrightarrow \mathcal{W}^{s-m,\gamma-m}(\mathcal{M}), \quad s \in \mathbb{R},$$

*if and only if*

$$\sigma(P) = 0, \quad \sigma_\wedge(P) = 0.$$

◀ The "if" part is trivial, since quantization is unique modulo negligible operators, which are compact. The "only if" part can be proved in a standard way by contradiction: one assumes that one of the symbols is nonzero at some point and then constructs a sequence of functions that weakly converges to zero but is not taken by the operator to a sequence strongly convergent to zero. We omit these technicalities. ▶

Let us now state the main theorems of the calculus of edge-degenerate pseudodifferential operators.

THEOREM 3.68 (on the composition of edge-degenerate operators). *The product of operators induces a bilinear mapping*

$$\mathrm{PSD}^{l}_{\gamma-m}(\mathcal{M}) \times \mathrm{PSD}^{m}_{\gamma}(\mathcal{M}) \longrightarrow \mathrm{PSD}^{l+m}_{\gamma}(\mathcal{M})$$

*for any $l$ and $m$. The symbols of the product are given by the formulas*

$$\sigma(D_2D_1) = \sigma(D_2)\sigma(D_1), \quad \sigma_\wedge(D_2D_1) = \sigma_\wedge(D_2)\sigma_\wedge(D_1). \tag{3.63}$$

The last assertion can also be represented in the following form. If $a \in \mathcal{A}^{l}_{\gamma-m}$ and $b \in \mathcal{A}^{m}_{\gamma}$, then

$$\widehat{(ab)} = \widehat{a}\,\widehat{b} \mod J\,\mathrm{Op}^{m+l}_{\gamma}.$$

THEOREM 3.69 (on the adjoint operator). *The passage to the adjoint operator with respect to the inner product in $\mathcal{W}^{0,0}(\mathcal{M})$ induces an antilinear mapping*

$$\mathrm{PSD}^{m}_{\gamma}(\mathcal{M}) \longrightarrow \mathrm{PSD}^{m}_{m-\gamma}(\mathcal{M}).$$

*The symbols of the adjoint operator are given by the formulas*

$$\sigma(D^*) = \sigma(D)^*, \quad \sigma_\wedge(D^*) = \sigma_\wedge(D)^*. \tag{3.64}$$

*The proof* of both theorems is by a straightforward computation. With some technical effort, it can be obtained from the composition theorems for usual pseudodifferential operators and for pseudodifferential operators with operator-valued symbols (Section 3.3). □

## 3.5. Ellipticity and Finiteness Theorems

### 3.5.1. The abstract finiteness theorem

Let us summarize the properties possessed by the $\psi$DO calculus in all cases considered in the preceding sections. To simplify the statements, we restrict ourselves to zero-order operators[4] and use an abstract algebraic language.

[4]As a rule, once the calculus of zero-order $\psi$DO has been constructed, the general case follows by order reduction.

A general algebra $\mathcal{A}$ of zero-order $\psi$DO is a subalgebra of the algebra $\mathcal{B}(\mathcal{H})$ of bounded operators in a Hilbert space $\mathcal{H}$. There is a *symbol mapping* defined on this algebra, i.e., a homomorphism $\sigma : \mathcal{A} \longrightarrow \mathcal{S}$ into a unital topological algebra $\mathcal{S}$ such that

$$\boxed{\sigma(A) = 0 \textit{ if and only if } A \textit{ belongs to the ideal } \mathcal{K} \textit{ of compact operators.}}$$

In other words, the homomorphism $\sigma$ generates a well-defined *monomorphism*

$$\widetilde{\sigma} : \mathcal{A}/(\mathcal{K} \cap \mathcal{A}) \longrightarrow \mathcal{S}$$

of the *Calkin algebra* $\mathcal{A}/(\mathcal{K}\cap\mathcal{A})$ into $\mathcal{S}$. Next, there is a continuous linear mapping

$$Q : \mathcal{S} \longrightarrow \mathcal{A}$$

such that

$$\boxed{\sigma Q = \mathrm{id} : \mathcal{S} \longrightarrow \mathcal{S} \textit{ (i.e., } Q \textit{ is the right inverse of } \sigma\textit{).}}$$

DEFINITION 3.70. Under the above-mentioned conditions, we refer to $\mathcal{S}$ as the *algebra of* (*principal*) *symbols* of elements of the algebra $\mathcal{A}$. The element $\sigma(A)$, where $A \in \mathcal{A}$, is called the *symbol* of the operator $A$. The mapping $Q$ is called *quantization*.

DEFINITION 3.71. An element $A \in \mathcal{A}$ is said to be *elliptic* if its symbol $\sigma(A)$ is an invertible element of the algebra $\mathcal{S}$. Invertible symbols are also called elliptic.

The main analytic fact of elliptic theory is the relationship between ellipticity and the Fredholm property: an operator is Fredholm provided that its symbol is invertible. Assertions of this kind are known as *finiteness theorems*. The following (entirely trivial) assertion shows that the finiteness theorem is valid in the abstract framework of Definition 3.70.

PROPOSITION 3.72. 1) *Under the above-mentioned conditions, the mapping $Q$ is a left almost inverse of $\sigma$ and an almost algebra homomorphism, i.e., a homomorphism modulo the ideal of compact operators*:

$$Q(\sigma(A)) - A \in \mathcal{K}, \quad Q(a)Q(b) - Q(ab) \in \mathcal{K} \quad \textit{for any } A \in \mathcal{A},\ a, b \in \mathcal{S}. \tag{3.65}$$

2) *If an operator $A \in \mathcal{A}$ is elliptic* (*i.e., if its symbol $\sigma(A)$ is invertible*), *then $A$ is Fredholm.*

*Proof.* We have

$$\sigma(Q(\sigma(A)) - A) = \sigma(A) - \sigma(A) = 0,$$

so that $Q(\sigma(A)) - A$ is compact. Likewise,

$$\sigma(Q(a)Q(b) - Q(ab)) = ab - ab = 0,$$

and $Q(a)Q(b) - Q(ab) \in \mathcal{K}$. This proves 1). To prove 2), note that if $A \in \mathcal{A}$ is elliptic, then $Q(\sigma(A)^{-1})$ is a two-sided almost inverse of $A$, that is, an inverse modulo the ideal of compact operators, and hence $A$ is Fredholm. □

*Remark* 3.73. The converse of 2) is also true under some additional assumptions. For example, suppose that $\mathcal{A}$ is a $C^*$-algebra. (In applications to pseudodifferential operators, this usually means the passage to the norm closure.) If the operator $A$ is Fredholm, then the coset $[A] \in \mathcal{B}(\mathcal{H})/\mathcal{K}$ is invertible. The ellipticity of $A$ will follow once we prove the invertibility of the coset of $A$ in $\mathcal{A}/(\mathcal{K}\cap\mathcal{A})$. This is however trivial: we have the $C^*$-algebra monomorphism

$$\mathcal{A}/(\mathcal{K} \cap \mathcal{A}) \longrightarrow \mathcal{B}(\mathcal{H})/\mathcal{K},$$

and the desired assertion follows from the theorem saying that an element invertible in a $C^*$-algebra is invertible in any $C^*$-subalgebra containing it.

An alternative, weaker assumption, which often holds already for the original algebras, is that the algebra of pseudodifferential operators is a local $C^*$-algebra (Blackadar 1998), whereby the spectrum of an element in a subalgebra coincides with its spectrum in the whole algebra.

Interestingly enough, one can say something about index theory even in this humble abstract framework. Namely, the conventional invariance properties of the index hold.

PROPOSITION 3.74. 1) *The index of an elliptic element $A \in \mathcal{A}$ is completely determined by the principal symbol $\sigma(A)$.*

2) *Let $A_0, A_1 \in \mathcal{A}$ be elements such that their symbols $\sigma(A_0)$ and $\sigma(A_1)$ are homotopic in the class of elliptic symbols (i.e., can be joined by a continuous curve $\sigma_t$, $t \in [0,1]$, consisting of elliptic symbols). Then*

$$\operatorname{ind} A_0 = \operatorname{ind} A_1.$$

*Proof.* Indeed, the operator $A - Q(\sigma(A))$ is compact by Proposition 3.72, 1), so that

$$\operatorname{ind} A = \operatorname{ind} Q(\sigma(A)),$$

whence assertion 1) follows. To prove assertion 2), it suffices to join $A_0$ and $A_1$ by a continuous curve $A_t$ consisting of Fredholm operators, since the index of a Fredholm operator is homotopy invariant. In view of 1), we can assume that $A_j = Q(\sigma(A_j))$, $j = 0, 1$, and then the desired curve can be taken in the form $A_t = Q(\sigma_t)$. □

Thus the solution of the index problem for a $\psi$DO algebra $\mathcal{A}$ can be carried out, for example, as follows:

1. One obtains a homotopy classification of the set of elliptic symbols.
2. One constructs an "index functional" $\operatorname{ind}_t$ on the set of homotopy classes of elliptic symbols such that the index formula

$$\operatorname{ind} A = \operatorname{ind}_t[\sigma(A)]$$

holds, where $[\sigma(A)]$ is the homotopy class of an elliptic symbol $\sigma(A)$.

*Remark* 3.75. In fact, we have somewhat simplified the situation: the genuine homotopy classification is usually intractable, and one deals with a *stable homotopy classification*, where the equivalence relation on symbols is generated by homotopies *and* by the addition of "trivial" direct summands (for which the index of the corresponding operators is a priori zero). In various implementations of this scheme, the class of trivial symbols is described in different ways.

The abstract finiteness theorem stated in this subsection implies finiteness theorems in specific situation as its special cases. For the reader's convenience, in the subsequent subsections we list the cases which are of interest to us; sometimes, the assertion of the finiteness theorem for a specific case is refined (i.e., made stronger).

### 3.5.2. Agranovich–Vishik families

DEFINITION 3.76. A classical parameter-dependent pseudodifferential operator $A$ is said to be *Agranovich–Vishik elliptic* if the principal symbol $\sigma(A)(x,\xi,p)$ is invertible on $T^*X \times V$ outside the zero section $\{\xi = 0, p = 0\}$.

Agranovich–Vishik ellipticity implies usual ellipticity for each value of the parameter $p$ but is much stronger: for an Agranovich–Vishik elliptic parameter-dependent $\psi$DO $A(p)$, one can construct a parameter-dependent pseudodifferential regularizer $R(p)$ (simply by setting $\sigma(R) = \sigma(A)^{-1}$), and then the remainders $Q(p)$ and $Q'(p)$ in the formulas

$$A(p)R(p) = 1 + Q(p), \quad R(p)A(p) = 1 + Q'(p)$$

are parameter-dependent $\psi$DO of order $\le -1$ and hence have small norms for large values of $|p|$. It follows that the individual operators are *not only Fredholm but also invertible* for large $|p|$.

### 3.5.3. Operators in sections of Hilbert bundles

The composition formula modulo compact operators given by Proposition 3.33 permits one to give a natural definition of ellipticity and prove the finiteness theorem.

Since we do not impose any homogeneity condition on the symbols, we must be careful in the definition of ellipticity.

DEFINITION 3.77. A symbol $p \in S^0_{CV}(T^*_0X, \mathfrak{H}_1, \mathfrak{H}_2)$ is said to be *elliptic* if there exists an $R > 0$ such that $p(x,\xi)$ is invertible for $|\xi| > R$ and the inverse satisfies

$$\psi(\xi)p(x,\xi)^{-1} \in S^0_{CV}(T^*_0X, \mathfrak{H}_2, \mathfrak{H}_1), \tag{3.66}$$

where $\psi(\xi)$ is a smooth function vanishing for $|\xi| \le R$ and equal to unity at infinity.

This is of course equivalent to saying that

$$\left\| p(x,\xi)^{-1} : H_{2\xi} \longrightarrow H_{1\xi} \right\|$$

is uniformly bounded for large $\xi$.

*Remark* 3.78. Note that if a symbol $p$ is elliptic, then it is Fredholm for all $(x,\xi) \in T^*_0X$ and its arbitrary fill-up $\widetilde{p}$ is Fredholm on $T^*X$. Indeed, this follows from the fact that the addition of a compact operator to a Fredholm operator does not destroy the Fredholm property.

PROPOSITION 3.79. *A symbol $p \in S^0_{CV}(T^*_0X, \mathfrak{H}_1, \mathfrak{H}_2)$ is elliptic if and only if there exists a symbol $q \in S^0_{CV}(T^*X, \mathfrak{H}_2, \mathfrak{H}_1)$ such that the symbols $q\widetilde{p} - 1$ and $\widetilde{p}q - 1$ are compactly supported and compact-valued for any fill-up $\widetilde{p} \in S^0_{CV}(T^*X)$ of the symbol $p$.*

*Proof.* We define $q(x,\xi)$ for $|\xi| > R$ by the formula

$$q(x,\xi) = p(x,\xi)^{-1}.$$

Outside the balls of radius $R$, this symbol satisfies the estimates (3.27) and has a compact fiber variation. We extend it into the interior of the above-mentioned balls using the construction in the proof of Lemma 3.27 with an excision function $\chi(t)$ equal to zero for $t \le R$ and with the cosphere bundle of radius $R$. By construction, the resulting symbol lies in $S^0_{CV}(T^*X)$; we again denote it by $q(x,\xi)$. The fact that $q\widetilde{p} - 1$ and $\widetilde{p}q - 1$ are compactly supported and compact-valued is then obvious. □

*Remark* 3.80. This shows that the ellipticity of a symbol $p \in S^0_{CV}(T^*_0X, \mathfrak{H}_1, \mathfrak{H}_2)$ is equivalent to the invertibility of the corresponding "principal symbol," i.e., the element generated by $p$ in the quotient space $S^0_{CV}(T^*_0X, \mathfrak{H}_1, \mathfrak{H}_2)/J^{-1}_K(T^*X, \mathfrak{H}_1, \mathfrak{H}_2)$.

THEOREM 3.81 (the finiteness theorem). *If*

$$p \in S^0_{CV}(T^*_0X, \mathfrak{H}_1, \mathfrak{H}_2)$$

*is an elliptic symbol, then the operator*

$$P = p(x, -i\partial/\partial x) : \mathcal{H}_1 \longrightarrow \mathcal{H}_2$$

*is Fredholm.*

*Proof.* This is a special case of Proposition 3.72. □

### 3.5.4. Operators on singular manifolds

***Cone-degenerate operators.*** The composition Theorem 3.44 implies definition of ellipticity and the validity of the finiteness Theorem 2.13 in the standard way (cf. Proposition 3.72). Naturally, the theorem holds not only for differential but also for pseudodifferential operators. We give this statement, since it is refined compared with the general case: the smoothness of solutions is also proved.

DEFINITION 3.82. An operator $D \in \Psi_\gamma^m(M)$ on a manifold with conical singularities is said to be *elliptic* if its symbol $\mathfrak{s}(D) \in \Sigma_\gamma^m(M)$ is invertible.[5]

THEOREM 3.83. *If an operator $D \in \Psi_\gamma^m(M)$ is elliptic, then it is Fredholm. Moreover, its kernel and cokernel (and hence the index) are independent of the smoothness exponent $s$.*

◀ Only the independence of the kernel and cokernel on the smoothness exponent needs some clarification. We construct an almost inverse $R \in \Psi_{\gamma-m}^{-m}(M)$. Then the remainders $Q_j$, $j = 1, 2$, in the formulas

$$RD - 1 = Q_1, \quad DR - 1 = Q_2$$

are negligible. In particular, they raise smoothness by one by the definition of a negligible operator. Now if, say, $u$ is an element of $\operatorname{Ker} D$ in $H^{s,\gamma}(\mathcal{M})$, then

$$u = RDu - Q_1 u = -Q_1 u \in H^{s+1,\gamma}(\mathcal{M}),$$

and the desired assertion follows by induction. ▶

***Edge symbols.*** In view of the composition formula for edge symbols provided by Theorem 3.54, the finiteness theorem for them is an immediate consequence of the standard scheme of reasoning provided by Proposition 3.72.

THEOREM 3.84. *Let $D(x,\xi) \in \mathrm{Edge}_\gamma^m$ be a given edge symbol. If the interior symbol $\sigma(D)(\omega,\xi,p,q)$ is invertible for $(\xi,p,q) \neq 0$ and the cone symbol $\sigma_c(D)$ is invertible on the weight line $\mathcal{L}_\gamma$, then the operator family*

$$D(x,\xi) : \mathcal{K}^{s,\gamma}(K_\Omega) \longrightarrow \mathcal{K}^{s-m,\gamma-m}(K_\Omega) \tag{3.67}$$

*is Fredholm for all $s \in \mathbb{R}$, and its kernel, cokernel, and index are independent of $s$. Moreover, $D(x,\xi)$ has an almost inverse in $\mathrm{Edge}_{\gamma-m}^{-m}$. If $D(x,\xi)$ is invertible, then the inverse lies in $\mathrm{Edge}_{\gamma-m}^{-m}$.*

As before, the independence of the kernel and cokernel of the smoothness exponent is derived from the specific properties of remainders in the composition theorem.

[5]The inverse then automatically belongs to $\Sigma_{\gamma-m}^{-m}(M)$.

***Edge-degenerate operators.*** Since the composition theorem holds for edge-degenerate operators, the definition of ellipticity and the finiteness theorem follow by the standard scheme of Proposition 3.72. Let us give the statement of the corresponding theorem.

THEOREM 3.85. *Let $D \in \mathrm{PSD}^m_\gamma(\mathcal{M})$ be an elliptic operator in the sense that its interior symbol and edge symbol are invertible on $T^*_0\mathcal{M}$ and $T^*_0X$, respectively. Then $D$ is Fredholm in the spaces*

$$D : \mathcal{W}^{s,\gamma}(\mathcal{M}) \longrightarrow \mathcal{W}^{s-m,\gamma-m}(\mathcal{M}), \quad s \in \mathbb{R}.$$

*The kernel, cokernel, and index of $D$ are independent of $s$.*

### 3.5.5. Supplement. Proof of Theorem 2.27

We present the proof, even though it is technical, because it has not been published so far.

For brevity, we give the proof for the case of a one-dimensional edge. In the general case, the argument is similar. We deal with the operator family

$$A_h = \overset{2}{r}{}^{-1} a\left(\overset{2}{r}, ihr\overset{1}{\frac{\partial}{\partial r}} + ih\frac{k-1}{2}\right) : \mathcal{K}^{1,1}(K_\Omega) \longrightarrow \mathcal{K}^{0,0}(K_\Omega). \tag{3.68}$$

For the operator (3.68), we construct a two-sided almost inverse $R_h$ modulo $O(h)$ in operator norm.

By $L^k(\Omega) = L^k(\Omega, \overline{\mathbb{R}}_+ \times \mathbb{R})$, $k = 0, \pm 1$, we denote the space of classical pseudodifferential operators of order $k$ on $\Omega$ with parameters $(r,p) \in \overline{\mathbb{R}}_+ \times \mathbb{R}$ in the sense of Agranovich–Vishik. Then $a(r,p) \in L^1(\Omega)$, and under the assumptions of the theorem there exists an $\varepsilon_0 > 0$ such that the family $\varphi(r, p + i\varepsilon) = a(r, p + i\varepsilon)^{-1}$ is defined for all $(r,p) \in \mathbb{R}^2$ and $|\varepsilon| \le \varepsilon_0$; moreover, $\varphi(r, p + i\varepsilon) \in L^{-1}(\Omega)$ uniformly with respect to $\varepsilon \in [-\varepsilon_0, \varepsilon_0]$. We set

$$R_h = \overset{2}{r}\varphi\left(\overset{2}{r}, ihr\overset{1}{\frac{\partial}{\partial r}} + ih\frac{k+1}{2}\right). \tag{3.69}$$

We claim that

1) the operator (3.69) is well defined and bounded in the spaces

$$R_h : \mathcal{K}^{0,0}(K_\Omega) \longrightarrow \mathcal{K}^{1,1}(K_\Omega) \tag{3.70}$$

2) the relations

$$A_h R_h = 1 + Q_h, \quad R_h A_h = 1 + \widetilde{Q}_h \tag{3.71}$$

hold, where

$$\begin{aligned} \left\| Q_h : \mathcal{K}^{0,0}(K_\Omega) \longrightarrow \mathcal{K}^{0,0}(K_\Omega) \right\| &= O(h), \\ \left\| \widetilde{Q}_h : \mathcal{K}^{1,1}(K_\Omega) \longrightarrow \mathcal{K}^{1,1}(K_\Omega) \right\| &= O(h) \end{aligned} \tag{3.72}$$

Now the standard technique of Neumann series implies the assertion of the theorem, so that it remains to prove items 1) and 2).

We make use of the following lemma (see Nazaikinskii, Savin, Schulze and Sternin 2004*a*).

LEMMA 3.86. *Let $f(r,p) \in L^0(\Omega)$. Then the operator*

$$f\left(\overset{2}{r}, ihr\overset{1}{\frac{\partial}{\partial r}} + ih\frac{k+1}{2}\right) : \mathcal{K}^{0,0}(K_\Omega) \longrightarrow \mathcal{K}^{0,0}(K_\Omega) \tag{3.73}$$

*is bounded uniformly with respect to $h \in (0,1]$.*

The lemma implies that the operator $R_h$ is well defined and bounded in $\mathcal{K}^{0,0}(K_\Omega)$, since $r\varphi(r,p) \in L^0(\Omega)$. To show that $R_h$ is actually bounded in the spaces (3.70) as well, we denote the norms $\mathcal{K}^{0,0}(K_\Omega)$ and $\mathcal{K}^{1,1}(K_\Omega)$ by $\|\cdot\|$ and $\|\cdot\|_1$, respectively, and recall that[6]

$$\mathcal{K}^{0,0}(K_\Omega) = L^2(K_\Omega, r^k dr d\omega),$$
$$\|u\|_1 \simeq \|u\| + \|r^{-1}u\| + \|\partial u/\partial r\| + \|r^{-1}\partial u/\partial \omega\|.$$

Thus it suffices to show that the operators

$$r^{-1} \circ R_h, \quad \partial/\partial r \circ R_h, \quad r^{-1}\partial/\partial\omega \circ R_h$$

are continuous in $\mathcal{K}^{0,0}(K_\Omega)$. Each of these operators can be represented as a function of operators in the form $\Phi(\overset{2}{r}, ir\overset{1}{\partial/\partial r} + i(k+1)/2)$, where the symbol $\Phi(r,p)$ is given by the formula

$$\Phi(r,p) = \begin{cases} \varphi(r,p), \\ \left(1 - ip + r\dfrac{\partial}{\partial r}\right)\varphi(r,p), \\ \dfrac{\partial}{\partial\omega} \circ \varphi(r,p) \end{cases}$$

in each of the cases, respectively, and belongs to $L^0(\Omega)$, so that it remains to refer to the lemma once more.

Let us verify (3.71) and (3.72). We compute the first composition in (3.71). Omitting the argument $\omega$ of the coefficients and the numbers over the operator arguments in $\varphi$ and denoting $\widehat{p} := ir\partial/\partial r$ and $\widehat{q} := -i\partial/\partial\omega$, we obtain

$$\begin{aligned} A_h R_h &= r^{-1}\big[a_0 + a_1 r + a_2(\widehat{p} + ih(k-1)/2) + a_3\widehat{q}\big] r\varphi(r, \widehat{p} + ih(k+1)/2) \\ &= r^{-1}r\big[a_0 + a_1 r + a_2(\widehat{p} + ih(k+1)/2) + a_3\widehat{q}\big]\varphi(r, \widehat{p} + ih(k+1)/2) \\ &= F_h(r, \widehat{p} + ih(k+1)/2), \end{aligned}$$

where

$$F_h(r,p) = a(r,p)\varphi(r,p) + iha_2 r\frac{\partial\varphi}{\partial r}(r,p) = 1 + iha_2 r\frac{\partial\varphi}{\partial r}(r,p).$$

---

[6]More rigorously, one should replace the last term in the expression for $\|u\|_1$ by a sum over local charts on $\Omega$ involving a partition of unity.

Since

$$a_2 r \frac{\partial \varphi}{\partial r}(r,p) \in L^{-1}(\Omega) \subset L^0(\Omega),$$

we see that the estimate $O(h)$ for the remainder

$$Q_h = iha_2 r \frac{\partial \varphi}{\partial r}(r,\widehat{p})$$

follows directly from the lemma.

Now let us compute the composition $R_h A_h$:

$$\begin{aligned} R_h A_h &= r\varphi(r,\widehat{p}+ih(k+1)/2)r^{-1}\big[a_0 + a_1 r + a_2(\widehat{p}+ih(k-1)/2) + a_3\widehat{q}\big] \\ &= rr^{-1}\varphi(r,\widehat{p}+ih(k-1)/2)\big[a_0 + a_1 r + a_2(\widehat{p}+ih(k-1)/2) + a_3\widehat{q}\big] \\ &= G_h(r,\widehat{p}+ih(k-1)/2), \end{aligned}$$

where

$$G_h(r,p) = \varphi(r,p)a(r,p) + [\varphi(r,p+ih) - \varphi(r,p)]a_1 r = 1 + ih\Psi_h(r,p),$$

and moreover,

$$\Psi_h(r,p) = \left[\int_0^1 \frac{\partial \varphi}{\partial p}(r,p+ih\lambda)d\lambda\right] a_1 r,$$

so that

$$\Psi_h(r,p+i\varepsilon) \in L^{-1}(\Omega) \subset L^0(\Omega)$$

uniformly with respect to $\varepsilon \in [-\varepsilon_0,\varepsilon_0]$ and $h \in (0,\varepsilon_0]$. Computations similar to the preceding show that the operator

$$\Psi_h\left(\overset{2}{r}, ihr\overset{1}{\frac{\partial}{\partial r}} + ih\frac{k-1}{2}\right) : \mathcal{K}^{1,1}(K_\Omega) \longrightarrow \mathcal{K}^{1,1}(K_\Omega)$$

is bounded uniformly with respect to $h \in (0,\varepsilon_0]$, which completes the proof.

## 3.6. Index Theorems on Smooth Closed Manifolds

This section deals with index theorems on smooth closed manifolds. First of all, this is the classical Atiyah–Singer theorem (which we recall in the first subsection). We also consider Luke's index theorem for operators in Hilbert bundles. Both theorems will be applied in Part III in the computation of the index on singular manifolds.

### 3.6.1. The classical Atiyah–Singer index theorem

We have already noted that the index of a classical elliptic $\psi$DO on a smooth compact manifold is completely determined by its principal symbol and is a homotopy invariant of the latter. More precisely, since the index of an operator with unit principal symbol is zero, it follows that the index of an elliptic operator is a stable homotopy invariant of the principal symbol. In more detail, we say that two elliptic symbols $\sigma_1$ and $\sigma_2$ over a manifold $M$ are *stably homotopic* if for some nonnegative numbers $N_1$ and $N_2$ the symbols $\sigma_1 \oplus \mathbf{1}_{N_1}$ and $\sigma_2 \oplus \mathbf{1}_{N_2}$ are homotopic in the class of elliptic symbols. The set of equivalence classes of elliptic symbols with respect to this equivalence relation is denoted by $\operatorname{Ell}(M)$. It is an abelian group with respect to the operation induced by the direct sum of elliptic operators.

One can give a different description of the group $\operatorname{Ell}(M)$ in terms of vector bundles and $K$-theory. An elliptic symbol

$$\sigma : \pi^*E \longrightarrow \pi^*F, \quad \text{where } \pi : T^*M \longrightarrow M,$$

over a manifold $M$ specifies an isomorphism of the bundles $\pi^*E$ and $\pi^*F$ over $T^*M$ outside the zero section of $T^*M$. In other words, we are given a triple $(\pi^*E, \pi^*F, \sigma)$ consisting of two bundles over the space $T^*M$ and an isomorphism of these bundles outside a compact set. By definition, such a triple specifies an element of the *$K$-group with compact supports* of the space $T^*M$. We denote this element by $[\sigma] \in K_c(T^*M)$.

◀ Recall that the $K$-group with compact supports of a space $X$ is defined as the set of equivalence classes of triples $(E, F, \sigma)$, where $E$ and $F$ are bundles over $X$ and $\sigma$ is an isomorphism of these bundles outside some compact subset of $X$, by the following equivalence relation: two triples $(E_1, F_1, \sigma_1)$ and $(E_2, F_2, \sigma_2)$ are said to be *equivalent* if they become isomorphic after the addition of some trivial triples as direct summands. Moreover, a triple $(E, F, \sigma)$ is said to be *trivial* if $\sigma$ can be extended to an isomorphism defined everywhere on $X$. ▶

THEOREM 3.87 (Atiyah and Singer 1968). *The mapping $\sigma \longmapsto [\sigma]$ is a well-defined abelian group isomorphism*

$$\chi : \operatorname{Ell}(M) \longrightarrow K_c(T^*M). \tag{3.74}$$

The homomorphism $\chi$ is called the *difference construction*, and the element $[\sigma]$ is referred to as the difference element corresponding to the elliptic symbol $\sigma$. Informally speaking, the word "difference" means here that the element $[\sigma]$ is represented as the difference $\pi^*E - \pi^*F$ of some vector bundles over $T^*X$ isomorphic (i.e., cancelling in some sense) at infinity.

Thus the difference construction takes each symbol of an elliptic operator on $M$ to an element of the $K$-group with compact supports of the cotangent bundle of $M$.

It is clear from the preceding that the index of an elliptic operator $A$ is determined by the *difference element* $[\sigma(A)]$ of its symbol.

Atiyah and Singer presented a homomorphism that computes the index in these terms. Namely, if $p : X \longrightarrow Y$ is a smooth mapping of closed manifolds, then in $K$-theory there is a well-defined *direct image homomorphism* (Atiyah 1989)

$$p_! : K_c(T^*X) \longrightarrow K_c(T^*Y).$$

In particular, if $Y = pt$ is a point and $p$ is a constant mapping, we obtain the homomorphism

$$p_! : K_c(T^*X) \longrightarrow K_c(pt) \equiv \mathbb{Z}.$$

THEOREM 3.88 (the Atiyah–Singer index theorem). *For elliptic operators on $M$, the index formula*

$$\operatorname{ind} A = p_!([\sigma(A)]) \tag{3.75}$$

*is valid.*

### 3.6.2. Luke's index theorem

In this subsection, we show how the Atiyah–Singer index theorem expressing the index of an elliptic operator on a smooth manifold can be generalized to the case in which the operator acts in spaces of sections of Hilbert bundles. This generalization was originally carried out in (Luke 1972).

***The difference construction for operator-valued symbols.*** Let $p$ be an operator-valued elliptic symbol of compact variation on $T_0^*X$. It turns out that, just as in the finite-dimensional case, to this symbol there corresponds a well-defined difference element $[p] \in K_c(T^*X)$ of the $K$-group with compact supports of the cotangent bundle. It is defined as follows. The symbol $p$ (more precisely, its fill-up) is a family of Fredholm operators invertible outside a compact set on the cotangent bundle $T^*X$. This family has a well-defined index

$$\operatorname{ind} p \in K_c(T^*X)$$

(*families index*), and one sets $[p] \stackrel{\text{def}}{=} \operatorname{ind} p$.

Let us recall the construction of the families index. The index of a single Fredholm operator is an integer. Now if we have a family $T_y$ of Fredholm operators parametrized by a compact set $Y$, then the "index" of this family is naturally defined as an element of the $K$-group of the parameter space. Indeed, if $\dim \operatorname{Ker} T_y = \text{const}$ (which can always be achieved by a small perturbation of the family), then the kernels and cokernels of operators in the family form vector bundles over the parameter space, and the *index of the family* $T$ is defined as the difference of these two bundles:

$$\operatorname{ind} T = [\operatorname{Ker} T] - [\operatorname{Coker} T] \in K(Y). \tag{3.76}$$

The relationship with the ordinary integer-valued index is obvious: for the singleton $Y = pt$, we have $K(pt) = \mathbb{Z}$ and $\operatorname{ind} T = \dim \operatorname{Ker} T - \dim \operatorname{Coker} T$.

It is known that the families index in the general case is not only independent of a perturbation making the family of kernels a vector bundle but also remains unchanged under continuous deformations of the family. Let $[Y, \mathrm{Fred}]$ be the set of homotopy classes of Fredholm families. The famous Atiyah–Jänich theorem (Atiyah 1989) states that the families index defines an isomorphism

$$[Y, \mathrm{Fred}] \overset{\mathrm{ind}}{\simeq} K(Y)$$

of this set with the $K$-group.

For a noncompact parameter space, it is natural to consider the $K$-group $K_c(Y)$ with compact supports. It turns out that if a Fredholm family is invertible for parameter values outside a compact set, then it also has an index that is an element of the $K$-group with compact supports:

$$\operatorname{ind} T \in K_c(Y).$$

To define this element, one uses an appropriate perturbation of the original family by a family of finite-dimensional operators that vanish at infinity.

◀ More precisely, the perturbation is added to the direct sum of the original family with the zero endomorphism of the trivial bundle $\mathbb{C}^N$ of sufficiently large dimension $N$. Since the perturbation is compactly supported, it follows that both the kernel and the cokernel of the perturbed family at infinity coincide with $\mathbb{C}^N$ and hence we have an isomorphism between them outside a compact set. If, moreover, the kernel and the cokernel are vector bundles (which is guaranteed by the choice of the perturbation), then we have a triple defining an element of the $K$-group with compact supports. ▶

One can readily establish the following properties of the families index.

1) For a compact parameter space, the index coincides with the expression (3.76).

2) The index of a family of operators in finite-dimensional spaces coincides with the difference construction earlier in this subsection.

3) The index is invariant under deformations: if $T_{y,t}$ is a homotopy of Fredholm families, then $\operatorname{ind} T_t = \operatorname{ind} T_0 \in K_c(Y)$; here it is assumed that $T_{y,t}$ is uniformly continuous on some compact set $K \subset Y$ and is always invertible outside $K$.

***The index theorem for operators in sections of Hilbert bundles.*** We shall consider only the case in which the tempered families of norms occurring in the definition of section spaces of Hilbert bundles and symbol classes are constant:

$$\|u\|_\xi \equiv \|u\|_H, \quad H_\xi \equiv H.$$

Then the corresponding section spaces have the simple form

$$\mathcal{H} \equiv \mathcal{F}^{-1} \int H_\xi \, d\xi = L^2(X, H).$$

From now on until the end of the section, we denote the corresponding classes of symbols of compact variation by $S^0_{CV}(T^*_0 X)$, omitting the indication of spaces for brevity.

THEOREM 3.89. *Let $p \in S^0_{CV}(T^*_0 X)$ be an elliptic symbol. Then the index of the corresponding elliptic operator is expressed as the direct image in $K$-theory:*

$$\operatorname{ind} p\left(x, -i\frac{\partial}{\partial x}\right) = \pi_! \operatorname{ind} p, \tag{3.77}$$

*where $\pi : X \longrightarrow pt$ is the mapping of the manifold $X$ into a point.*

*Proof.* Without loss of generality, we can assume from the very beginning that $p \in S^0_{CV}(T^*X)$ (i.e., the fill-up has already been chosen). The proof of the index theorem splits into two stages.

1) *A special case: symbols homogeneous for large* $|\xi|$.

Suppose that the symbol $p$ satisfies the condition

$$p(x, t\xi) = p(x, \xi) \qquad \text{for} \quad \xi \geq R \quad \text{and} \quad t > 1. \tag{3.78}$$

The proof for this case is given in (Luke 1972). The main idea is to deform the operator to the direct sum of an operator induced by a vector bundle isomorphism and an elliptic operator with finite-dimensional symbol.

With this homotopy, it suffices to verify the index formula for the resulting direct sum. The invertible operator does not contribute to (3.77), and for the finite-dimensional symbol the desired formula coincides with the Atiyah–Singer formula.

2) *The general case: reduction to homogeneous symbols.*

In this case, we use a method due to Hörmander (1985*a*, Theorem 19.2.3) to reduce general elliptic symbols to homogeneous symbols. Hörmander's theorem gives the desired reduction for symbols acting in finite-dimensional bundles, and our main task is to verify that all constructions remain valid in the infinite-dimensional case.

The reduction is based on the following assertion, which we give in a special case convenient to us.

PROPOSITION 3.90 (Hörmander 1985*a*, Theorem 19.1.10). *Consider strongly continuous operator families $T_\varepsilon, S_\varepsilon \in \mathcal{B}(\mathcal{H})$, $\varepsilon \in [0, \varepsilon_0]$. If the families $S_\varepsilon T_\varepsilon - 1$ and $T_\varepsilon S_\varepsilon - 1$ are uniformly compact, then* $\operatorname{ind} S_\varepsilon = -\operatorname{ind} T_\varepsilon$ *is independent of* $\varepsilon$.

◀ An operator family $Q_\varepsilon \in \mathcal{B}(\mathcal{H})$ is said to be *uniformly compact* if the union of images of the unit ball under the action of all $Q_\varepsilon$ is precompact, i.e., has a compact closure. ▶

Let $q \in S^0_{CV}(T^*X)$ be a symbol such that $pq = qp = 1$ for $|\xi| > R$.
Let $\psi(t)$, $t \geq 0$, be a smooth real function such that

$$\psi(t) = \begin{cases} 1 & \text{for } t < 1, \\ 1/t & \text{for } t > 2. \end{cases} \tag{3.79}$$

Consider the symbols

$$p_\varepsilon(x,\xi) = p(x, \xi\psi(\varepsilon\xi)), \quad q_\varepsilon(x,\xi) = q(x, \xi\psi(\varepsilon\xi)). \tag{3.80}$$

They have the following properties:

1) $p_0 = p$ and $q_0 = q$

2) $p_\varepsilon$ and $q_\varepsilon$ are uniformly bounded in $S^0_{CV}(T^*X)$ for $\varepsilon \in [0,1]$

3) for $\varepsilon > 0$, the symbols $p_\varepsilon$ and $q_\varepsilon$ satisfy condition (3.78), and for sufficiently small $\varepsilon$ they are elliptic

4) for sufficiently small $\varepsilon > 0$, the compactly supported compact-valued symbols

$$r_1 = p_\varepsilon q_\varepsilon - 1, \quad r_2 = q_\varepsilon p_\varepsilon - 1 \tag{3.81}$$

are independent of $\varepsilon$

The proof coincides word for word with the one given by Hörmander (1985*a*, Theorem 19.2.3) for the finite-dimensional case.

We define operators

$$P_\varepsilon = p_\varepsilon\left(x, -i\frac{\partial}{\partial x}\right), \quad Q_\varepsilon = q_\varepsilon\left(x, -i\frac{\partial}{\partial x}\right) \tag{3.82}$$

by the formulas (3.36)–(3.37) with coordinate neighborhoods and partition of unity independent of $\varepsilon$. We claim that (for sufficiently small $\varepsilon$)

(a) $\operatorname{ind} p_\varepsilon \in K_c(T^*X)$ is independent of $\varepsilon$

(b) the families $P_\varepsilon$ and $Q_\varepsilon$ satisfy the assumptions of Proposition 3.90

This implies the index theorem, since, on the one hand, the symbol $p_\varepsilon$ is homogeneous at infinity for $\varepsilon > 0$, so that $\operatorname{ind} P_\varepsilon = \pi_! \operatorname{ind} p_\varepsilon$ (by the first part of the proof), and on the other hand, the passage to the limit as $\varepsilon \to 0$ is possible by Proposition 3.90. Hence it suffices to prove assertions (a) and (b).

Assertion (a) follows from the homotopy invariance of the difference element, since for sufficiently small $\varepsilon$ the symbol $p_\varepsilon$ varies with $\varepsilon$ only outside a sufficiently large ball $\{|\xi| > R\}$, where $R \simeq 1/\varepsilon$, and remains invertible in the exterior of the ball.

To prove assertion (b), we note that

1) The families $P_\varepsilon$ and $Q_\varepsilon$ are strongly continuous, since they are uniformly bounded and each term occurring in their definition via the sum (3.36) over coordinate charts on $X$ is strongly continuous on the set of functions whose Fourier transform is compactly supported.

2) The operators $P_\varepsilon Q_\varepsilon - 1$ and $Q_\varepsilon P_\varepsilon - 1$ are compact and continuously depend on $\varepsilon$.

Indeed, compactness is obvious, while continuity follows from the fact that their complete symbols in local coordinate systems, as well as their derivatives, continuously depend on $\varepsilon$ on compact subsets of values of $\xi$, are uniformly bounded and uniformly decay as $\xi \to \infty$, and hence are uniformly continuous in $\varepsilon$ for all $\xi$.

It follows from 2) that the families $P_\varepsilon Q_\varepsilon - 1$ and $Q_\varepsilon P_\varepsilon - 1$ are uniformly compact.

Thus the proof of the theorem is complete. □

### 3.6.3. Application of Luke's theorem to singular manifolds

***The index of operators with unit interior symbol on a manifold with edges.*** In this and the next subsections, we consider some simplest index formulas on manifolds with edges.

Let $\mathcal{M}$ be a manifold with edge $X$, and let $M$ be the corresponding stretched manifold.

We define a *Mellin operator* as a zero-order edge-degenerate operator whose interior symbol is zero. It follows that the edge symbol of a Mellin operator treated as an operator function on the cotangent bundle $T^*X$ of the edge has a compact fiber variation.

Thus we can apply Luke's formula to give an index formula for operators of the form $1 + \mathfrak{M}$, where $\mathfrak{M}$ is a Mellin operator. The edge symbol of $\mathfrak{M}$ will be denoted by $m$. The ellipticity condition in this case means that the operator function $1 + m(x, \xi)$ is invertible for $(x, \xi) \in T_0^*X$. Such a symbol determines a difference element in the $K$-group with compact supports of the cotangent bundle. This element is denoted by

$$[1 + m] \in K_c(T^*X),$$

and Luke's formula is valid.

PROPOSITION 3.91. *The index of an elliptic edge-degenerate operator $1+\mathfrak{M}$ with a Mellin operator $\mathfrak{M}$ is equal to the direct image of the difference element determined by the edge symbol of the operator:*

$$\operatorname{ind}(1 + \mathfrak{M}) = p_![1 + m], \tag{3.83}$$

*where $p_! : K_c(T^*X) \to K_c(pt) = \mathbb{Z}$ is the direct image map induced by the projection into a point.*

*Proof.* The action of $\mathfrak{M}$ far from the edge can be assumed to be zero. Thus the index of $1+\mathfrak{M}$ is equal to the index of an operator with the same interior and edge symbols on the bundle with fiber the infinite cone $K_\Omega$. The latter operator on the bundle of infinite cones can be treated as a pseudodifferential operator on $X$ with operator-valued symbol $1+m(x,\xi)$ (see Section 3.3). Hence our index formula is a special case of Luke's formula (3.77). □

*Remark* 3.92. If we additionally assume that not only the interior symbol of the Mellin operator is zero but also the conormal symbol $\sigma_c(m)$ of the edge symbol is zero (in this case, we speak of *Green operators*), then the edge symbol will be compact-valued, and hence $m$ can be approximated by a finite-dimensional symbol in such a way that the sum $1+m$ remains elliptic. Up to this approximation, the computation of the index in this case is reduced to the ordinary Atiyah–Singer formula.

***The index on bundles with smooth base and fiber having conical points.*** In the preceding item, we have given an index formula for elliptic edge operators with unit interior symbols. Geometrically, the key point in the computation is the fact that essentially it was carried out on a special manifold with edges, namely, on an infinite wedge, which is globally a bundle over the edge $X$, and the Mellin operator itself is represented as a pseudodifferential operator acting in sections of infinite-dimensional vector bundles on the edge.

Now we apply the same technique for the case in which the manifold $\mathcal{M}$ itself is globally represented as a bundle

$$\pi : \mathcal{M} \longrightarrow X$$

whose fiber is a closed manifold $\mathcal{N}$ with isolated conical singularities. Consider the simplest example.

EXAMPLE 3.93. Let $\pi : E \to X$ be a smooth bundle with closed base $X$ and fiber $\Omega$. Then there is an associated *suspension bundle*

$$\widetilde{\pi} : \Sigma E \longrightarrow X$$

whose fiber is the suspension over $\Omega$ (see Example 1.22),

$$\Sigma\Omega = \{\{\Omega \times [0,2]\}/\{\Omega \times \{0\}\}\}/\{\Omega \times \{2\}\},$$

obtained from the finite cylinder $\Omega \times [0,2]$ with base $\Omega$ by shrinking both bases into points (which are conical singularities of the resulting manifold). The total space $\Sigma E$ is a manifold with edge that is the disjoint union of two copies of the manifold $X$.

Let $\mathbf{D}$ be an elliptic edge operator on $\mathcal{M}$. For brevity, we restrict ourselves to the case of a zero-order scalar operator. The reader can readily generalize this to operators of arbitrary order acting in sections of vector bundles.

PROPOSITION 3.94. *An elliptic zero-order edge-degenerate operator* $\mathbf{D}$ *can be represented modulo compact operators as an operator*

$$\mathbf{D} = D\left(x, -i\frac{\partial}{\partial x}\right) \tag{3.84}$$

*on the edge* $X$ *with elliptic*[7] *operator-valued symbol*

$$D(x,\xi) \in S^0_{CV}(T^*_0 X)$$

*with compact fiber variation ranging in the space of cone-degenerate pseudodifferential operators on the fiber* $\mathcal{N}$.

*In particular, the index is given by the Luke formula*

$$\operatorname{ind} \mathbf{D} = p_![D], \qquad \text{where } [D] \in K_c(T^*X). \tag{3.85}$$

*The proof* follows directly from definitions. □

### 3.6.4. Supplement. Order reduction

In subsequent chapters, it is often convenient to deal with zero-order operators instead of the originally given ones. The passage to zero-order operators is known as *order reduction.* It can be done as follows. Suppose that we are given an $m$th-order elliptic edge-degenerate operator

$$D : \mathcal{W}^{s,\gamma}(\mathcal{M}, E) \longrightarrow \mathcal{W}^{s-m,\gamma-m}(\mathcal{M}, F),$$

where $E$ and $F$ are vector bundles over $\mathcal{M}$. We know from the preceding that the ellipticity condition for $D$, as well as the kernel and cokernel of $D$, are independent of the smoothness exponent $s$. Hence all questions of interest in elliptic theory such as the homotopy classification or the index are also independent of $s$. Without loss of generality, we can assume then that $s = \gamma$, i.e., the operator $D$ acts in the spaces

$$D : \mathcal{W}^{\gamma,\gamma}(\mathcal{M}, E) \longrightarrow \mathcal{W}^{\gamma-m,\gamma-m}(\mathcal{M}, F).$$

Consider the edge-degenerate pseudodifferential operators

$$\begin{aligned} D_0 &: \mathcal{W}^{\gamma-m,\gamma-m}(\mathcal{M}, F) \longrightarrow \mathcal{W}^{0,0}(\mathcal{M}, F), \\ D_1 &: \mathcal{W}^{0,0}(\mathcal{M}, E) \longrightarrow \mathcal{W}^{\gamma,\gamma}(\mathcal{M}, E) \end{aligned}$$

of orders $\gamma - m$ and $-\gamma$, respectively, such that the interior symbols of these operators are equal to $1_F$ and $1_E$ on the unit spheres in $T^*_0\mathcal{M}$ and the edge symbols are equal to

$$\begin{aligned} \sigma_\wedge(D_0) &= \left(1 - \Delta_{K_\Omega} + \frac{1}{r^2}\right)^{(\gamma-m)/2} \otimes 1_F, \\ \sigma_\wedge(D_1) &= \left(1 - \Delta_{K_\Omega} + \frac{1}{r^2}\right)^{-\gamma/2} \otimes 1_E \end{aligned}$$

[7]Recall that an operator-valued symbol defined on $T^*_0 X$ is said to be elliptic if it is Fredholm everywhere and uniformly invertible outside a compact set.

on the unit spheres in $T_0^*X$. This is well defined, since the operator

$$1 - \triangle_{K_\Omega} + \frac{1}{r^2}$$

is positive definite and self-adjoint in $\mathcal{K}^{0,0}(K_\Omega)$. Consequently, its arbitrary powers are well defined, and one can prove that they themselves are families of cone-degenerate pseudodifferential operators on $K_\Omega$ with interior and cone symbols being the corresponding powers of the interior and cone symbols of the original operator. The operators $D_0$ and $D_1$ are elliptic operators of index zero with unit interior symbol on the unit spheres. Then the operator

$$D_0 D D_1 = \mathcal{W}^{0,0}(\mathcal{M}, E) \longrightarrow \mathcal{W}^{0,0}(\mathcal{M}, F)$$

is the order reduction of $D$.

# Chapter 4

# Localization (Surgery) in Elliptic Theory

The aim of this chapter is to describe the surgery method in index theory, which is one of the main tools in our investigation of topological problems of elliptic theory on singular manifolds in Part III.

## 4.1. The Index Locality Principle

### 4.1.1. What is locality?

The index of an elliptic operator $D$ on a smooth compact manifold $M$ without boundary does not change if lower-order terms are added to the operator (since the index of a Fredholm operator is stable under compact perturbations) and hence depends only on the principal symbol $\sigma(D)$ of the operator $D$. In topological terms, the index is expressed via the principal symbol by the Atiyah–Singer formula (Atiyah and Singer 1968)

$$\operatorname{ind} D = p_![\sigma(D)],$$

where $[\sigma(D)]$ is the element of the $K$-group with compact supports of the cotangent bundle of $M$ determined via the principal symbol by the difference construction and $p_!$ is the direct image in $K$-theory induced by the mapping $p : M \longrightarrow \{pt\}$ of $M$ into a point. How can one describe, however, the *analytic* structure of the dependence of the index on the principal symbol? Apparently, the most detailed answer is given by the so-called "local index formula" (e.g., Gilkey 1995): the index of an elliptic operator $D$ on a closed manifold $M$ is expressed by the formula

$$\operatorname{ind} D = \int_M \alpha(x), \tag{4.1}$$

where the "local density" $\alpha(x)$ at a point $x \in M$ depends only on the principal symbol $\sigma(D)$ and its derivatives in the fiber $T_x^*M$ of the cotangent bundle over $x$, and moreover, one can write out explicit formulas for this density.

It is rather difficult to extend such a locality property of the index to more general situations (e.g., to the case of the index of elliptic operators on manifolds with singularities): in these situations, there are far more complicated (noncommutative) principal symbol structures, for which formulas like (4.1) are no longer valid.

Hence the following coarser locality property (which directly follows from (4.1) in the case of elliptic operators on a smooth compact manifold without boundary) will be of interest to us.

> *If one changes the operator $D$ (within the class of elliptic operators) on some open subset $V \subset M$, then the index changes by a number that depends only on the structure of the original and new operators on $V$ and is independent of the structure of $D$ outside $V$.*

◀ Indeed, by (4.1) the variation of the index is equal to

$$\int_V \alpha(x) - \int_V \alpha'(x),$$

where $\alpha'(x)$ is the density corresponding to the new operator, for the integrals over $M \setminus V$ cancel out. ▶

This argument shows that the variation of the operator can be interpreted rather widely: for example, we can change not only the operator itself but also the bundles in whose sections it acts and even the manifold $M$ itself. (More precisely, we can replace $V$ by some other set $V'$, leaving $M \setminus V$ unchanged.)

In the next subsection, we show this locality property in the simplest case directly, without resorting to the local index formula (4.1). The idea in this example will help us later on to state the general index locality principle as well.

### 4.1.2. A pilot example

Let

$$D, D' : C^\infty(M, E) \longrightarrow C^\infty(M, F) \tag{4.2}$$

be two elliptic pseudodifferential operators ($\psi$DO) acting on the same smooth compact manifold $M$ without boundary in sections of the same bundles and coinciding everywhere outside a compact subset $V \subset M$. In this situation, the index locality property can readily be proved by elementary means.

◀ For example, one can argue as follows. The difference of indices of $D$ and $D'$ is equal to

$$\operatorname{ind} D - \operatorname{ind} D' = \operatorname{ind} D(D')^{-1}, \tag{4.3}$$

where $(D')^{-1}$ is an almost inverse[1] of $D'$. In turn, the index of the elliptic $\psi$DO $D(D')^{-1}$ is completely determined by its principal symbol $\sigma = \sigma(D)\sigma(D')^{-1}$, which is equal to unity over $M \setminus V$ and hence actually depends only on the values of the principal symbols $\sigma(D)$ and $\sigma(D')$ over $V$. ▶

[1] An inverse modulo compact operators.

What is the essence of the argument given in this example? Obviously, the key point is that, for elliptic $\psi$DO, taking the product and passing to an almost inverse are local operations (modulo compact operators, which do not affect the index anyway). This follows from the (pseudo)locality property of (pseudo)differential operators. It is the locality[2] of these operations that implies that the operator $D(D')^{-1}$ is the identity operator outside $V$ and is determined in $V$ only by the operators $D$ and $D'$ also in $V$. Moreover, we actually use only the locality with respect to the pair of sets $V$ and $M \setminus V$: were the operators nonlocal in $V$ and in $M \setminus V$ separately, our conclusions would remain in force.

Thus, the pilot example essentially shows that we deal with the implication

$$\boxed{\text{locality of elliptic } \psi\text{DO}} \implies \boxed{\text{locality of the index}}$$

The abstract index locality principle given in forthcoming subsections is just a generalization of this implication. To speak of local operators in the abstract case, we need the following.

- There should be an analog of the notion of *support* for elements of spaces where the operators will act.
- For such spaces, we should define an analog of the notion of an *operator whose integral kernel is localized near the diagonal.*

Such spaces, which will be called *collar spaces*, are introduced in the following subsection, and then we define an appropriate class of operators, referred to as *proper operators*, in such spaces.

### 4.1.3. Collar spaces

The support of a smooth section $u \in C^\infty(M, E)$ of a vector bundle $E$ over a manifold $M$ is defined as the closure of the set of points $x \in M$ such that $u(x) \neq 0$. It is important in this definition that $u$ is a mapping defined on $M$. However, there is an equivalent definition that does not rely on this fact explicitly. Namely, since $C^\infty(M, E)$ is a module over the algebra $C^\infty(M)$ of smooth functions on $M$, one can define the support $\operatorname{supp} u$ as the intersection of zero sets of all functions $\varphi \in C^\infty(M)$ that annihilate $u$. The latter definition is valid for an arbitrary $C^\infty(M)$-module regardless of its nature (i.e., whether elements of this module are defined on $M$).

Collar spaces are just defined as modules over a function algebra, and so their supports are described by the above-mentioned construction. To state the index locality principle, we need not consider any general function algebras; it suffices to consider functions on the closed interval $[-1, 1]$. (Accordingly, the supports of elements of collar spaces will be subsets of this interval.)

[2]We omit the prefix "pseudo" in what follows.

Consider the unital topological algebra $C^\infty([-1,1])$ of smooth functions $\varphi(t)$, $t \in [-1,1]$, on the interval $[-1,1]$.

DEFINITION 4.1. A *collar space* is a separable Hilbert space $H$ equipped with the structure of a module over $C^\infty([-1,1])$ such that the action of $C^\infty([-1,1])$ is continuous in the uniform operator topology and the unit function $\mathbf{1} \in C^\infty([-1,1])$ acts as the identity operator in $H$.

Let us give one of the main examples of collar spaces.

EXAMPLE 4.2. Let $M$ be a compact $C^\infty$ manifold without boundary, and let

$$\chi : M \longrightarrow [-1,1]$$

be a smooth mapping. Then each Sobolev space $H^s(M)$ can be made a collar space if we define a natural action of $C^\infty([-1,1])$ on $H^s(M)$ by the formula

$$(\varphi f)(x) = \varphi(\chi(x))f(x), \quad x \in M,$$

for any $\varphi \in C^\infty([-1,1])$ and $f \in H^s(M)$. In this and similar more general cases, the subset

$$U = \overline{\chi^{-1}(-1,1)} \subset M,$$

of the manifold $M$, where the bar stands for the closure, will be referred to as the *collar*. We represent this graphically in Fig. 4.1, where the collar $U$ is dashed and the function $\chi$ takes the values $\chi = -1$ to the left of the collar (i.e., in $M_-$) and $\chi = +1$ to the right of the collar (i.e., in $M_+$).

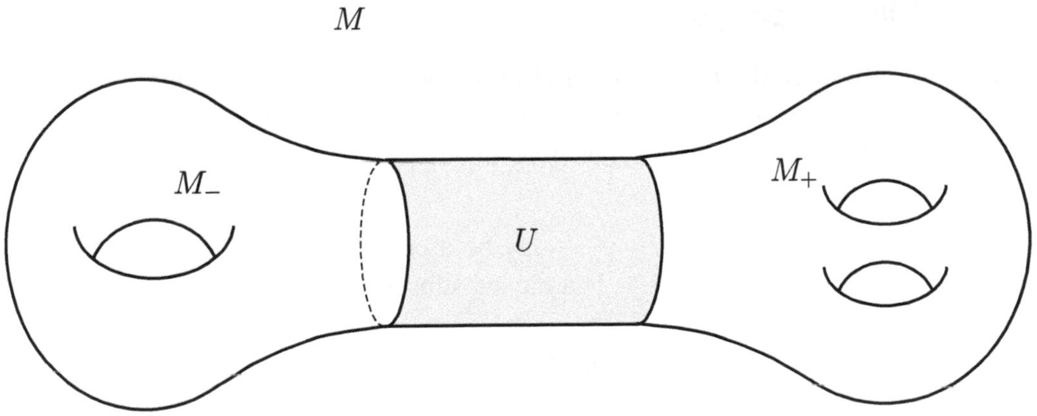


Figure 4.1. A manifold with a collar.

For any element $h \in H$ of a collar space $H$, there is a naturally defined notion of *support*.

DEFINITION 4.3. The *support* of an element $h \in H$ is the subset $\operatorname{supp} h$ of the interval $[-1,1]$ given by the formula

$$\operatorname{supp} h = \bigcap \varphi^{-1}(0),$$

where the intersection is taken over all $\varphi \in C^\infty([-1,1])$ such that $\varphi h = 0$.

The supports of elements of a collar space have all the natural properties. Just to list a few, $\operatorname{supp} h$ is closed; $\operatorname{supp} h$ is empty if and only if $h = 0$;

$$\operatorname{supp}(h_1 + h_2) \subset \operatorname{supp} h_1 \cup \operatorname{supp} h_2.$$

Note that in the situation of Example 4.2 the set $\operatorname{supp} h$ is just the closure of the image of the ordinary support of $h$ under the mapping $\chi$. For example, if $f \in H^s(M)$ is supported in $M_-$, then $\operatorname{supp} f = \{-1\}$.

One can single out subspaces of a collar space $H$ by imposing conditions on the supports of elements. Let $F \subset [-1,1]$ be a given subset. Then the set

$$\overset{\circ}{H}_F = \{h \in H \mid \operatorname{supp} h \subset F\}$$

is a linear manifold (lineal) in $H$, which is closed provided that $F$ is closed. In the general case, we define a subspace $H_F \subset H$ as the *closure* of $\overset{\circ}{H}_F$. Note that it is not true in general that $H_F = H_{\bar F}$.

Just as in the case of ordinary supports, functions with disjoint supports are linearly independent. More precisely, if $F_1, \dots, F_m \subset [-1,1]$ are subsets such that $\bar F_j \cap \bar F_k = \varnothing$ for $j \neq k$, then

$$H_{F_1 \cup \dots \cup F_m} = H_{F_1} \oplus \dots \oplus H_{F_m},$$

where the sum is direct but not necessarily orthogonal.

### 4.1.4. Elliptic operators

Here we introduce the notion of an elliptic operator in a collar space. This notion inherits two characteristic properties of elliptic $\psi$DO, namely, pseudolocality and the Fredholm property. First, let us give an adequate statement of pseudolocality in the context of collar spaces.

DEFINITION 4.4. Let $A : H_1 \longrightarrow H_2$ be a continuous linear operator in collar spaces and $K \subset [-1,1] \times [-1,1]$ a closed subset. We say that *the support of $A$ is contained in $K$* if

$$\operatorname{supp} Ah \subset K(\operatorname{supp} h) \tag{4.4}$$

for every $h \in H$, where $K$ is treated as a multimapping of the interval $[-1,1]$ into itself:

$$Kx \overset{\text{def}}{=} \{y \in [-1,1] \mid (x,y) \in K\}.$$

The *support* $\operatorname{supp} A$ of the operator $A$ is the intersection of all closed sets $K$ with property (4.4).

It readily follows from the definition that if $A$ and $B$ are operators in collar spaces, then

$$\operatorname{supp}(BA) \subset \operatorname{supp} B \circ \operatorname{supp} A,$$

where the right-hand side is understood as the composition of multimappings.

In the theory of $\psi$DO, pseudolocality is understood as the fact that the integral kernel of a $\psi$DO is supported modulo smooth functions in an arbitrarily small neighborhood of the diagonal. Hence, using smooth cutoff functions concentrated near the diagonal, one can include an elliptic $\psi$DO in a continuous family of elliptic operators with the same symbol and with kernels supported in a shrinking neighborhood of the diagonal as the parameter of the family tends to zero. Let us transfer this property to the case of collar spaces. Let

$$\Delta = \{(x,x) \mid x \in [-1,1]\} \subset [-1,1] \times [-1,1]$$

be the diagonal, and let

$$\Delta_\varepsilon = \{(x,y) \in [-1,1] \times [-1,1] \mid |x-y| < \varepsilon\} \tag{4.5}$$

be its $\varepsilon$-neighborhood.

DEFINITION 4.5. A *proper operator* in collar spaces $H_1$ and $H_2$ is a family of continuous linear operators

$$A_\delta : H_1 \longrightarrow H_2$$

with parameter $\delta > 0$ such that

(i) $A_\delta$ continuously depends on $\delta$ in the uniform operator topology

(ii) for each $\varepsilon > 0$, there exists a $\delta_0 > 0$ such that

$$\operatorname{supp} A_\delta \subset \Delta_\varepsilon \quad \text{for} \quad \delta < \delta_0 \tag{4.6}$$

In the situation of Example 4.2, $\psi$DO in Sobolev spaces can naturally be viewed as proper operators (more precisely, included in appropriate families).

◀ Indeed, as was indicated above, a $\psi$DO can be included in a continuous family of $\psi$DO with the same symbol and with supports of the integral kernels shrinking to the diagonal (in $M \times M$). It remains to note that the preimage of the $\varepsilon$-neighborhood of the diagonal in $M \times M$ under the mapping $\chi \times \chi$ is contained in $\Delta_\varepsilon$. ▶

Now we can give the definition of elliptic operators in collar spaces.

DEFINITION 4.6. An *elliptic operator* in collar spaces $H$ and $G$ is a proper operator

$$D_\delta : H \longrightarrow G$$

that is Fredholm for each $\delta$ and has an almost inverse $D_\delta^{-1}$ that is also a proper operator.

Here, as usual, an almost inverse of a bounded operator $A$ is defined as an operator $A^{-1}$ such that the products $AA^{-1}$ and $A^{-1}A$ differ from the identity operators by compact operators in the corresponding spaces.

*Remark* 4.7. The class of elliptic operators in collar spaces considered here is wider than Atiyah's (1969) class of *abstract* elliptic operators that consists of Fredholm operators $A : H \longrightarrow G$ acting in Hilbert spaces $H$ and $G$ equipped with the structure of $C(X)$-modules (where $C(X)$ is the $C^*$-algebra of continuous functions on a compact set $X$) and commuting with the action of $C(X)$ modulo compact operators: the operator

$$\varphi A - A\varphi : H \longrightarrow G$$

is compact for every $\varphi \in C(X)$.

More precisely, every $C([-1,1])$-module is a collar space by virtue of the continuous embedding $C^\infty([-1,1]) \subset C([-1,1])$, and each abstract elliptic operator $A$ on $[-1,1]$ can be embedded in a family of abstract elliptic operators that is an elliptic operator in collar spaces. However, the opposite is not true.

◀ Indeed, define the structure of a collar space on $L_2(\mathbb{R})$ via the mapping

$$\chi : \mathbb{R} \longrightarrow [-1,1], \quad \chi(t) = \frac{2}{\pi} \tan^{-1} t.$$

The operator

$$\begin{aligned} A_\delta : L_2(\mathbb{R}) &\longrightarrow L_2(\mathbb{R}) \\ f(t) &\longmapsto f(t+\delta) \end{aligned}$$

is an elliptic operator in collar spaces but is not an abstract elliptic operator for any $\delta > 0$. ▶

### 4.1.5. Surgery and the relative index theorem

The relative index theorem stated in this subsection expresses the index locality principle at the abstract level of collar spaces. The *relative index* of two Fredholm operators $D'$ and $D$ is the difference

$$\operatorname{ind}(D', D) = \operatorname{ind} D' - \operatorname{ind} D.$$

In the locality property considered in Section 4.1.2 for the case of a smooth compact manifold without boundary, the operator $D'$ was obtained from an elliptic operator $D$ on a manifold $M$ by some change on a subset $V \subset M$. (Moreover, $V$ itself could change, but $M \setminus V$ remained intact.) At the abstract level, we should first make it clear what changes of operators are to be considered. Admissible changes will be referred to as *surgeries*; they include surgeries of collar spaces themselves (which corresponds to the replacement of $V$ by $V'$) and associated surgeries of operators.

Let $H_1$ and $H_2$ be collar spaces.

DEFINITION 4.8. If for some $F \subset [-1,1]$ there is a given isomorphism (not necessarily isometric)

$$j : H_{1F} \cong H_{2F}$$

of $C^\infty([-1,1])$-modules, then we say that $H_1$ and $H_2$ *coincide* on $F$ (or *are obtained from each other by surgery on* $[-1,1]\backslash F$). In this case, we write

$$H_1 \overset{F}{=} H_2 \quad \text{or} \quad H_1 \overset{[-1,1]\backslash F}{\longleftrightarrow} H_2.$$

We point out that the *specific* isomorphism is important here (rather than the existence of *some* isomorphism).

Let us now extend the notion of surgery to operators. Since proper (and, in particular, elliptic) operators in collar spaces do not change the supports of elements "too much," one can reasonably speak of coincidence of such operators on a subset of $[-1,1]$ provided that the spaces themselves where the operators act coincide on the subset.

DEFINITION 4.9. Let $F \subset [-1,1]$ be a subset open in $[-1,1]$, and let $H_1 \overset{F}{=} H_2$ and $G_1 \overset{F}{=} G_2$ be collar spaces. We say that proper operators

$$A_1 : H_1 \to G_1, \qquad A_2 : H_2 \to G_2 \tag{4.7}$$

*coincide on* $F$ if for each compact subset $K \subset F$ the following condition is satisfied: there exists a number $\delta_0 = \delta_0(K) > 0$ such that

$$A_{1\delta} h = A_{2\delta} h \tag{4.8}$$

whenever $\delta < \delta_0$ and $\operatorname{supp} h \subset K$. Under this condition, we also say that $A_1$ is obtained from $A_2$ by *surgery* on $[-1,1]\backslash F$ and write

$$A_1 \overset{F}{=} A_2 \quad \text{or} \quad A_1 \overset{[-1,1]\backslash F}{\longleftrightarrow} A_2.$$

We note that (4.8) is well defined, since $h \in H_{1K} \cong H_{2K}$ and for small $\delta$ one has $A_{1\delta}h$, $A_{2\delta}h \in G_{1F} \cong G_{2F}$; the latter inclusion is valid since $F$ is open.

In the following, we deal with *surgery diagrams* rather than individual surgeries, more precisely, with squares of the form

$$\begin{array}{ccc} H_1 & \overset{B}{\longleftrightarrow} & H_2 \\ {\scriptstyle A}\updownarrow & & \updownarrow{\scriptstyle C} \\ H_3 & \underset{D}{\longleftrightarrow} & H_4, \end{array}$$

where the $H_i$ are collar spaces and $A$, $B$, $C$, $D \subset [-1, 1]$. This square is said to *commute* if the diagram

$$\begin{array}{ccc} H_{1F} & \approx & H_{2F} \\ \wr & & \wr \\ H_{3F} & \approx & H_{4F}, \end{array} \qquad F = [-1,1]\backslash\{A \cup B \cup C \cup D\},$$

of isomorphisms commutes. (These homomorphisms are the restrictions to the relevant subspaces of the corresponding isomorphisms occurring in Definition 4.8.) A similar square of surgeries for *operators* is said to commute if the underlying squares of surgeries of collar spaces commute.

Let us now state the relative index theorem, which is the main result of this section.

THEOREM 4.10. *Let the surgery diagram*

$$\begin{array}{ccc} D & \xleftrightarrow{-1} & D_- \\ {\scriptstyle 1}\updownarrow & & \updownarrow{\scriptstyle 1} \\ D_+ & \xleftrightarrow[-1]{} & D_\pm \end{array} \tag{4.9}$$

*of elliptic operators in collar spaces commute. Then the relative indices of the operators occurring in the diagram satisfy the relation*

$$\operatorname{ind}(D, D_-) = \operatorname{ind}(D_+, D_\pm).$$

*The proof* can be found in Nazaikinskii and Sternin (2001).

The following sections provide examples of applications of the index locality principle.

## 4.2. Localization in Index Theory on Smooth Manifolds

### 4.2.1. The index excision property

The surgery for smooth manifolds can be visualized as the *index excision property*, which is described in what follows.

Let $S$ be a two-sided hypersurface in a smooth closed compact manifold $M$. We take a tubular neighborhood $S \times [-1, 1] \subset M$. Then we can define two smooth compact manifolds (see Fig. 4.2).

- The first manifold $M \setminus S \times (-1, 1)$ is obtained by cutting $S \times (-1, 1)$ away from $M$ and then by gluing the components $S \times \{\pm 1\}$ of the boundary of the resulting manifold together along the identity diffeomorphism.[3]

[3]Thus we obtain a manifold diffeomorphic to $M$.

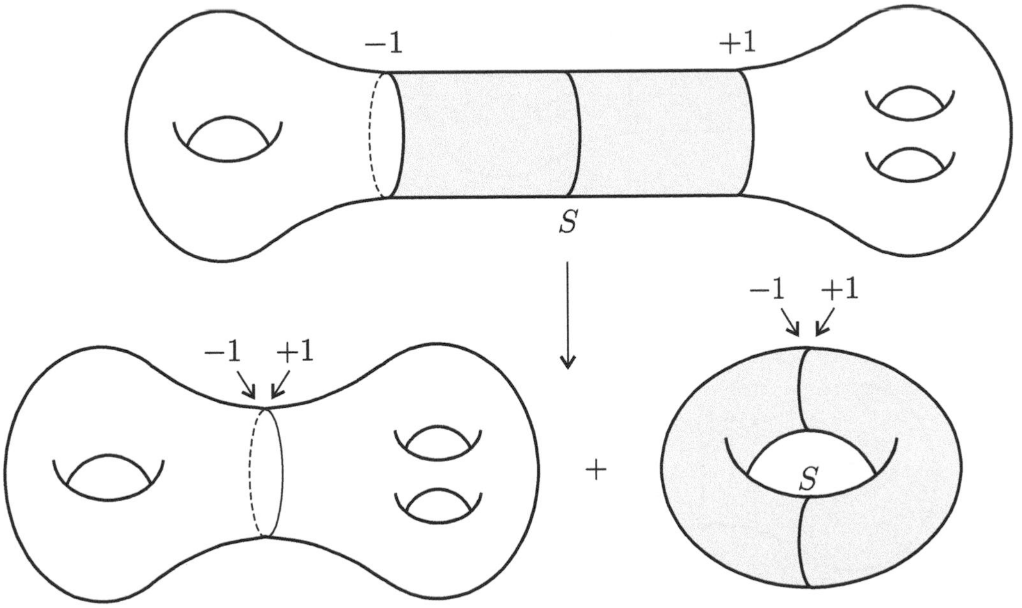


Figure 4.2. Cutting away a neighborhood of a hypersurface.

- The second manifold is the torus $S \times \mathbb{S}^1$ obtained by gluing together the boundary components of the deleted part $S \times [-1, 1]$ along the identity diffeomorphism.

Let us carry out a similar surgery in elliptic theory. Let $D_M$ be an elliptic operator on $M$ such that its principal symbol is the same on both components of the boundary of the tubular neighborhood:

$$\sigma(D_M)|_{S\times\{-1\}} = \sigma(D_M)|_{S\times\{1\}}.$$

Under this condition, the symbol $\sigma(D_M)$ naturally determines symbols on the above-mentioned manifolds $M \setminus S \times (-1, 1)$ and $S \times \mathbb{S}^1$. We denote the operators corresponding to the symbols by $D_{M\setminus S\times(-1,1)}$ and $D_{S\times\mathbb{S}^1}$.

INDEX EXCISION PROPERTY.

$$\operatorname{ind} D_M = \operatorname{ind} D_{M\setminus S\times(-1,1)} + \operatorname{ind} D_{S\times\mathbb{S}^1}. \tag{4.10}$$

*Proof.* See the surgery diagram in Fig. 4.3, where the index of the operator in the right bottom corner is zero. □

Note that the index excision property remains valid (together with the proof) even if $M$ has singularities. We should only assume that the hypersurface $S$ (together with a tubular neighborhood) lies in the smooth part of the manifold. A typical example of such a hypersurface is the submanifold $r = \varepsilon$ of a manifold with edge (a *section parallel to the edge*).

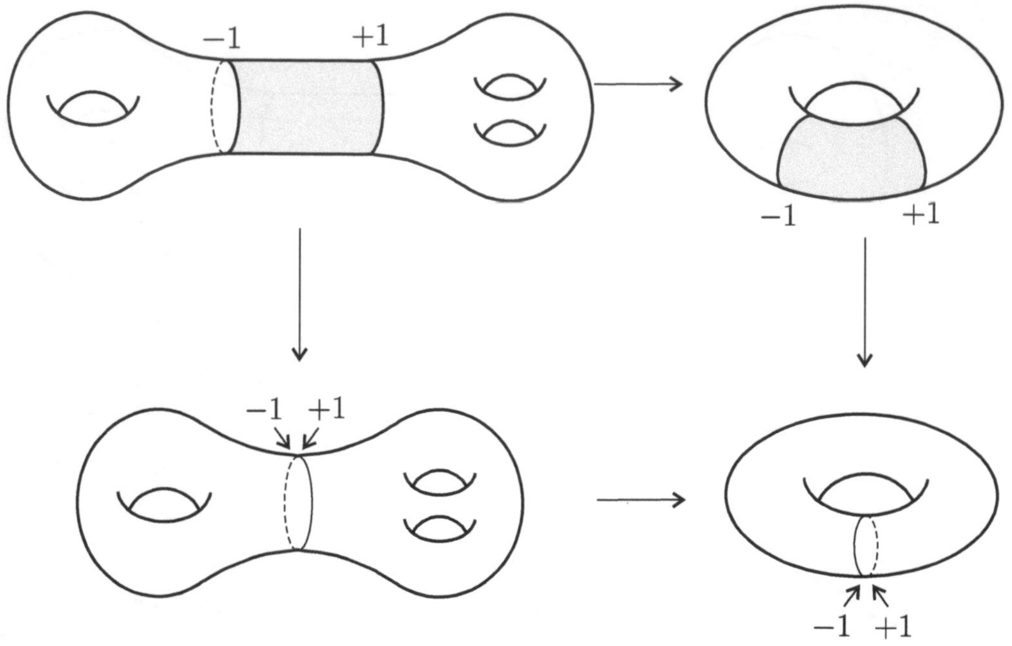


Figure 4.3. Surgery diagram proving the index excision property.

### 4.2.2. The Booß–Wojciechowski theorem

In this subsection, we study how the index of an elliptic $\psi$DO on a manifold changes under surgery of the manifold where the operator is defined and some associated surgery of the bundles in whose sections it acts. The index increment formula can naturally be treated as a *relative index formula*, for the symbol before and after the surgery is essentially the same but is realized differently depending on how the manifold and the bundles have been glued from pieces.

***Surgery of manifolds and vector bundles.*** Let $M$ be a closed smooth manifold, and let $S \subset M$ be an embedded smooth compact two-sided submanifold of codimension 1. Next, let $U$ be a collar neighborhood of $S$. We take some trivialization $U = (-1, 1) \times S$ of $U$ and use the coordinates $(t, s)$, $t \in (-1, 1)$, $s \in S$. Let

$$g : S \longrightarrow S$$

be a diffeomorphism. We perform the following operation: we cut $M$ along $S$ and glue together again, identifying each point $(-0, s)$ on the left coast of the cut with the corresponding point $(+0, g(s))$ on the right coast. The resulting smooth manifold (the smooth structure is well defined, since we have chosen and fixed the trivialization) will be denoted by $M_g$ and called the *surgery of $M$ via $g$*.

Let $E$ be a vector bundle over $M$. Suppose that we are given a vector bundle isomorphism

$$\mu : E|_S \longrightarrow g^* (E|_S) .$$

Then over $M_g$ there is a naturally defined vector bundle $E_\mu$ (by clutching along $S$ with the help of $\mu$), which will be called the *surgery of $E$ via the pair $(g,\mu)$*.

***Surgery of elliptic symbols.*** Now let $E$ and $F$ be vector bundles over $M$, and let

$$a : \pi^*E \longrightarrow \pi^*F,$$

where $\pi : T_0^*M \to T_0^*M$ is the natural projection, be an elliptic symbol of some order $m$. We can assume that $a$ is independent of the coordinate $t$ in a sufficiently small neighborhood of $S$ (that is, $a \equiv a_0$ in that neighborhood).

◀ Indeed, we can achieve this by choosing the representation

$$E|_U = \widetilde{\pi}^{-1}(E|_S), \qquad F|_U = \widetilde{\pi}^{-1}(F|_S),$$

of the bundles $E, F$ over $U = [-1,1] \times S$, where $\widetilde{\pi}\colon [-1,1] \times S \to S$ is the natural projection, and by passing to a homotopic symbol if necessary. ▶

Consider the restriction of this symbol to $S$ (denoted by the same letter):

$$a_0 \overset{\text{def}}{=} a|_{\pi^{-1}S}\colon \pi^*E|_{\pi^{-1}S} \longrightarrow \pi^*F|_{\pi^{-1}S}. \tag{4.11}$$

With regard to the trivialization chosen, this mapping can be rewritten in the form

$$a_0(p,s,\xi)\colon E_s \longrightarrow F_s, \qquad p^2 + |\xi|^2 \neq 0, \quad s \in S,$$

where $p$ is the dual variable of $t$ and $\xi$ is a point in the fiber of $T^*S$ over $s$.

Suppose that a surgery $g$ of $M$ and associated surgeries $\mu_E$ and $\mu_F$ of the bundles $E$ and $F$ are given. If the diagram

$$\begin{array}{ccc} E_s & \xrightarrow{\;a_0(p,s,\xi)\;} & F_s \\ {\scriptstyle \mu_E(s)}\big\downarrow & & \big\downarrow{\scriptstyle \mu_F(s)} \\ E_{g(s)} & \xrightarrow[\;a_0(p,g(s),{}^tg_s(s)^{-1}\xi)\;]{} & F_{g(s)}, \end{array} \tag{4.12}$$

where ${}^tg_s(s)$ is the transposed Jacobi matrix of the mapping $g$ at the point $s$, commutes, then the surgery gives a new (smooth) symbol $\widetilde{a}$ on the cotangent bundle $T_0^*M_g$. (The smoothness of the newly obtained symbol is guaranteed by the independence of $a$ on the coordinate $t$ in a neighborhood of $S$.)

***The relative index.*** Our task is to establish how the surgery affects the index of the corresponding $\psi$DO. By the locality principle, this index increment (the relative index) depends only on the surgery on $S$, and so we use the relative index theorem to pass to a simpler model.

Thus let $A$ and $\widetilde{A}$ be operators with principal symbols $a$ and $\widetilde{a}$ on the manifolds $M$ and $M_g$, respectively, obtained from each other by the above-mentioned surgery. The problem is to find the relative index $\operatorname{ind}\widetilde{A} - \operatorname{ind} A$.

It follows from Theorem 4.10 that the relative index is independent of the structure of the manifolds and operators in question outside a small neighborhood of $S$. Hence we can use the simplest model for the computations. Namely, consider the manifold $M = S \times \mathbb{S}^1$ and the elliptic pseudodifferential operator

$$A_0 : H^s(M, E) \to H^{s-m}(M, F)$$

of order $m$ with principal symbol $a_0$ independent of $\varphi \in \mathbb{S}^1$. (The bundles $E$ and $F$ are lifted to $M$ with the help of the natural projection $M \longrightarrow S$.) Next, let $M_g$ be the surgery of $M$ with the help of $g$, let $E_{\mu_E}$ and $F_{g,\mu_F}$ be the associated surgeries of the bundles $E$ and $F$, and let

$$\widetilde{A}_0 : H^s(M_g, E_{\mu_E}) \to H^{s-m}(M_g, F_{\mu_F})$$

be the new elliptic pseudodifferential operator with principal symbol $\widetilde{a}_0$ coinciding with $A_0$ outside a neighborhood of the set $S$, where the surgery is done.

The index of $A_0$ is zero, since its symbol is independent of $\varphi \in \mathbb{S}^1$. Hence in this model only one term in the expression for the relative index is nontrivial. We arrive at the following expression for the relative index of $A$ and $\widetilde{A}$.

THEOREM 4.11 (Booß-Bavnbek and Wojciechowski 1993).

$$\operatorname{ind} \widetilde{A} - \operatorname{ind} A = \operatorname{ind} \widetilde{A}_0.$$

Thus the relative index of surgery along a two-sided hypersurface is expressed as the index of an elliptic operator on the mapping torus of $g$.

### 4.2.3. The Gromov–Lawson theorem

The locality principle in index theory on noncompact manifolds was apparently obtained for the first time (for the special case of Dirac operators on noncompact Riemannian manifolds) by Gromov and Lawson (1983), who obtained the corresponding relative index theorem. (Later, Anghel (1993) generalized their result to arbitrary self-adjoint elliptic first-order operators on a complete Riemannian manifold.) In this subsection, we briefly describe the result due to Gromov and Lawson and its relationship to the general index locality principle.

***The geometric setup.*** Let $X_0$ and $X_1$ be complete even-dimensional Riemannian manifolds, and let $D_0$ and $D_1$ be generalized Dirac operators on $X_0$ and $X_1$, respectively, acting on sections of vector bundles $S_0$ and $S_1$. We say that $D_0$ and $D_1$ *coincide at infinity* if there exist compact subsets $K_0 \subset X_0$ and $K_1 \subset X_1$, an isometry

$$F : X_0 \backslash K_0 \xrightarrow{\approx} X_1 \backslash K_1,$$

and an isometry

$$\widetilde{F} : S_0\big|_{X_0 \backslash K_0} \longrightarrow S_1\big|_{X_1 \backslash K_1}$$

of vector bundles such that

$$D_1 = \widetilde{F} \circ D_0 \circ \widetilde{F}^{-1} \quad \text{on} \quad X_1 \backslash K_1.$$

To simplify the notation, we identify $X_0 \backslash K_0$ with $X_1 \backslash K_1$ and write

$$D_0 = D_1 \quad \text{on} \quad \Omega = X_0 \backslash K_0 \cong X_1 \backslash K_1.$$

***The relative index.*** In this situation, we can define the *topological relative index* $\operatorname{ind}_t(D_1^+, D_0^+)$ of the operators

$$D_1^+ : \Gamma(S_1^+) \longrightarrow \Gamma(S_1^-) \quad \text{and} \quad D_0^+ : \Gamma(S_0^+) \longrightarrow \Gamma(S_0^-)$$

as follows. If $X_0$ and $X_1$ are compact, then we simply set

$$\operatorname{ind}_t(D_1^+, D_0^+) = \operatorname{ind}(D_1^+) - \operatorname{ind}(D_0^+).$$

If $X_0$ (and hence $X_1$) is noncompact, then we use the following procedure. We cut the manifolds $X_0$ and $X_1$ along some compact hypersurface $H \subset \Omega$ and compactify them by attaching some compact manifold with boundary $H$. The operators $D_0^+$ and $D_1^+$ can be extended to elliptic operators $\widetilde{D}_0^+$ and $\widetilde{D}_1^+$ on the compact manifolds thus obtained. Now we set

$$\operatorname{ind}_t(D_1^+, D_0^+) = \operatorname{ind}(\widetilde{D}_1^+) - \operatorname{ind}(\widetilde{D}_0^+). \tag{4.13}$$

It follows from the relative index Theorem 4.10 that the right-hand side of (4.13) is independent of the arbitrariness in the above construction.

***The Gromov–Lawson theorem.*** Next, let the operators $D_0^+$ and $D_1^+$ be *positive at infinity*. (The precise definition is given in Gromov and Lawson 1983; roughly speaking, this condition means that the free terms in the operators $(D_0^+)^* D_0^+$ and $(D_1^+)^* D_1^+$ expressed via covariant derivatives are positive.) Then the operators $D_0^+$ and $D_1^+$ are Fredholm, and one can define the *analytical relative index*

$$\operatorname{ind}_a(D_1^+, D_0^+) = \operatorname{ind}_a(D_1^+) - \operatorname{ind}_a(D_0^+). \tag{4.14}$$

The Gromov–Lawson relative index theorem states that *the topological and analytical relative indices coincide*:

$$\operatorname{ind}_a(D_1^+, D_0^+) = \operatorname{ind}_t(D_1^+, D_0^+). \tag{4.15}$$

In Gromov and Lawson (1983), one can also find a more general theorem pertaining to the case in which the operators $D_0$ and $D_1$ coincide only on some of the "ends" of $X_0$ and $X_1$ at infinity. In this case, one again has a formula like (4.15), where the right-hand side is no longer the "topological relative index," but it is rather the analytical index of some elliptic Fredholm operator on a (generally speaking, noncompact) manifold obtained from $X_0$ and $X_1$ by cutting away the "common" ends along some hypersurface followed by gluing along that hypersurface. The proof uses the same technique.

We can conclude (as is easily seen from the second theorem) that the *topological index* actually has nothing to do with the Gromov–Lawson relative index theorem: this theorem states the equality of the *analytical* relative indices for two pairs of operators obtained from each other by simultaneous surgery on a part of the manifold where they coincide; the topological index occurs in the answer only if the newly obtained operators fall within the scope of the Atiyah–Singer theorem. (On the other hand, naturally, the *applications* of theorems of that type are just related to transforming the original operators to new operators such that the Atiyah–Singer theorem or any other theorem expressing the index in topological terms can be used.) The coincidence of analytic relative indices readily follows from Theorem 4.10.

## 4.3. Surgery for the Index of Elliptic Operators on Singular Manifolds

On singular manifolds, we shall use surgeries of the following type.

***Surgery of manifolds and operators.*** Let $M$ be a manifold with singularities and/or cylindrical ends. (The case in which $M$ is a smooth compact manifold is not excluded.) Let $S \subset M$ be a closed compact hypersurface that does not meet the singularities and divides $M$ into two parts $M_+$ and $M_-$:

$$M_+ \cup M_- = M, \quad M_+ \cap M_- = S.$$

Next, let $M'$, $\widetilde{M}$, and $\widetilde{M}'$ be manifolds with singularities and/or cylindrical ends obtained from $M$ by the replacement of one or both of the parts $M_+$ and $M_-$ by some other parts $N_+$ and $N_-$:

$$\begin{aligned} M &= M_+ \bigcup_S M_-, & M' &= N_+ \bigcup_S M_-, \\ \widetilde{M} &= M_+ \bigcup_S N_-, & \widetilde{M}' &= N_+ \bigcup_S N_-. \end{aligned} \tag{4.16}$$

Finally, let $D$, $D'$, $\widetilde{D}$, and $\widetilde{D}'$ be elliptic operators on the respective manifolds $M$, $M'$, $\widetilde{M}$, and $\widetilde{M}'$ such that

$$\begin{aligned} D &= D' \quad \text{on } M_-, & D &= \widetilde{D} \quad \text{on } M_+, \\ \widetilde{D} &= \widetilde{D}' \quad \text{on } N_-, & D' &= \widetilde{D}' \quad \text{on } N_+. \end{aligned} \tag{4.17}$$

*Remark* 4.12. The meaning of the equalities in (4.17) is clear for differential operators, since they are local. For pseudodifferential operators, we require the coincidence of the interior symbols over the corresponding part and the coincidence of other symbol components pertaining to the singularities lying in the corresponding part.

***The index locality principle.*** We represent relations (4.16), (4.17) schematically by the *surgery diagram*

$$\begin{array}{ccc} D & \overset{+}{\longleftrightarrow} & D' \\ {\scriptstyle -}\Big\updownarrow & & \Big\updownarrow{\scriptstyle -} \\ \widetilde{D} & \underset{+}{\longleftrightarrow} & \widetilde{D}', \end{array} \tag{4.18}$$

where the plus and minus signs on the arrows indicate which parts of the manifold and the operator undergo surgery, and say that this diagram *commutes*.

The following theorem describes the behavior of index under surgeries.

THEOREM 4.13. *Let the surgery diagram* (4.18) *commute. Then*

$$\operatorname{ind} D - \operatorname{ind} D' = \operatorname{ind} \widetilde{D} - \operatorname{ind} \widetilde{D}'. \tag{4.19}$$

*Proof.* This is of course a special case of Theorem 4.10. □

*Remark* 4.14. This means that index increments resulting from modifications of an elliptic operator on disjoint parts of the manifold are independent. We shall sometimes refer to (4.19) as the *index locality principle* for singular manifolds.

## 4.4. Relative Index Formulas on Manifolds with Isolated Singularities

In this section, we use surgery to obtain index formulas on manifolds with singularities in some of the simplest situations.

We shall assume that all operators in question are differential. Although all results remain valid for pseudodifferential operators, the proofs in some cases require more complicated techniques (e.g., in Proposition 4.17). For brevity, we consider only the conical case. Using the cylindrical representation (see Definition 1.30), the reader can readily extend all the results to other types of isolated singularities, e.g., cusps.

Let us discuss properties of the index which are due to the singular points and have no analogs on smooth manifolds. Recall that the index on smooth closed manifolds is invariant under perturbations (deformations) of the operator. More precisely, the index remains unchanged under the following transformations:

- an arbitrary variation in lower-order terms of an elliptic operator

- a continuous deformation (preserving the ellipticity) of the coefficients of an elliptic operator

These assertions follow from standard theorems of functional analysis stating the invariance of the index under compact perturbations and continuous deformations of a Fredholm operator.

Needless to say, these two properties remain valid on manifolds with singularities (by the same theorems of functional analysis).[4]

However, if we weaken the condition and require the ellipticity of the interior symbol alone, then both assertions about the invariance of the index (as well as the assertion that the operator is Fredholm) fail. Formulas describing the variation of the index under the above-mentioned transformations are called *relative index formulas*. They can be obtained in this situation by the general surgery method.

To obtain relative index formulas, we need to compute the index in a model situation. This will be done in the next subsection.

We shall consider cone-degenerate operators

$$D = D\left(\omega, r, -i\frac{\partial}{\partial\omega}, ir\frac{\partial}{\partial r}\right) : H^{s,\gamma}(\mathcal{M}, E) \longrightarrow H^{s-m,\gamma}(\mathcal{M}, F) \qquad (4.20)$$

of order $m$ without the factor $r^{-m}$.

The index of a cone-degenerate operator is independent of the smoothness exponent $s$ of the Sobolev space, just as in the smooth case. However, the choice of the weight $\gamma$ affects the index. The index of the operator $D$ in the spaces (4.20) will be denoted by $\operatorname{ind}_\gamma D$.

### 4.4.1. The index of invariant operators on the cylinder

First, we consider a pilot example.

EXAMPLE 4.15. The *Cauchy–Riemann operator*

$$\frac{\partial}{\partial t} + i\frac{\partial}{\partial\varphi}$$

on the infinite cylinder $\mathbb{R} \times \mathbb{S}^1$ can be viewed as a cone-degenerate operator. To see this, it suffices to make logarithmic changes of variables as $t \to \pm\infty$. (For example, for $t \to +\infty$ one sets $t = -\log r$.)

PROPOSITION 4.16. *The operator*

$$\frac{\partial}{\partial t} + i\frac{\partial}{\partial\varphi} : H^{s,\gamma_1,\gamma_2}(\mathbb{R} \times \mathbb{S}^1) \longrightarrow H^{s-m,\gamma_1,\gamma_2}(\mathbb{R} \times \mathbb{S}^1)$$

[4]We note that *lower-order terms* in the singular case can be treated as the coefficients of the operator which occur in neither the interior symbol nor the conormal symbol.

*(where $\gamma_{1,2}$ are the weights at $\mp\infty$) is Fredholm provided that $\gamma_1, \gamma_2 \notin \mathbb{Z}$, and its index is given by the expression*

$$\operatorname{ind}_{\gamma_1,\gamma_2}\left(\frac{\partial}{\partial t}+i\frac{\partial}{\partial\varphi}\right)=[-\gamma_2]-[\gamma_1],$$

*where $[\,\cdot\,]$ is the integer part of a number. In particular, by varying the weight at infinity, one obtains an arbitrary prescribed value of the index.*

◀ *Hint.* Use separation of variables to give a closed-form description of the kernel and cokernel of the operator in the weighted spaces: for $\gamma_1 \le -\gamma_2$, the cokernel is trivial and the kernel is spanned by the exponentials $e^{n(t+i\varphi)}$ with $\gamma_1 < n < -\gamma_2$. In the computations, it is convenient to use the fact that the norm in $H^{0,\gamma_1,\gamma_2}$ is equivalent to the norm given by the square root of the integral

$$\int_{\mathbb{R}} \|u(t)\|^2_{L^2(\Omega)}\varphi(t)dt, \text{ where } \varphi(t)=\begin{cases} e^{-2\gamma_1 t} & \text{at } -\infty, \\ e^{2\gamma_2 t} & \text{at } +\infty. \end{cases}$$

For $\gamma_1 > -\gamma_2$, the kernel and cokernel change places. ▶

***The general case.*** Consider the cylinder $\mathbb{R}\times\Omega$ with smooth closed base $\Omega$. We represent the cylinder as the smooth part of a manifold with two conical singular points corresponding to two infinite parts of the cylinder. The corresponding com-

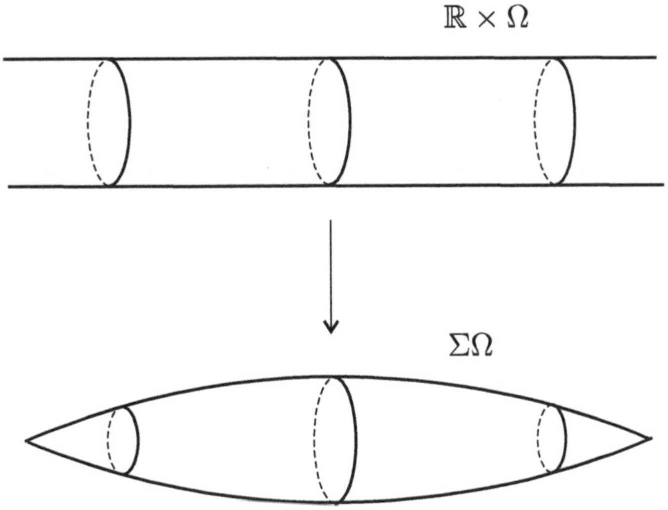


Figure 4.4. An infinite cylinder and the suspension.

pact manifold with singularities is known in topology as the *suspension* over $\Omega$ and is denoted by $\Sigma\Omega$ (see Fig. 4.4). The suspension can also be represented as the space obtained by gluing two identical cones over $\Omega$ along their bases.

On the cylinder, we consider the operator

$$D\left(-i\frac{\partial}{\partial t}\right) : H^{s,\gamma_1,\gamma_2}(\mathbb{R}\times\Omega) \longrightarrow H^{s-m,\gamma_1,\gamma_2}(\mathbb{R}\times\Omega)$$

obtained from a parameter-dependent elliptic operator $D(p)$ on the base of the cylinder by the substitution $p \mapsto -i\partial/\partial t$. This operator is invariant under the translations $t \mapsto t+a$, and the index of the corresponding operator on the closed smooth manifold $\mathbb{S}^1 \times \Omega$ is zero. However, in the noncompact case the index may be nonzero, and we give a closed-form expression for it in the following assertion. Suppose that the operator $D(-i\partial/\partial t)$ is elliptic with weight $\gamma_1$ at $-\infty$ and weight $\gamma_2$ at $+\infty$ (cf. Proposition 4.16).

PROPOSITION 4.17. *For $\gamma_1 < -\gamma_2$, the index is equal to the sum of multiplicities* (*see Appendix A*) *of singular points of the family $D(p)$ in the strip between the weight lines*:

$$\operatorname{ind}_{\gamma_1,\gamma_2} D\left(-i\frac{\partial}{\partial t}\right) = \sum_{\gamma_1<\operatorname{Im} p_j<-\gamma_2} m_D(p_j). \tag{4.21}$$

*If $\gamma_1 > -\gamma_2$, then the sign on the sum is opposite* (*and the singular points are taken in the strip $-\gamma_2 < \operatorname{Im} p_j < \gamma_1$*).

*Proof.* For simplicity, we compute the index for the case in which all zeros of the family $D(p)$ are simple (i.e., have the unit multiplicity). By applying the Euler method to the homogeneous equation

$$D\left(-i\frac{\partial}{\partial t}\right)u = 0,$$

we find that its solutions are represented as linear combinations of functions of the form

$$e^{it\lambda}u_\lambda, \tag{4.22}$$

where $\lambda \in \mathbb{C}$ is a zero of the family $D(p)$ and $u_\lambda$ is a nontrivial solution of the corresponding equation $D(\lambda)u_\lambda = 0$. Furthermore, the condition that the solution (4.22) must belong to the weighted space $H^{s,\gamma_1,\gamma_2}$ results in the additional requirement $\gamma_1 < \operatorname{Im}\lambda < -\gamma_2$. Thus if $\gamma_1 < -\gamma_2$, then the dimension of the kernel of the operator is equal to the number of zeros of $D(p)$ in the strip. If $\gamma_1 > -\gamma_2$, we conclude that the kernel is trivial.

Considering the adjoint operator, we see that the situation with the cokernel is opposite: the cokernel is trivial if the kernel is nontrivial, and vice versa. Thus we have proved the formula (4.21). □

*Remark* 4.18. The proof of this assertion for general differential operators is given in (Sternin 1972).

We shall show in the next subsection that for arbitrary operators on the cylinder the index formula can also be written in terms of multiplicities of zeros of a family in a way similar to the index formula (4.21). The corresponding (more complicated) expression is called the *spectral flow* for families of parameter-dependent elliptic operators. This notion is explained in Appendix A.

### 4.4.2. Relative index formulas

Consider the following operations over operators on manifolds with conical singularities:

1) change of the weight exponent $\gamma$ of the Sobolev space $H^{s,\gamma}$

2) change of the conormal symbol

3) homotopy of the operator

These operations can violate the ellipticity. Hence the index is no longer preserved. The aim of this subsection is to obtain formulas for the variation of the index of elliptic operators under these operations.

THEOREM 4.19 (relative index formulas). 1) (change of the weight exponent). *For an operator $D$ elliptic with respect to two weights $\gamma_1 < \gamma_2$, the index variation under the passage from one weight to the other is equal to the sum of multiplicities of singular points of the conormal symbol $\sigma_c(D)$ in the strip between the weight lines*:

$$\operatorname{ind}_{\gamma_1} D - \operatorname{ind}_{\gamma_2} D = \sum_{\gamma_1 < \operatorname{Im} p_j < \gamma_2} m_{\sigma_c(D)}(p_j). \tag{4.23}$$

2) (change of the conormal symbol). *Let $D_1$ and $D_2$ be elliptic operators with the same interior symbol. Then the difference of their indices is equal to the spectral flow (see Appendix A) across the weight line $\operatorname{Im} p = \gamma$ of the linear homotopy $t\sigma_c(D_1) + (1-t)\sigma_c(D_2)$ with parameter $t$ joining the conormal symbols*:

$$\operatorname{ind}_\gamma D_2 - \operatorname{ind}_\gamma D_1 = \operatorname{sf}_\gamma\{t\sigma_c(D_1) + (1-t)\sigma_c(D_2)\}_{t\in[0,1]}. \tag{4.24}$$

3) (homotopy of operators). *Let $\{D_a\}_{a\in[0,1]}$ be a continuous homotopy of cone-degenerate operators with elliptic interior symbols. If the operators at the beginning and the end of the homotopy are elliptic with respect to some weight $\gamma$, then the index variation is equal to the spectral flow across the weight line $\operatorname{Im} p = \gamma$ of the family of conormal symbols*

$$\operatorname{ind}_\gamma D_0 - \operatorname{ind}_\gamma D_1 = \operatorname{sf}_\gamma\{\sigma_c(D_a)\}_{a\in[0,1]}. \tag{4.25}$$

*Proof.* Although the homotopy formula (4.25) implies all preceding formulas as simple corollaries, it will be convenient to us to prove first item 1, then item 3, and finally the formula in item 2.

1. Let us prove (4.23). Consider an operator $D$ elliptic with respect to two weights $\gamma_1$ and $\gamma_2$. By the index locality principle (see Section 4.3), the variation in the index of the operator under the change of weight is independent of the interior symbol of the operator outside the singular point. In other words, the difference of the indices is determined by the conormal symbol. To compute the difference, we use the surgery technique described in Section 4.3.

We deal with a neighborhood of the conical singularity in the cylindrical representation. Consider the surgery diagram shown in Fig. 4.5. Geometrically, the "vertical" surgeries correspond to a change of the weight at the conical point (from $\gamma_2$ to $\gamma_1$) and the horizontal surgeries correspond to the replacement of the smooth part of the manifold by a conical point with weight $-\gamma_1$. Then on the cylinder we obtain the translation-invariant operator $\sigma_c(D)\,(-i\partial/\partial t)$. An application of the relative index theorem of Section 4.3 to this commutative diagram of surgeries gives an expression for the difference of indices of the operator $D$ with respect to different weights:

$$\operatorname{ind}_{\gamma_2} D - \operatorname{ind}_{\gamma_1} D = \operatorname{ind}_{-\gamma_1,\gamma_2} \sigma_c(D)\left(-i\frac{\partial}{\partial t}\right). \tag{4.26}$$

The index of the resulting invariant operator on the cylinder is just minus the sum of multiplicities of the poles in (4.23) (see Proposition 4.17).

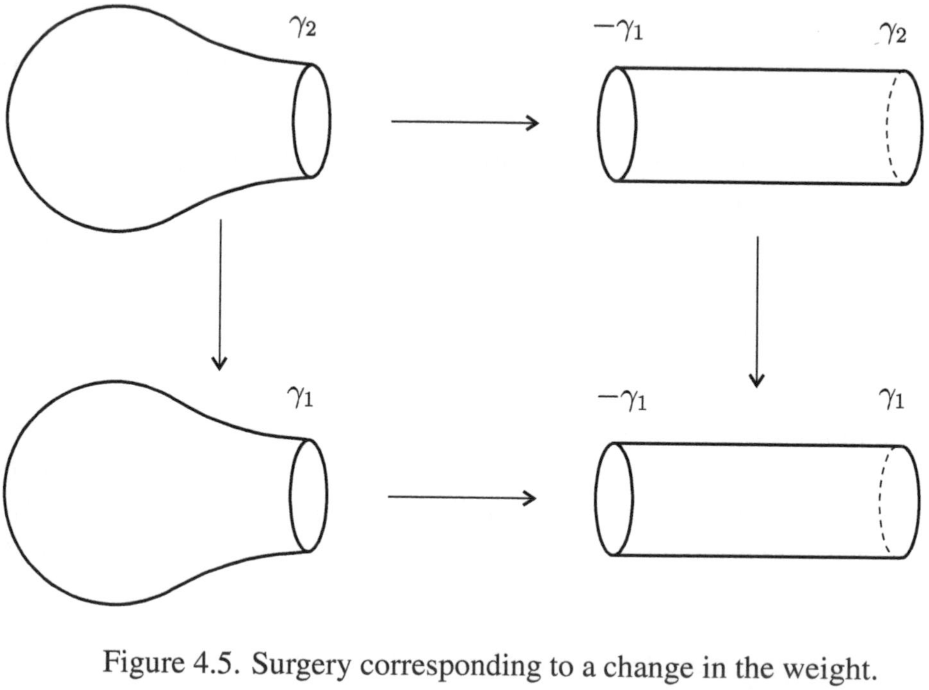


Figure 4.5. Surgery corresponding to a change in the weight.

2. Now let us obtain a formula for the change in the index under homotopies. To this end, we show that the index and the spectral flow vary by the same rule.

Consider a homotopy $D_a$ of operators. For some parameter values, these operators fail to be Fredholm. (The conormal symbol $\sigma_c(D_a)$ may be noninvertible on the weight line $\operatorname{Im} p = \gamma$.) To obtain a family of *Fredholm* operators, just as in the definition of the spectral flow (see Appendix A), for the family of conormal symbols $\sigma_c(D_a)$ we take some *admissible broken line* with vertices $(a_i, \lambda_i)$ such that

$$a_0 = 0, \lambda_0 = \gamma, \qquad a_N = 1, \lambda_N = \gamma.$$

(Recall that a broken line is said to be admissible for a family $\sigma_c(D_a)$ if for all parameter values on the interval $a_i \le a \le a_{i+1}$ there are no singular points of the family on the weight line $\operatorname{Im} p = \lambda_i$.)

Using an admissible broken line, we can readily construct a family of Fredholm operators: we choose the weight $\lambda_i$ for the operator $D_a$ on the half-open interval $a_i \le a < a_{i+1}$.

We know that only the vertices of the broken line contribute to the spectral flow. The contribution at $a = a_i$ is the sum of multiplicities of singular points of the family in the strip between the weight lines corresponding to $\lambda_i$ and $\lambda_{i+1}$.

Thus to prove (4.25) it suffices to verify that

- the index $\operatorname{ind}_{\lambda_i} D_a$ is constant on each interval $a_i \le a \le a_{i+1}$
- the variation of the index as the parameter $a$ passes through some vertex $a_i$ is equal to the sum of multiplicities of zeros of the family in the strip between the weight lines (this corresponds to a change of weight at the singular point)

In the first case, the index is preserved by virtue of the continuity and ellipticity of the family on the intervals, and the index variation in the second case is described by the formula (4.23) proved earlier.

Thus we arrive at (4.25).

3. Finally, the formula (4.24) is obtained by the substitution of the linear homotopy $(1-t)D_0 + tD_1$ of operators on the manifold $\mathcal{M}$ into (4.25). □

Relative index formulas can be used to compute the index in some special classes of operators. Examples of such applications will be considered in the next two subsections.

*Remark* 4.20. For operators on manifolds with conical points, one also has an analog of the Agranovich formula: if two elliptic operators $D_1$ and $D_2$ coincide in a neighborhood of the singularity, then the difference of their indices is determined by the interior symbols outside the neighborhood and

$$\operatorname{ind}_\gamma D_1 - \operatorname{ind}_\gamma D_2 = \operatorname{ind} D_1 D_2^{-1}, \tag{4.27}$$

where $D_2^{-1}$ is an almost inverse operator. The product $D_1 D_2^{-1}$ on the right-hand side of the formula acts as the identity operator in a neighborhood of the singular point. Hence its index can be computed by the ordinary Atiyah–Singer formula.[5] The formula (4.27) follows from the logarithmic property of the index.

Note that this construction remains valid if we require only that the conormal symbols of the two operators coincide.

### 4.4.3. The index of operators with unit interior symbol on a manifold with conical singularities.

By a *Mellin operator* we mean a cone-degenerate operator whose interior symbol is zero. Our aim is to compute the index of elliptic operators of the form $1+\mathfrak{M}$, where $\mathfrak{M}$ is a Mellin operator. Such operators arise, say, as quotients of elliptic differential operators with coinciding interior symbols. The conormal symbol of the Mellin operator will be denoted by $m = \sigma_c(\mathfrak{M})$. The ellipticity condition means in this case that the function $1+m$ is invertible on the corresponding weight line.

The elliptic symbol $1+m$ defines a topological invariant. Indeed, the conormal symbol $m$ of the Mellin operator is compact-valued and tends to zero at infinity. Thus the symbol $1+m$ can be treated as a continuous function on the circle $\mathbb{S}^1 = \mathcal{L}_\gamma \cup \{\infty\}$ obtained as the one-point compactification of the weight line. By

$$w_\gamma(1+m) \in \mathbb{Z}$$

we denote the *winding number* of this invertible operator function.

Recall that the space of invertible operators of the form $1+K$, where $K$ is compact, has the fundamental group isomorphic to $\mathbb{Z}$, the isomorphism just being given by the winding number.

If the symbol of the Mellin operator is of finite rank, then the winding number can be computed by the well-known formula

$$w_\gamma(1+m) = \frac{1}{2\pi i} \int_{\operatorname{Im} p=\gamma} \operatorname{Tr}((1+m)^{-1} dm).$$

PROPOSITION 4.21. *The index of the operator* $1+\mathfrak{M}$ *is equal to the winding number of its conormal symbol* $1+m$:

$$\operatorname{ind}_\gamma(1+\mathfrak{M}) = w_\gamma(1+m).$$

[5]The Atiyah–Singer theorem remains valid for manifolds with boundary if in a neighborhood of the boundary the operator is induced by a bundle isomorphism, i.e., "does not contain differentiations."

*Proof.* The operator $1+\mathfrak{M}$ is linearly homotopic to the identity operator. By the relative index formula (4.24), the index of $1+\mathfrak{M}$ is equal to the spectral flow of the homotopy $1+tm$ as $t$ varies from zero to unity. The relation

$$\operatorname{sf}_\gamma\{1+tm\}_{t\in[0,1]} = w_\gamma(1+m) \tag{4.28}$$

can be verified by standard methods:

1. First, using the homotopy invariance of both sides, we can replace $m$ by a finite-dimensional family.

2. For finite-dimensional families, the relation for a family of diagonal matrices with the only nonunit diagonal entry equal to $(p+i)/(p-i)$ can be verified directly.

3. The preceding computation suffices for establishing the formula in the general case, since finite-dimensional invertible families modulo homotopy form the fundamental group $\pi_1(\mathbf{GL}(N)) \simeq \mathbb{Z}$ of the group of invertible matrices, the generator being the family indicated in item 2. □

### 4.4.4. The index of general operators on the cylinder

As an application of relative index formulas, we give a formula for the index of an arbitrary elliptic operator $D$ stabilizing at infinity on the infinite cylinder $\mathbb{R}\times\Omega$. In particular, the stabilization condition holds for cone-degenerate operators corresponding to the suspension $\Sigma\Omega$. In contrast to the case considered in Section 4.4.1, here we do not assume that the operator is translation-invariant with respect to the variable $t$. Without loss of generality, we can assume that the interior symbol of the operator is independent of $t$ for large $t$.

EXERCISE 4.22. Show that the last condition can be ensured by a small deformation and addition of compact operators. Moreover, the deformation can also be chosen in such a way that the conormal symbol remains constant.

◄ *Hint.* Use the fact that the interior symbol of a stabilizing operator on the infinite cylinder can be viewed as the ordinary symbol over the stretched manifold, which is the compact manifold $\Omega\times[0,1]$ with boundary. The latter symbol can readily be deformed to a symbol constant with respect to the normal variable in a neighborhood of the boundary. ►

It turns out that for the case of the infinite cylinder the interior symbol and the conormal symbols at $\pm\infty$ can be considered together. To this end, we treat the operator $D$ as an operator on the line $\mathbb{R}$ with an operator-valued symbol (in the sense of Section 3.3) acting in function spaces on $\Omega$. The operator-valued symbol, which we denote by $D_t(p)$, is a homotopy of parameter-dependent elliptic operators. Here the parameter of the family is $p\in\mathcal{L}_\gamma$, and the homotopy parameter is

$t' \in (-\infty, +\infty)$. Moreover, the values of the homotopy at infinity are determined by the conormal symbols:

$$D_{t'} = \begin{cases} \sigma_c(D)|_{t\to-\infty} & \text{for } t' \le -N, \\ \sigma_c(D)|_{t\to+\infty} & \text{for } t' \ge N. \end{cases}$$

EXERCISE 4.23. Show that, conversely, to the family $D_t(p), t \in [-N, N]$ one can assign an operator on the infinite cylinder. Moreover, the correspondence

$$\text{(operator on the cylinder)} \leftrightarrow \text{(homotopy of families with parameter)}$$

is one-to-one if operators on the cylinder are considered modulo compact operators and homotopies modulo compact-valued homotopies vanishing for $|t| \ge N$.

◀ *Hint.* It suffices to observe that a family of parameter-dependent operators is uniquely determined by its principal symbol modulo families ranging in the set of compact operators. ▶

PROPOSITION 4.24. *The index of an operator on the infinite cylinder is equal to the spectral flow of the corresponding family of parameter-dependent elliptic operators*:

$$\operatorname{ind}_{\gamma,-\gamma} D = -\operatorname{sf}\{D_t\}_{t\in[-N,N]},$$

*where* sf *stands for the spectral flow across the weight line* $\operatorname{Im} p = \gamma$.

*Proof.* The idea of the proof is to use the fact that the cylinder $\Omega \times [-N, N]$ can be retracted to its base and thus deform $D$ to an operator with coefficients independent of $t$.

Specifically, we equip the family $D_t$ with an additional parameter $a$ by the formula

$$D_{t,a} = D_{-N+a(t+N)}.$$

This deformation of families continuously joins the homotopy $D_t$ corresponding to the original operator $D$ (for $a = 1$) with the constant family $D_0$ ($a = 0$), which corresponds to an operator with coefficients independent of $t$; the index of the latter operator is zero (see Section 4.4.1). Hence the index of $D$ is equal to the index variation under the homotopy and coincides with the spectral flow of the corresponding family of conormal symbols at $+\infty$, where the conormal symbol varies from $D_{-N}$ to $D_N$. (The conormal symbol at $-\infty$ does not vary.) The proof is complete. □

# Part III

# Topological Problems

# Chapter 5

# Index Theory

In this chapter, we consider the index problem for elliptic operators on a compact singular manifold $\mathcal{M}$. We deal only with manifolds with cone or edge singularities, the former being formally a special case of the latter if the edge $X$ is zero-dimensional. To avoid unnecessary wordiness and parallel statements, we adopt the convention that for $\dim X = 0$ the term "edge symbol," as well as the corresponding notation $\sigma_\wedge$, refers to the conormal symbol $\sigma_c$ whenever our argument pertains to both edge and conical cases. If some results are valid (or make sense) only for one of the two cases, we always indicate this explicitly. The operators are always assumed to act in the weighted Sobolev spaces $H^{s,\gamma}(\mathcal{M})$ or $\mathcal{W}^{s,\gamma}(\mathcal{M})$. Unless specified otherwise, $\gamma$ is fixed (and can safely be assumed to be zero).

## 5.1. Statement of the Problem

### 5.1.1. What does one expect of an index formula?

Most generally, the index problem is the problem of computing the index of an elliptic operator $D$ on a manifold $\mathcal{M}$ in terms of the principal symbol $\mathfrak{s}(D)$, that is, of finding a functional $f$ (necessarily homotopy invariant) such that

$$\operatorname{ind} D = f(\mathfrak{s}(D)). \tag{5.1}$$

This abstract setting does not take account of any additional structures that might be present on $\mathcal{M}$ and can be refined if they are known. Specifically, we deal with singular manifolds $\mathcal{M}$, and these can be viewed as *stratified manifolds* with two smooth strata,[1] the *singular stratum* $X$ and the *interior stratum* $M^\circ = \mathcal{M} \setminus X$.

In topology, there is a well-known phenomenon of splitting of invariants into parts corresponding to strata. For example, the Euler characteristic of $\mathcal{M}$ splits,

$$\chi(\mathcal{M}) = \chi(\mathcal{M} \setminus X) + \chi(X),$$

if $\dim \mathcal{M}$ is even. It is reasonable to ask whether this phenomenon occurs also in index theory, i.e., whether the functional $f$ in (5.1) can be represented as the sum $f = f_1 + f_2$ of "contributions" of the strata. Note that the principal symbol

[1] This is valid for cone or edge singularities; there are more strata for higher singularities.

$\mathfrak{s}(D) = (\sigma(D), \sigma_\wedge(D))$ itself consists of two components associated with the stratification, and so we can rephrase our problem as follows:

*Does there exist a stratified index formula*

$$\operatorname{ind} D = f_1(\sigma(D)) + f_2(\sigma_\wedge(D)), \tag{5.2}$$

*where $f_1$ and $f_2$ are homotopy invariant functionals of the respective symbols? How should we compute the functionals $f_1$ and $f_2$?*

We address both questions in this chapter. Namely, we find the obstruction to the existence of such formulas, describe classes of operators in which the obstruction is zero, and write out explicit expressions for $f_1$ and $f_2$ in a number of cases.

*Remark* 5.1. Index formulas of the form (5.2) are especially important in elliptic theory on singular manifolds in that they separate homotopy invariant index contributions of finite and infinite-dimensional nature.

### 5.1.2. When do stratified index formulas exist?

Our first task is to study whether stratified index formulas like (5.2) with homotopy invariant functionals $f_1$ and $f_2$ exist at all. We do so by trying to construct a counterexample. Namely, consider an elliptic operator $D$ on a singular manifold $\mathcal{M}$. Using periodic homotopies of the boundary symbol

$$\sigma_\partial(D) = \sigma(D)\big|_{\partial T_0^*\mathcal{M}}$$

of $D$, we shall construct elliptic operators $D'$ such that $\sigma_\wedge(D') = \sigma_\wedge(D)$ and $\sigma(D')$ is homotopic to $\sigma(D)$ in the class of elliptic interior symbols. If this method produces an operator $D'$ such that $\operatorname{ind} D' \neq \operatorname{ind} D$, then the formula (5.2) obviously cannot be true.

Thus let us embed the boundary symbol $\sigma_\partial(D)$ in a continuous periodic family $\mu$ of elliptic boundary symbols:

$$\mu = \{\mu_\varphi\}, \quad \varphi \in [0, 2\pi]; \qquad \mu_0 = \mu_{2\pi} = \sigma_\partial(D).$$

This family can be treated as an elliptic symbol on $T_0^*(\partial M \times [0, 2\pi])$.

◄ Indeed, boundary symbols live on $\partial T_0^*\mathcal{M} = (T^*\partial M \times \mathbb{R})_0$, and hence $\mu$ is defined on $(T^*\partial M \times \mathbb{R})_0 \times [0, 2\pi] = T_0^*(\partial M \times [0, 2\pi])$. ►

By periodicity, we obtain an elliptic symbol (which we also denote by $\mu$) on $T_0^*(\partial M \times \mathbb{S}^1)$ and hence an elliptic operator $\widehat{\mu}$ on the "torus" $\partial M \times \mathbb{S}^1$.

***The excision trick.*** Without loss of generality, we can assume that the interior symbol of $D$ is independent of the variable $r$ in a collar neighborhood of $\partial M$. We cut $M$ along the submanifold $\{r = \varepsilon\} \simeq \partial M$ and paste the cylinder $\partial M \times [0, 2\pi]$ (with the symbol $\mu$ defined over it) between the coasts of the cut.

The resulting manifold is diffeomorphic to $M$, and the new interior symbol $\sigma'$ obtained from $\sigma(D)$ by this procedure is homotopic to $\sigma(D)$. (The homotopy can readily be obtained from $\mu$ itself.) Let $D'$ be an elliptic operator on $\mathcal{M}$ with

$$\sigma(D') = \sigma', \quad \sigma_\wedge(D') = \sigma_\wedge(D).$$

By the index excision property (4.10), we have

$$\operatorname{ind} D' = \operatorname{ind} D + \operatorname{ind} \widehat{\mu}. \tag{5.3}$$

We arrive at the following conclusion.

> *If there exists a periodic homotopy $\mu$ of elliptic boundary symbols such that the index* $\operatorname{ind} \widehat{\mu}$ *of the associated elliptic operator $\widehat{\mu}$ on the torus is nonzero, then the stratified index formula cannot hold.*

◀ The reader might ask how one produces a counterexample starting from such a homotopy, having in mind that the boundary symbol $\mu_0$ need not be the boundary symbol of any elliptic operator $D$. This is true indeed, but we can always proceed to the new homotopy

$$\widetilde{\mu}_\varphi = \mu_\varphi \mu_0^{-1}, \qquad \widetilde{\mu}_0 = \widetilde{\mu}_{2\pi} = 1.$$

Then $\operatorname{ind} \widehat{\widetilde{\mu}} = \operatorname{ind} \widehat{\mu} \neq 0$, and we can take $D = 1$. ▶

Unfortunately, there always exists a homotopy of this sort. (In the simplest case $\partial M = \mathbb{S}^1$, this follows from Example A.9 in Appendix A in conjunction with Proposition 5.6 below.)

> ***It follows that there is no stratified index formula with homotopy invariant terms in the class of all elliptic operators on $\mathcal{M}$.***

Thus if we insist on stratified index formulas, we have to sacrifice generality by reducing the class of elliptic operators considered, or the class of homotopies with respect to which we expect homotopy invariance, or both. How do we do that? Note that our counterexamples are based solely on boundary symbols. Hence it seems reasonable to reduce the class of operators by imposing some conditions on their boundary symbols.

***Subclasses of elliptic operators.*** Suppose that $\mathcal{M}$ is a singular manifold and $\Sigma \subset \mathcal{O}^*(\partial T_0^*\mathcal{M})$ is a set of elliptic boundary symbols. We assume that $\Sigma$ contains only symbols corresponding to elliptic theory on $\mathcal{M}$, that is, each element of $\Sigma$ can be realized as the boundary symbol of some elliptic operator $D$. This is equivalent to saying that for each $\sigma_\partial \in \Sigma$ there exists

(i) an invertible interior symbol $\sigma$ on $T_0^*\mathcal{M}$ with

$$\sigma \big|_{\partial T_0^*\mathcal{M}} = \sigma_\partial$$

(ii) an invertible edge symbol $\sigma_\wedge$ with

$$\sigma(\sigma_\wedge) = \sigma_\partial$$

We denote the sets of all such symbols (where $\sigma_\partial$ ranges over $\Sigma$) by $\mathfrak{S}$ and $\mathfrak{S}_\wedge$, respectively. Next, by $\mathrm{Ell}_\Sigma$ we denote the set of elliptic operators $D$ on $\mathcal{M}$ such that $\sigma_\partial(D) \in \Sigma$.

DEFINITION 5.2. We say that the class $\mathrm{Ell}_\Sigma$ admits a *stratified index formula* if there exist two homotopy invariant functionals

$$f_1 : \mathfrak{S} \longrightarrow \mathbb{R}, \qquad f_2 : \mathfrak{S}_\wedge \longrightarrow \mathbb{R}$$

such that relation (5.2) holds for each operator $D \in \mathrm{Ell}_\Sigma$.

***The obstruction.*** It is clear from the preceding that the class $\mathrm{Ell}_\Sigma$ cannot admit stratified index formulas if some periodic family of elements of $\Sigma$ defines an operator with nonzero index on the torus. The following theorem shows that this is in fact the only obstruction.

THEOREM 5.3. *The class* $\mathrm{Ell}_\Sigma$ *admits a stratified index formula if and only if for each periodic family* $\mu = \{\mu_\varphi\}$, $\varphi \in \mathbb{S}^1$, *of boundary symbols in* $\Sigma$ *the index of the corresponding operator on the torus* $\partial M \times \mathbb{S}^1$ *is zero*:

$$\mu = \{\mu_\varphi\} \subset \Sigma \quad \Longrightarrow \quad \operatorname{ind} \widehat{\mu} = 0. \tag{5.4}$$

*Proof.* We only need to prove the sufficiency of condition (5.4). Suppose that this condition is satisfied; we shall construct $f_1$ and $f_2$.

1. *Auxiliary construction.* For each arcwise connected component $\widetilde{\Sigma} \subset \Sigma$, we pick up an elliptic edge symbol $a_\wedge \in \mathfrak{S}_\wedge$ such that $\sigma(a_\wedge) \in \widetilde{\Sigma}$.

2. *Construction of the functional* $f_1$. Now let $\sigma \in \mathfrak{S}$ be an elliptic interior symbol such that $\sigma_\partial = \sigma|_{\partial T^*\mathcal{M}} \in \widetilde{\Sigma}$. We construct an auxiliary elliptic operator $A$ as follows. Take some homotopy $\{\nu_t\} \in \Sigma$ joining $\sigma(a_\wedge)$ with $\sigma_\partial$:

$$\nu_0 = \sigma(a_\wedge), \qquad \nu_1 = \sigma_\partial. \tag{5.5}$$

The interior symbol $\sigma(A)$ is obtained from $\sigma$ by attaching the homotopy $\nu$ at the collar, which we write down symbolically as

$$\sigma(A) = \nu \sqcup \sigma.$$

Next, we set $\sigma_\wedge(A) = a_\wedge$; this is possible, since the compatibility condition is satisfied: $\sigma_\partial(A) = \nu_0 = \sigma(a_\wedge)$. This uniquely determines $A$ modulo compact operators, and we set

$$f_1(\sigma) = \operatorname{ind} A. \tag{5.6}$$

LEMMA 5.4. *This is well defined: the index of $A$ is independent of the choice of $\nu$ and hence determined by $\sigma$ alone.*

◀ Indeed, let $\widetilde{\nu}$ be another homotopy with the property (5.5), and let $\widetilde{A}$ be the corresponding operator. Then $\sigma(\widetilde{A}) = \widetilde{\nu} \sqcup \sigma$ is homotopic to $\nu \sqcup \nu^{[-1]} \sqcup \widetilde{\nu} \sqcup \sigma$, where

$$\nu_t^{[-1]} = \nu_{1-t},$$

via a homotopy with constant boundary symbol, and hence $\widetilde{A}$ is homotopic to the operator $A'$ with

$$\sigma(A') = \nu \sqcup \nu^{[-1]} \sqcup \widetilde{\nu} \sqcup \sigma, \qquad \sigma_\wedge(A') = a_\wedge,$$

so that $\operatorname{ind} \widetilde{A} = \operatorname{ind} A'$. Note that $\mu = \nu^{[-1]} \sqcup \widetilde{\nu} \subset \Sigma$ is a periodic homotopy, and hence

$$\operatorname{ind} A' = \operatorname{ind} A + \operatorname{ind} \widehat{\mu}$$

by the index excision property. It remains to note that $\operatorname{ind} \widehat{\mu} = 0$ by virtue of condition (5.4). ▶

Now it is easily seen from the construction that $f_1$ is homotopy invariant.

3. *Construction of the functional $f_2$.* If $\sigma_\wedge \in \mathfrak{S}_\wedge$, then we take an arbitrary $D \in \operatorname{Ell}_\Sigma$ with $\sigma_\wedge(D) = \sigma_\wedge$ and set

$$f_2(\sigma_\wedge) = \operatorname{ind} D - f_1(\sigma(D)). \tag{5.7}$$

This is again well defined.

◀ Indeed, $f_1(\sigma(D)) = \operatorname{ind} D'$, where $\sigma_\wedge(D') = a_\wedge$ is the distinguished element in the connected component of $\sigma_\wedge(D)$ and $\sigma(D') = \nu \sqcup \sigma(D)$, $\nu$ being a homotopy that joins $\sigma(a_\wedge)$ with $\sigma_\partial(D)$. Now if we take another operator, $\widetilde{D}$, in (5.7), then $f_1(\sigma(\widetilde{D})) = \operatorname{ind} \widetilde{D}'$, where

$$\sigma_\wedge(\widetilde{D}') = a_\wedge, \qquad \sigma(\widetilde{D}') = \nu \sqcup \sigma(\widetilde{D}).$$

We arrive at the surgery diagram

$$\begin{array}{ccc} D & \longleftrightarrow & D' \\ \updownarrow & & \updownarrow \\ \widetilde{D} & \longleftrightarrow & \widetilde{D}', \end{array}$$

where the horizontal arrows represent the surgery that attaches $\nu$ and replaces the edge symbol by $a$ and the vertical arrows represent the surgery that replaces $\sigma(D)$ by $\sigma(D')$. By the relative index Theorem 4.13,

$$\operatorname{ind} D - \operatorname{ind} D' = \operatorname{ind} \widetilde{D} - \operatorname{ind} \widetilde{D}',$$

which exactly means that the right-hand side of (5.7) is independent of the choice of $D$. ►

Now the formula (5.2) holds by construction, and it remains to check that $f_2$ is homotopy invariant, but this is actually trivial, since a homotopy of elliptic edge symbols can obviously be lifted to a homotopy of elliptic operators, and then the desired homotopy invariance of $f_2$ follows from that of the index and $f_1$. □

*Remark* 5.5. One can also consider the (apparently more general) problem on the existence of index formulas of the form

$$\operatorname{ind} D = f\big([\sigma(D)], [\sigma_\wedge(D)]\big), \tag{5.8}$$

where $f$ is a functional defined on pairs of homotopy classes of elliptic symbols (the interior symbol and the edge symbol taken separately). However, the obstruction is the same. Indeed, the proof of *necessity* of condition (5.4) does not use the linearity of (5.2) and hence remains valid for (5.8); on the other hand, (5.2) is a special case of (5.8), whence the desired equivalence follows.

***Representation of the obstruction via the spectral flow.*** It turns out that the obstruction to the existence of a stratified index formula can also be expressed in terms of the spectral flow (we recall the definition of the spectral flow in Appendix A). Specifically, let $\mu = \{\mu_\varphi\}$, $\varphi \in \mathbb{S}^1$, be a periodic family of boundary symbols. Take some periodic family $\{D_\varphi(p)\}$ of elliptic operators with parameter $p \in \mathbb{R}$ (in the sense of Agranovich and Vishik 1964) on $\partial M$ such that

$$\sigma(D_\varphi) = \mu_\varphi.$$

Then the operator $\widehat{\mu}$ on the torus $\partial M \times \mathbb{S}^1$ can be represented modulo compact operators as the operator on $\mathbb{S}^1$ with operator-valued symbol $D_\varphi(p)$:

$$\widehat{\mu} = D_\varphi\left(-i\frac{\partial}{\partial\varphi}\right).$$

Suppose that, in addition, $D_0(p)$ is invertible for all $p \in \mathbb{R}$ (which can always be ensured by the addition of a lower-order operator). Then the following assertion holds.

PROPOSITION 5.6. *One has*

$$\operatorname{ind} D_\varphi\left(-i\frac{\partial}{\partial\varphi}\right) = -\operatorname{sf}\{D_\varphi\}_{\varphi=0}^{2\pi}. \tag{5.9}$$

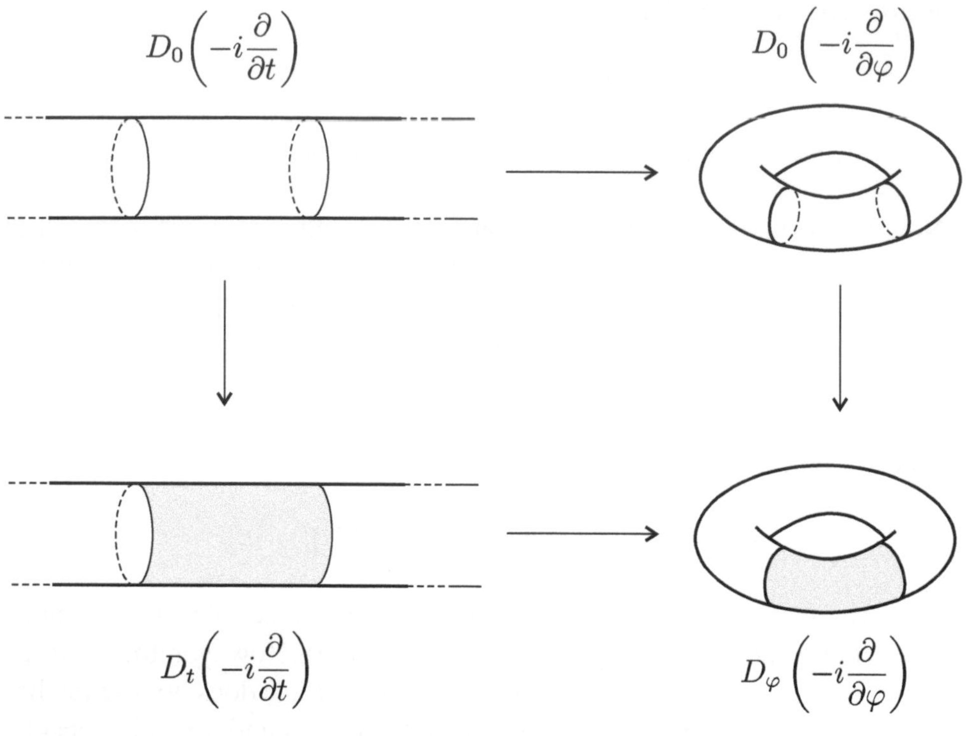


Figure 5.1. Surgery on the cylinder and the torus.

*Proof.* We represent the spectral flow of the family $D_\varphi, 0 \leq \varphi \leq 2\pi$, as minus the index of an elliptic operator $D$ on the infinite cylinder (see Proposition 4.24) with coordinate $t$. Now we use the periodicity of $D_\varphi$. Consider the surgery diagram shown in Fig. 5.1. The upper left corner shows the cylinder with the operator $D$ on it. The horizontal arrows stand for the surgery in which neighborhoods of infinity are cut away and the corresponding bases of the cylinder are glued together to form a torus. The vertical arrows stand for the surgery in which the operator is replaced on the dashed cylindrical part by an operator with coefficients independent of $t$. The indices of the operators in the upper row are zero (they are translation invariant), and so the relative index Theorem 4.13 says that the indices of the operators in the lower row are equal, as desired. □

Thus when speaking of operators on manifolds with conical singularities, we can say that

***the obstruction to the existence of stratified index formulas is the spectral flow of periodic families of conormal symbols whose principal symbols lie in $\Sigma$.***

### 5.1.3. Classes of equivariant symbols

Theorem 5.3 exactly describes the obstruction but does not provide any useful stratified index formulas, since it actually proves only the *existence* of the functionals $f_1$ and $f_2$. This is not surprising, since here we deal with arbitrary sets $\Sigma$. In what follows, we shall describe specific classes $\Sigma$ of boundary symbols satisfying the necessary and sufficient condition (5.4) and give specific index formulas. How to find such classes?

Here a (very coarse) analogy with physics may be of help. As we have seen, condition (5.4) says that the "spectral flow through the boundary $\partial M$" should be zero for periodic families. In physical language, this condition resembles a conservation law. (The variable conserved is of course $f_1$, the "index contribution" of the interior symbol.) Where do physicists get their conservation laws from? The answer is well known, it is perhaps one of the most general principles of physics:

| |
|---|
| *If you are after conservation laws, you should look for symmetries.* |

We use this wisdom in our situation and consider boundary symbols satisfying certain symmetry properties.

***The simplest symmetries.*** 1) Possibly the simplest symmetry (and also very natural from the physical viewpoint) is the *self-adjointness condition*

$$\sigma_\partial^* = \sigma_\partial. \tag{5.10}$$

(Here we should assume that $\sigma_\partial$ is an endomorphism of some Hermitian vector bundle $E$.)

Let $\Sigma_{sa}$ be the set of elliptic boundary symbols satisfying condition (5.10). Then the class $\mathrm{Ell}_{\Sigma_{sa}}$ admits stratified index formulas. Indeed, if $\mu = \{\mu_\varphi\} \subset \Sigma_{sa}$ is a periodic family, then the operator $\widehat{\mu}$ on the torus $\partial M \times \mathbb{S}^1$ satisfies

$$\widehat{\mu} = \widehat{\mu}^* + \text{ a compact operator,}$$

and hence $\operatorname{ind}\widehat{\mu} = 0$, so that Theorem 5.3 applies.

2) Another very important, though simple, symmetry is as follows. Recall that a boundary symbol is a family of elliptic symbols with parameter $p \in \mathbb{R}$ on $T^*\partial M$:

$$\sigma_\partial = \sigma_\partial(p), \quad p \in \mathbb{R}.$$

We consider the following *symmetry condition with respect to the conormal variable*:

$$\sigma_\partial(-p) = \sigma_\partial(p). \tag{5.11}$$

Physically, this condition can be interpreted as "invariance with respect to time inversion." (Recall that in the cylindrical representation $p$ is the momentum dual to the cylindrical variable $t$.)

This condition will appear independently in the next section, when we apply surgery techniques to the index problem.

Let $\Sigma_{ti}$ be the set of elliptic boundary symbols satisfying condition (5.11).

PROPOSITION 5.7. *The class* $\mathrm{Ell}_{\Sigma_{ti}}$ *admits stratified index formulas.*

*Proof.* Indeed, let $\mu = \{\mu_\varphi\} \subset \Sigma_{ti}$ be a continuous periodic family. We should prove that $\operatorname{ind}\widehat{\mu} = 0$ for the corresponding operator on the torus. To this end, consider the family $\mu'$ given by the formula

$$\mu'(p) = \mu^{[-1]}(-p).$$

(Thus we simultaneously change the sign of the conormal variable and of the parameter of the family.) Then $\widehat{\mu}'$ and $\widehat{\mu}$ are related by the change of variable $\varphi \longmapsto -\varphi$ on the torus and hence

$$\operatorname{ind}\widehat{\mu} = \operatorname{ind}\widehat{\mu}'.$$

On the other hand, it follows from condition (5.11) for $\mu$ that $\nu = \mu \sqcup \mu'$ is also a *continuous* (and periodic) family, and the index excision property implies that

$$\operatorname{ind}\widehat{\nu} = \operatorname{ind}\widehat{\mu} + \operatorname{ind}\widehat{\mu}' = 2\operatorname{ind}\widehat{\mu}.$$

The family $\nu$ is homotopic to a family independent of $\varphi$, and so $\operatorname{ind}\widehat{\nu} = 0$ (for example, this follows from Proposition 5.6), and we arrive at the desired conclusion that $\operatorname{ind}\widehat{\mu} = 0$. □

EXAMPLE 5.8. The Beltrami–Laplace operator belongs to both $\mathrm{Ell}_{\Sigma_{sa}}$ and $\mathrm{Ell}_{\Sigma_{ti}}$.

Unfortunately, these simplest symmetry conditions are very restrictive and it is not so easy to give further meaningful examples of operators whose boundary symbol satisfies one of these conditions. For example, there is no first-order operator satisfying condition (5.11). Thus we shall look for more general conditions.

***Twisted symmetries.*** It turns out that the desired more general conditions can be obtained from the basic conditions (5.10) and (5.11) if we apply some "twisting." Twisting can affect both interior degrees of freedom (bundles where symbols act) and exterior degrees of freedom (the variables on which symbols depend). In other words, we shall modify the basic symmetry conditions by using some vector bundle isomorphisms and diffeomorphisms of $\partial T_0^*\mathcal{M}$.

We restrict ourselves to the case of fiberwise linear mappings of $\partial T_0^*\mathcal{M}$. Let

$$g : \partial M \longrightarrow \partial M$$

be a diffeomorphism, and let

$$\alpha : \partial T^*\mathcal{M} \longrightarrow \partial T^*\mathcal{M}$$

be a vector bundle isomorphism over $g$. This means that the diagram

$$\begin{array}{ccc} \partial T^*\mathcal{M} & \xrightarrow{\alpha} & \partial T^*\mathcal{M} \\ {\scriptstyle\pi}\downarrow & & \downarrow{\scriptstyle\pi} \\ \partial M & \xrightarrow[g]{} & \partial M, \end{array}$$

where $\pi : \partial T^*\mathcal{M} \longrightarrow \partial M$ is the natural projection, commutes and the restriction

$$\alpha_y : T_y^*\mathcal{M} \longrightarrow T_{g(y)}^*\mathcal{M}$$

of $\alpha$ to the fiber over an arbitrary point $y \in \partial M$ is linear.

◀ Just in case, note that the mapping $\alpha$ is not assumed to preserve the symplectic form. ▶

Depending on whether $\alpha$ is orientation-preserving or orientation-reversing, we shall consider the following conditions, generalizing self-adjointness (5.10) and invariance (5.11), respectively.

1. **The anti-equivariance condition**

$$\alpha^*\sigma_\partial = f\sigma_\partial^* e \quad (\alpha \text{ is orientation-preserving}), \tag{5.12}$$

where

$$e : g^*E \longrightarrow F^*, \qquad f : E^* \longrightarrow g^*F$$

are vector bundle isomorphisms.

2. **The equivariance condition**

$$\alpha^*\sigma_\partial = f\sigma_\partial e \quad (\alpha \text{ is orientation-reversing}), \tag{5.13}$$

where

$$e : g^*E \longrightarrow E, \qquad f : F \longrightarrow g^*F$$

are vector bundle isomorphisms.

*Remark* 5.9. In the case of the invariance condition (5.13), the mapping $\alpha$ is the composition of an orientation-preserving mapping with the inversion $p \longmapsto -p$. Thus we see that each of our twisted symmetries is obtained from one of the two basic ones by twisting with an orientation-preserving mapping of $\partial T_0^*\mathcal{M}$ and with some bundle isomorphisms.

PROPOSITION 5.10. *The class of operators whose boundary symbols satisfy the equivariance condition* (*or the anti-equivariance condition*) *admits stratified index formulas.*

*Remark* 5.11. Here a remark is in order. The notion of an (anti-)equivariant symbol involves the choice of vector bundle isomorphisms $e$ and $f$. Hence a *continuous homotopy of* (*anti-*)*equivariant symbols* should be understood as a homotopy such that these isomorphisms also vary continuously. Strictly speaking, we have not stated the theorem on the obstruction to stratified index formulas in the case of symbols with this additional structure. The reader can however verify that Theorem 5.3 (together with the proof) can be transferred to this case without any modifications.

The proof of Proposition 5.10 can be obtained by a straightforward computation of the index of the operator on the torus in Theorem 5.3 with the use of the Atiyah–Singer theorem. However, we prefer a simpler proof, which we postpone until the end of Section 5.2.2.

Instead of (5.12), we shall usually deal with the following modification of the anti-equivariance condition, suitable for zero-order symbols:

$$\alpha^*\sigma_\partial = f\sigma_\partial^{-1}e \quad (\alpha \text{ is orientation-preserving}). \tag{5.14}$$

Needless to say, the vector bundle isomorphisms in condition (5.14) act in the bundles

$$e : g^*E \longrightarrow F, \qquad f : E \longrightarrow g^*F.$$

Proposition 5.10 remains valid for condition (5.14), since $\sigma_\partial^*$ and $\sigma_\partial^{-1}$ are homotopic.

DEFINITION 5.12. By $\Sigma_\alpha$ we denote the class of boundary symbols satisfying the above-mentioned (anti-)equivariance condition, and by $\mathrm{Ell}_\alpha$ we denote the class of elliptic operators on $\mathcal{M}$ whose boundary symbols belong to $\Sigma_\alpha$.

It always will be clear from the context which of the three conditions is actually meant. In accordance with the above remark, elements of $\Sigma_a$ are actually triples $(\sigma_\partial, e, f)$, where $e$ and $f$ are the bundle isomorphisms occurring in the corresponding condition.

*Remark* 5.13. The (anti-)equivariance conditions (5.13) and (5.14) can be represented in the form

$$G^*\sigma_\partial = \sigma_\partial \tag{5.15}$$

where

$$G^*\sigma_\partial = f^{-1}(\alpha^*\sigma_\partial)e^{-1}, \quad (5.16)$$

$$G^*\sigma_\partial = e(\alpha^*\sigma_\partial)^{-1}f, \quad (5.17)$$

respectively. This is often convenient.

### 5.1.4. Examples

Let us give some examples of operators satisfying the above-mentioned equivariance conditions. For simplicity, we consider the case of manifolds with boundary (i.e., edges with $\Omega = \{pt\}$).

***The Euler operator.*** The Euler operator is an important elliptic operator satisfying the symmetry condition (5.13). Let us briefly recall the definitions; details can be found in Lawson and Michelsohn (1989) and Palais (1965).

Let $M$ be an even-dimensional Riemannian manifold with boundary. We also assume that $M$ is oriented. The Euler operator

$$D = d + d^* : \Lambda^{ev}(M) \longrightarrow \Lambda^{odd}(M)$$

acts from the space of differential forms of even degree into the space of forms of odd degree and is defined in terms of the exterior differential $d$ and the formally adjoint operator $d^*$.

The adjoint operator is taken with respect to the inner product $\langle \cdot, \cdot \rangle$ of forms defined as the integral

$$\langle \omega_1, \omega_2 \rangle \stackrel{\text{def}}{=} \int_M \omega_1 \wedge *\omega_2$$

over the manifold, where

$$* : \Lambda^k(M) \to \Lambda^{n-k}(M), \quad n = \dim M,$$

is the Hodge star operator. This linear operator is uniquely determined by the property

$$*(e_1 \wedge e_2 \wedge \ldots \wedge e_k) = e_{k+1} \wedge \ldots \wedge e_n,$$

where $e_1, \ldots, e_n$ is an arbitrary positive (consistent with the orientation) orthonormal basis of covectors.

Using integration by parts, one can express $d^*$ via $d$ and the Hodge operator:

$$d^* = (-1)^{n(k+1)+1} * d*, \quad d^* : \Lambda^k(M) \to \Lambda^{k-1}(M).$$

Let us study the Euler operator in a collar neighborhood of the boundary of the manifold. In this neighborhood, differential forms can be represented as sums

$$\omega = \omega_1 + dt \wedge \omega_2,$$

where $\omega_{1,2}$ do not contain the differential $dt$ of the normal coordinate $t$. This expansion induces expansions of the spaces of even (odd) forms on $M$ in terms of forms on the boundary:

$$\Lambda^{ev/odd}(M)|_{\partial M} \simeq \Lambda^{ev/odd}(\partial M) \oplus \Lambda^{odd/ev}(\partial M). \tag{5.18}$$

Let us equip the neighborhood of the boundary with a product metric $dt^2 + g$, where $g$ is some metric on $\partial M$. Then the Euler operator can be represented in the following matrix form in the expansion (5.18) in a neighborhood of the boundary:

$$d + d^* = \begin{pmatrix} d' + d'^* & -\dfrac{\partial}{\partial t} \\ \dfrac{\partial}{\partial t} & -d' - d'^* \end{pmatrix}, \tag{5.19}$$

where $d'$ is the exterior differential on $\partial M$. This formula can be proved by a routine computation.

Starting from the representation (5.19), one can readily write out a symmetry of the form (5.13) for the Euler operator, where $\alpha$ is the inversion $p \longmapsto -p$ (with diffeomorphism $g = \mathrm{id}$). It is given by the matrix identity

$$\begin{pmatrix} d' + d'^* & \frac{\partial}{\partial t} \\ -\frac{\partial}{\partial t} & -d' - d'^* \end{pmatrix} = \begin{pmatrix} 1 & 0 \\ 0 & -1 \end{pmatrix} \begin{pmatrix} d' + d'^* & -\frac{\partial}{\partial t} \\ \frac{\partial}{\partial t} & -d' - d'^* \end{pmatrix} \begin{pmatrix} 1 & 0 \\ 0 & -1 \end{pmatrix}.$$

We point out that geometrically the bundle isomorphism with diagonal matrix $\mathrm{diag}(1, -1)$ merely corresponds to the change $dt \mapsto -dt$ of sign of the differential.

*Remark* 5.14. Using (5.19), we can show that the anti-equivariance condition (5.12) with $\alpha = \mathrm{id}$ is also valid for the Euler operator.

***The Cauchy–Riemann operator.*** The Cauchy–Riemann operator on the cylinder $[0, 1] \times \mathbb{S}^1$ has the form

$$-i\frac{\partial}{\partial t} + \frac{\partial}{\partial \varphi}.$$

Its principal symbol is equal to $p + i\xi$. This operator has a symmetry corresponding to the diffeomorphism $g(\varphi) = -\varphi$. It has the form

$$(g^*\sigma_\partial)(-p) = -\sigma_\partial(p),$$

where $\sigma_\partial$ is the boundary symbol of the Cauchy–Riemann operator.

One should note that the Cauchy–Riemann operator does not have symmetries with respect to the reflection $p \mapsto -p$. The reason for this is topological: this reflection gives the symbol $-p + i\xi$, which is neither equal nor even homotopic to the original symbol.[2] Hence the Cauchy–Riemann operator has no symmetries (5.13) with underlying diffeomorphism $\alpha : p \mapsto -p$.

[2]These symbols, viewed as functions on the circle $\{p^2 + \xi^2 = 1\}$, have distinct winding numbers.

The absence of local symmetries with respect to $p \mapsto -p$ is typical of geometric operators (the Dirac operator, the signature operator, and the Dolbeault operator). In what follows, the signature operator plays an important role, and hence we study symmetries only for this operator.

***The signature operator.*** Just as the Euler operator, the *signature operator* on an oriented $2k$-dimensional manifold $M$ is determined by the differential expression $d + d^*$ (see Palais 1965):

$$\mathcal{H} = d + d^* : \Lambda^+ (M) \longrightarrow \Lambda^- (M) .$$

However, it is considered in the subspaces of differential forms $\omega \in \Lambda^* (M)$ that are eigenvectors of the involution $\gamma$:

$$\omega \in \Lambda^\pm(M) \Longleftrightarrow \gamma\omega = \pm\omega, \quad \text{where } \gamma|_{\Lambda^p(M)} = i^{p(p-1)+k} * .$$

The forms belonging to these subspaces are said to be *self-dual* and *antiself-dual*, respectively.

The signature operator has no symmetries with respect to $p \mapsto -p$.

Moreover, just as with the Cauchy–Riemann operator, there are no symmetries even at a point. Indeed, the difference element of the symbol $\sigma(\mathcal{H})$ at a point $x \in M$ generates the group $K_c(T_x^* M) \otimes \mathbb{Q} \simeq \mathbb{Q}$ (see Palais 1965). Moreover, the change $p \leftrightarrow -p$ of sign of the covariable acts on the $K$-group as the change of sign of the generator. (The last assertion can readily be proved, since by virtue of the Bott periodicity it suffices to carry out the verification in the two-dimensional case for the group $K_c(\mathbb{R}^2) = K_c(\mathbb{C})$ with generator (Bott element) determined by the isomorphism $z$ of one-dimensional trivial bundles outside zero.) Thus the symbols $\sigma(\mathcal{H})(p,\xi)$ and $\sigma(\mathcal{H})(-p,\xi)$ determine distinct elements in $K$-theory. Hence the impossibility of a symmetry condition with respect to $p \mapsto -p$ is obvious.

To obtain a symmetry for the signature operator, let us write out this operator explicitly in a neighborhood of the boundary.

On the boundary of the manifold, one has the bundle isomorphisms

$$\Lambda^\pm (M)\big|_{\partial M} \simeq \Lambda^* (\partial M)$$

given by the formula $\omega \mapsto (\omega \pm \gamma\omega)/2$, $\omega \in \Lambda^*(\partial M)$. If we take a product metric in a collar neighborhood of the boundary, then the signature operator is represented modulo vector bundle isomorphisms in the form (Atiyah, Patodi and Singer 1975)

$$\frac{\partial}{\partial t} + A,$$

where the elliptic self-adjoint operator $A$ acting on the boundary is known as the *odd signature operator* and is determined by the expression

$$A : \Lambda^* (\partial M) \longrightarrow \Lambda^* (\partial M) , \quad A\omega = (\varepsilon d * + \varepsilon' * d)\, \omega. \tag{5.20}$$

Here $\varepsilon$ and $\varepsilon'$ are some unimodular complex numbers depending on the degree of the form and the dimension of the manifold. Since they are not important in what follows, we omit explicit expressions for these numbers.

We see from (5.20) that the operator $A$ anticommutes with a diffeomorphism $g : \partial M \longrightarrow \partial M$ provided that *$g$ reverses the orientation of the boundary*. In this case, the signature operator has the twisted symmetry property

$$g^* \left( \frac{\partial}{\partial t} + A \right) = - \left( -\frac{\partial}{\partial t} + A \right).$$

EXERCISE 5.15. Show that a symmetry with

$$\alpha = g^* \times \alpha_0 : T^*\partial M \times \mathbb{R} \longrightarrow T^*M,$$

where $g^*$ is the differential of a diffeomorphism $g : \partial M \to \partial M$ and $\alpha_0 : p \mapsto -p$, is possible for the signature operator only with an orientation-reversing diffeomorphism $g$ of the boundary.

## 5.2. Invariants of Interior Symbol and Symmetries

Consider some class $\Sigma_\alpha$ of boundary symbols satisfying the (anti-)equivariance condition. Then for operators $D \in \mathrm{Ell}_\alpha$ there exist stratified index formulas of the form (5.2), and the problem is to find homotopy invariants occurring in such formulas. In this section, we study homotopy invariants of the interior symbol. Thus the question is,

> *How should homotopy invariants be assigned to an elliptic symbol on a manifold with boundary?*

This question was studied rather comprehensively in elliptic theory on smooth manifolds with boundary.[3] Without going into detail, note that the most straightforward way in the construction of invariants is as follows:

- continue the operator to an elliptic operator on some closed manifold
- the index of the elliptic operator thus obtained can be taken as the desired homotopy invariant

This simple idea proves to be extremely useful in index theory on manifolds with singularities.

[3] That is, in the theory of boundary value problems for elliptic equations.

### 5.2.1. Example: Symmetry (5.11) in the conormal variable

The simplest way to construct a manifold without boundary from a manifold $M$ with boundary is to consider the *double*

$$2M \overset{\text{def}}{=} M \bigcup_{\partial M} M,$$

which is obtained by gluing two copies of $M$ along the boundary; see Fig. 5.2.

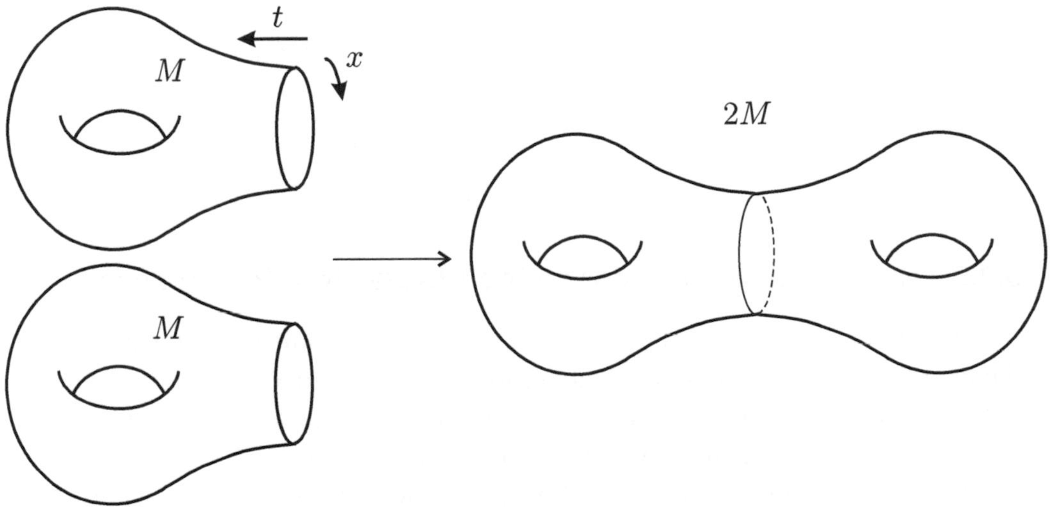


Figure 5.2. The double.

If we try to apply the same doubling and pasting procedure to elliptic operators, we notice that things are not so simple as with manifolds.

> *In general, operators cannot be pasted together along the boundary even at the level of principal symbols.*

Remarkably, it turns out that the condition of continuous pasting of the symbols on the double $2M$ coincides with the symmetry condition (5.11).

Let $D$ be an operator on $M$ with symbol $\sigma(D)$. We take two copies of $M$ and equip each of them with a copy of $D$. A neighborhood of the boundary in the first copy is pasted with the same neighborhood in the second copy with *inversion* of the normal direction rather than along the identity mapping (see Fig. 5.2):

$$(r, x) \longmapsto (-r, x),$$

where $(r, x)$ are the normal and tangential coordinates. Consequently, the symbols of two identical copies of $D$ coincide on the boundary under this pasting if and only if the restriction of the symbol to the boundary satisfies condition:

$$\sigma_\partial(D)(-p, \xi) = \sigma_\partial(D)(p, \xi). \tag{5.21}$$

Here $(p, \xi)$ are the dual momentum variables for $(r, x)$. We see that this gluing condition is the same as condition (5.11).

Thus if a symbol $\sigma(D)$ satisfying the symmetry condition is given on $M$, then on the double $2M$ we obtain a symbol[4] that, with some abuse of notation, will be denoted by $2\sigma(D)$. The corresponding operator will be denoted by $2D$. If the original symbol is elliptic, then, obviously, so is the continuation. Thus for the case of the simplest symmetry condition (5.11) we obtain an invariant of the interior symbol of $D$ in the form

$$f_1(\sigma(D)) = \frac{1}{2} \operatorname{ind} 2D. \tag{5.22}$$

### 5.2.2. The signature operator on the Atiyah–Bott–Patodi space

In the general (anti-)equivariance conditions, diffeomorphisms of (the boundary of) the phase space rather than the configuration space are involved, and so we cannot obtain invariants of the symbol by a method as simple as that in the preceding subsection. If we paste together two copies of $T^*\mathcal{M}$ along $\partial T^*\mathcal{M}$ using a general mapping $\alpha$ occurring in these conditions, then the resulting "double" of $T^*\mathcal{M}$ will not in general be the cotangent bundle of any closed manifold, and if we wish our invariant to be the index of some operator, we have to consider operators on the phase space.

In this subsection, we consider the passage from operators on the configuration space to operators on the phase space in the case of closed smooth manifolds. In the following subsections, we apply this technique to (anti-)equivariant symbols.

In index theory on closed manifolds, one often reduces the index of an arbitrary operator to the index of the Dirac operator with coefficients in some bundle determined by the symbol of the original operator.

This reduction is not always possible in our case, since the Dirac operator is based on the spinor structure of the manifold (see Lawson and Michelsohn 1989), while we admit diffeomorphisms $\alpha$ that only preserve (or invert) the orientation of manifolds. Hence we shall use the signature operator, in the construction of which it suffices to keep track of orientation. We also pass from the noncompact phase space to a compactification.

[4]Note that our symmetry condition ensures only that the symbol on the double is continuous. The reader can readily write out a condition ensuring smoothness. However, we automatically obtain a smooth continuation if (5.21) holds and the symbol is independent of $r$ in a neighborhood of the boundary.

Thus for an elliptic operator on the configuration space we construct a closed manifold (a compactification of the phase space) and a signature operator on it such that the index of the original operator is equal to the index of the signature operator with special coefficients.

***The compactification of the phase space.*** Consider a real vector bundle $E$ over a compact base. The total space of $E$ is always noncompact but can be compactified by adding the point at infinity to each fiber $E_x$. The resulting space is called the *Atiyah–Bott–Patodi space* of $E$ (Atiyah, Bott and Patodi 1973) and will be denoted by $DE$. The fiber of this space is a sphere of dimension $\dim E$. The Atiyah–Bott–Patodi space can also be defined by the formula

$$DE \overset{\text{def}}{=} S(E \oplus \mathbb{R}),$$

where $S$ is the bundle of unit spheres in a vector bundle (with respect to some metric).

Let us apply this operation to the cotangent bundle $T^*M$ of a smooth closed manifold $M$. Its Atiyah–Bott–Patodi space $D^*M$ is a bundle with base $M$ and fiber $\mathbb{S}^n$, $n = \dim M$, and is a smooth closed even-dimensional manifold. It can also be viewed as the result of gluing two copies of the unit disk bundle $B^*M$ in $T^*M$ along the unit spheres (see Fig. 5.3).

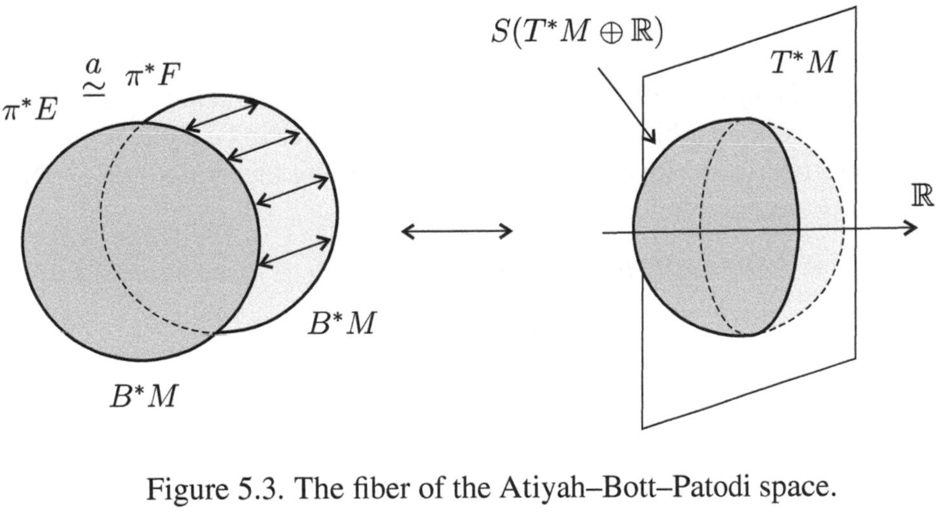


Figure 5.3. The fiber of the Atiyah–Bott–Patodi space.

***The coefficient bundle.*** To an elliptic symbol $a\colon \pi^*E \longrightarrow \pi^*F$ over $M$, there corresponds a vector bundle

$$[a] \in \operatorname{Vect}(D^*M)$$

on $D^*M$ defined as the clutching of the lifts of $E$ and $F$ to the unit disk bundles $B^*_+M$ and $B^*_-M$ along the isomorphism $a$ on the unit spheres (see Fig. 5.3).

***The signature operator.*** Recall the construction of an *operator with coefficients in a vector bundle*. If the symbol of an operator $D$ is $\sigma(D) : \pi^*E \longrightarrow \pi^*F$, then the symbol of the operator $D \otimes H$ with coefficients in a bundle $H \in \mathrm{Vect}(M)$ is by definition

$$\sigma(D) \otimes 1_H : \pi^*(E \otimes H) \longrightarrow \pi^*(F \otimes H).$$

Now we consider the signature operator

$$\mathcal{H}_a = \mathcal{H} \otimes (\Lambda^*(M) \otimes \mathbb{C})^{-1} \otimes [a] \tag{5.23}$$

on $D^*M$ with coefficients determined by the vector bundle $[a] \in \mathrm{Vect}(D^*M)$. Here $\mathcal{H}$ is the signature operator on $D^*M$, and the "inverse bundle" $\Lambda^*(M)^{-1}$ is understood as follows. The operation $E^{-1}$ for a vector bundle $E$ can be defined in terms of the formal series

$$E^{-1} \equiv \frac{1}{n-(n-E)} = \frac{1}{n}\left(1 + \sum_{l\geq 1}(1-n^{-1}E)^l\right) \in K(M) \otimes \mathbb{Q}, \tag{5.24}$$

where $n = \dim E$ is the dimension of the bundle. The result belongs to the $K$-group with rational coefficients. The series converges, since starting from some point it contains only zero terms: for sufficiently large $l$, the element $(\dim E - E)^l$ is zero in the $K$-group (Atiyah 1989).

Indeed, this result follows if we pass to cohomology with the help of the Chern character and use the fact that the Chern character of the difference $\dim E - E$ consists of cohomology classes of positive degrees.

***The coincidence of indices.*** The signature operator constructed here is important in that the index of an arbitrary operator $D$ on $M$ proves to be equal to the index of the signature operator with coefficients determined by the symbol of $D$. Namely, the following assertion holds.

PROPOSITION 5.16. *Let $D$ be an elliptic operator on a closed manifold $M$. Then*

$$\operatorname{ind} D = \operatorname{ind} \mathcal{H}_{\sigma(D)}. \tag{5.25}$$

*The proof* will be given in Supplement 5.7.

***Proof of Proposition 5.10.*** Using formula (5.25), we can now readily prove that the (anti-)invariance conditions indeed provide the existence of stratified index formulas. The proof is based on the invariance of the signature operator under orientation-preserving isometries.

By way of example, we carry out the proof for the case of the anti-equivariance condition (5.12); the proof for the other case is similar.

Thus let $\mu = \{\mu_\varphi\}$ be a periodic family satisfying condition (5.12):

$$\alpha^*\mu = f\mu^*e,$$

where $f$ and $e$ are periodic families of bundle isomorphisms. We can also interpret $f$ and $e$ as bundle isomorphisms on the torus $\partial M \times \mathbb{S}^1$. We have

$$\widehat{\alpha^*\mu} = \widehat{f\mu^* e} = f\widehat{\mu}^* e$$

and hence

$$\operatorname{ind}\widehat{\alpha^*\mu} = \operatorname{ind}\widehat{\mu}^* = -\operatorname{ind}\widehat{\mu}.$$

On the other hand, we express the index using the signature operator and obtain

$$\operatorname{ind}\widehat{\alpha^*\mu} = \operatorname{ind}\mathcal{H}_{\alpha^*\mu} = \operatorname{ind}\alpha\circ\mathcal{H}_\mu\circ\alpha^{-1} = \operatorname{ind}\mathcal{H}_\mu = \operatorname{ind}\widehat{\mu},$$

since $\alpha$ preserves the orientation. (Here $\alpha$ is defined on the Atiyah–Bott–Patodi space in a natural way.) We conclude that $\operatorname{ind}\widehat{\mu} = 0$, as desired.

### 5.2.3. The double of the signature operator for symmetries

Consider the class $\mathrm{Ell}_\alpha$ of operators whose boundary symbols satisfy the equivariance condition (5.13) with some orientation-reversing bundle isomorphism

$$\alpha : \partial T^*\mathcal{M} \to \partial T^*\mathcal{M}$$

over a diffeomorphism $g$ of $\partial M$.

***The twisted double of the phase space.*** We paste together two copies of $T^*\mathcal{M}$ along the common boundary with the use of the mapping $\alpha$ and obtain the manifold

$$2T^*\mathcal{M}_\alpha \overset{\text{def}}{=} T^*\mathcal{M}\bigcup_\alpha T^*\mathcal{M},$$

referred to as the *twisted double* of the phase space $T^*\mathcal{M}$. Similarly, we have the twisted double of the stretched manifold:

$$2M_g \overset{\text{def}}{=} M\bigcup_g M.$$

Note that $2T^*\mathcal{M}_\alpha$ is a vector bundle over $2M_g$:

$$2T^*\mathcal{M}_\alpha \in \operatorname{Vect}(2M_g).$$

***Continuation of the symbol to the twisted double.*** Let $D \in \mathrm{Ell}_\alpha$. Since the boundary symbol $\sigma_\partial(D)$ satisfies condition (5.13), it follows that two copies of the interior symbol $\sigma(D)$ can be pasted together to form a well-defined elliptic symbol

$$2\sigma_\alpha(D) : 2E_\alpha \longrightarrow 2F_\alpha$$

on the twisted double $2T^*\mathcal{M}_a$ without the zero section. The bundle $2E_\alpha$ is defined by clutching two copies of $\pi^*E$ at the boundary $\partial T^*\mathcal{M}$ with the help of the isomorphism $e$; the bundle $2F_\alpha$ is obtained in a similar way.

***The Atiyah–Bott–Patodi space.*** We add a point at infinity to each fiber of $2T^*\mathcal{M}_\alpha$ and obtain the Atiyah–Bott–Patodi space, which we denote by $2D^*\mathcal{M}_\alpha$. It is a bundle with closed base $2M_g$ and fiber the sphere of dimension $n = \dim M$ and is a smooth closed even-dimensional manifold. This manifold can also be obtained by pasting together two copies of the bundle $T^*\mathcal{M} \oplus \mathbb{R}$ along their common boundary $\partial T^*\mathcal{M} \oplus \mathbb{R}$ with the help of the isomorphism $\alpha \oplus 1$ and then taking the sphere bundle of the resulting vector bundle.

***The coefficient bundle.*** To the elliptic symbol $2\sigma_\alpha = 2\sigma_\alpha(D)$ on $2T^*\mathcal{M}_\alpha$, we assign the vector bundle

$$[2\sigma_\alpha] \in \mathrm{Vect}(2D^*\mathcal{M}_\alpha)$$

on the Atiyah–Bott–Patodi space in the usual manner (cf. the preceding subsection).

***The signature operator.*** Since we assume $\alpha$ to be orientation-reversing, we see that both the phase space $2T^*\mathcal{M}_\alpha$ and its Atiyah–Bott-Patodi space $2D^*\mathcal{M}_\alpha$ are oriented manifolds. Hence there is a well-defined signature operator $\mathcal{H}$ on the Atiyah-Bott–Patodi space $2D^*\mathcal{M}_\alpha$. We consider the operator $\mathcal{H}$ with coefficients:

$$\mathcal{H}_\sigma = \mathcal{H} \otimes (\Lambda^*(2M_g) \otimes \mathbb{C})^{-1} \otimes [2\sigma_\alpha]. \tag{5.26}$$

This is an elliptic operator on the closed compact manifold $2D^*\mathcal{M}_\alpha$.

### 5.2.4. The double of the signature operator for antisymmetries

In this subsection, we consider only condition (5.14). The case of condition (5.12) is similar to that of (5.14), and we leave it to the reader.

Thus consider the class $\mathrm{Ell}_\alpha$ of operators whose boundary symbols satisfy condition (5.14) with some orientation-preserving bundle isomorphism $\alpha$ over a diffeomorphism $g$ of $\partial M$.

***The Atiyah–Bott–Patodi space.*** If $\alpha$ is orientation-preserving, then the twisted double of the Atiyah–Bott–Patodi space constructed in the preceding subsection is not oriented and hence does not bear even the signature operator.

However, we can modify the construction. More precisely, we paste together two copies of $T^*\mathcal{M} \oplus \mathbb{R}$ along the diffeomorphism

$$\alpha \oplus -1 : \partial T^*\mathcal{M} \oplus \mathbb{R} \longrightarrow \partial T^*\mathcal{M} \oplus \mathbb{R}$$

of their boundaries. The resulting vector bundle is oriented. The associated sphere bundle can be viewed as the result of gluing two copies of the Atiyah–Bott–Patodi space $D^*\mathcal{M}$ along the diffeomorphism $\alpha \oplus -1$. In other words, in contrast to the equivariant case, we have an additional twisting that interchanges the hemispheres of the Atiyah–Bott–Patodi space. The resulting twisted Atiyah–Bott–Patodi space will be denoted by $2D^*\mathcal{M}'_\alpha$.

***The coefficient bundle.*** The reader can verify that the anti-equivariance condition for the symbol exactly coincides with the pasting condition for the two copies of the bundle

$$[\sigma] \in \mathrm{Vect}(D^*\mathcal{M})$$

under the action of the diffeomorphism $\alpha \oplus -1$. Thus for anti-equivariant symbols there is a well-defined vector bundle

$$[2\sigma_\alpha] \in \mathrm{Vect}(2D^*\mathcal{M}'_\alpha).$$

***The signature operator.*** Now there is a well-defined signature operator $\mathcal{H}$ on the Atiyah-Bott–Patodi space $2D^*\mathcal{M}'_\alpha$. We consider the operator $\mathcal{H}$ with coefficients:

$$\mathcal{H}_\sigma = \mathcal{H} \otimes (\Lambda^*(2M_g) \otimes \mathbb{C})^{-1} \otimes [2\sigma_\alpha]. \tag{5.27}$$

This is an elliptic operator on the closed compact manifold $2D^*\mathcal{M}'_\alpha$.

### 5.2.5. The invariant of the interior symbol

Now we are in a position to introduce the desired homotopy invariant of the interior symbol.

Let $\alpha$ be a bundle isomorphism as in one of the preceding two subsections, and let $\mathrm{Ell}_\alpha$ be the class of elliptic operators on $\mathcal{M}$ with boundary symbols satisfying condition (5.13) or (5.14), respectively.

DEFINITION 5.17. We set

$$f_1(\sigma) = \frac{1}{2} \operatorname{ind} \mathcal{H}_\sigma \tag{5.28}$$

for symbols $\sigma = \sigma(D)$ of elliptic operators $D \in \mathrm{Ell}_\alpha$, where $\mathcal{H}_\sigma$ is defined either by (5.26) or (5.27).

One can readily see that $f_1(\sigma)$ is indeed a homotopy invariant of $\sigma$ in the class of symbols whose boundary values lie in $\Sigma_\alpha$.

### 5.2.6. Does the invariant of the interior symbol depend on the choice of symmetry?

We have written out an invariant of the interior symbol under various symmetry conditions imposed on the symbol. Needless to say, the symbol can have several symmetries simultaneously. Hence we naturally arrive at the following question:

| |
|---|
| *Does the invariant of the interior symbol depend on the choice of symmetry?* |

We shall consider only the case of equivariance conditions (5.13). Our aim is to show that under certain circumstances the invariant of the interior symbol (determined by the operator (5.26)) is independent of the specific choice of symmetry. Before stating the result, we make an important remark.

Symmetries (5.13) are not symmetries of *operators* in general; i.e., they do not allow one to continue the original *operator* to some closed manifold (as in Section 5.2.1). However, even these symmetries can be treated as symmetries of the signature operator on the Atiyah–Bott–Patodi space.[5] Accordingly, we state the result of this subsection on the invariance of the contribution of the interior symbol only for symmetries of operators. In the sequel, this result will be used only for the signature operator with coefficients.

PROPOSITION 5.18. *Suppose that the interior symbol $\sigma$ of an elliptic operator $D$ on a manifold $M$ with boundary $\partial M$ is invariant with respect to two symmetries $G = (g, e, f)$ and $G' = (g', e', f')$, where $g, g' \colon \partial M \to \partial M$ are diffeomorphisms:*

$$[G^*\sigma](p) = \sigma(-p), \quad [(G')^*\sigma](p) = \sigma(-p).$$

*(Here $p$ is the conormal variable.) Then the indices of the corresponding operators on the manifolds $2M_g$ and $2M_{g'}$ obtained by doubling with the help of the respective symmetries are the same,*

$$\operatorname{ind}(2D_g) = \operatorname{ind}(2D_{g'}),$$

*provided that either the product $G'^{-1}G$ is of finite order or the symmetries commute: $GG' = G'G$.*

COROLLARY 5.19. *Under the assumptions of Proposition 5.18, the invariant $f_1(\sigma)$ of the interior symbol for two symmetries is the same.*

*Proof.* 1. Consider the difference $\operatorname{ind} 2D_g - \operatorname{ind} 2D_{g'}$. The operators $2D_g$ and $2D_{g'}$ are isomorphic outside some neighborhood of the hypersurface corresponding to $\partial M$.

Consider two surgeries of $2M_g$ and $2M_{g'}$. The first surgery replaces the diffeomorphism $g$ that determines the gluing by $g'$. The second surgery cuts away the complement of the tubular neighborhood of the hypersurface along which the doubles are glued and pastes the remaining components of the boundary along $g^{-1}$ (see Fig. 5.4). The diffeomorphisms in the right top corner cancel out (twisting first with $g$ and then with $g^{-1}$), and as a result we obtain the torus $\partial M \times \mathbb{S}^1$, whereas the lower row of the diagram contains the twisted torus obtained from the cylinder with base $\partial M$ by pasting the bases together along $g^{-1}g'$ (the mapping torus of $g^{-1}g'$).[6]

[5]Indeed, the signature operator (5.26) is obtained by gluing two copies of the signature operator on the halves $D^*M$.

[6]Recall that a diffeomorphism $f : X \to X$ defines the *mapping torus* $X \times [0,1]/\{(x,1) \sim (f(x),0)\}$, obtained from the cylinder by gluing the bases along $f$.

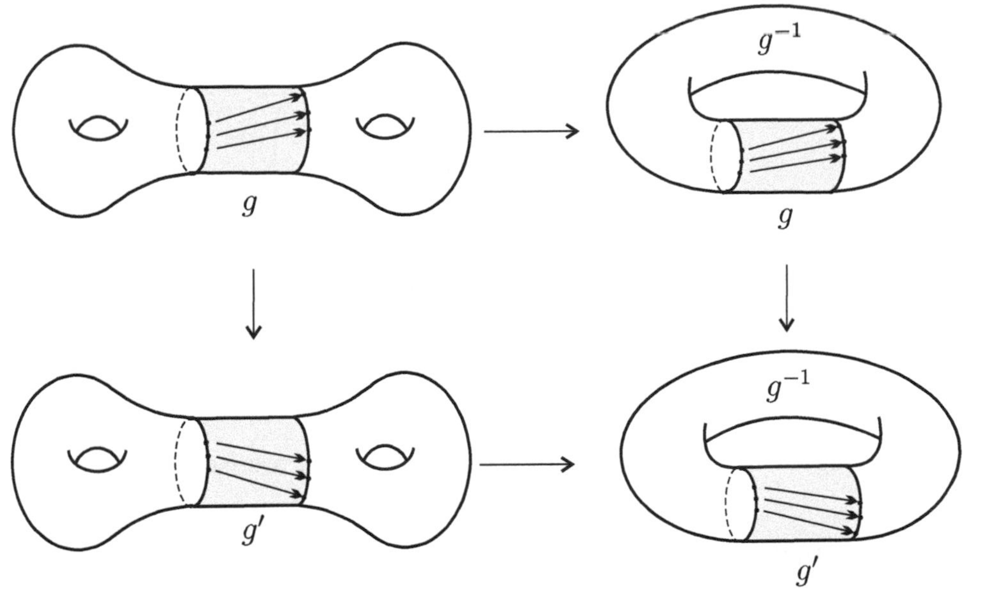


Figure 5.4. Surgery associated with a pair of symmetries.

At the level of elliptic operators, the surgery diagram shown in Fig. 5.4 has the form

$$\begin{array}{ccc} 2D_g & \xleftrightarrow{\text{cutting and pasting}} & T_{g^{-1}g} \\ {\scriptstyle g\leftrightarrow g'}\updownarrow & & \updownarrow{\scriptstyle g\leftrightarrow g'} \\ 2D_{g'} & \xleftrightarrow[\text{cutting and pasting}]{} & T_{g^{-1}g'}, \end{array} \tag{5.29}$$

where $T_f$ is an elliptic operator on the torus twisted by the diffeomorphism $f : \partial M \to \partial M$. The symbol of $T_f$ is independent of $t \in [0, 1]$ and is equal to $\sigma(D)|_{\partial T^*M}$.

The index of the operator in the upper right corner in (5.29) is zero, since its symbol is independent of $t$. By the relative index theorem (see Section 4.3), we obtain

$$\operatorname{ind} 2D_g - \operatorname{ind} 2D_{g'} = -\operatorname{ind} T_{g^{-1}g'}.$$

2. It remains to verify that the index of the operator on the twisted torus on the right-hand side is zero. In the first case (when $G'^{-1}G$ has finite order), this follows from the relation $\operatorname{ind} T_{f^2} = 2 \operatorname{ind} T_f$ and the finite order of the composition $G^{-1}G'$. If the symmetries $G$ and $G'$ commute, then the following relation is crucial:

$$\operatorname{ind} T_{g^{-1}g'} = \operatorname{ind} T_{g(g')^{-1}}.$$

This relation follows from the isomorphism of the operators in question (see Fig. 5.5): the twisted tori are obtained from two identical cylinders (each equipped with the symbol $\sigma(D)$), where the bases are pasted with the corresponding twisting. We obtain an isomorphism of operators by interchanging the cylinders.

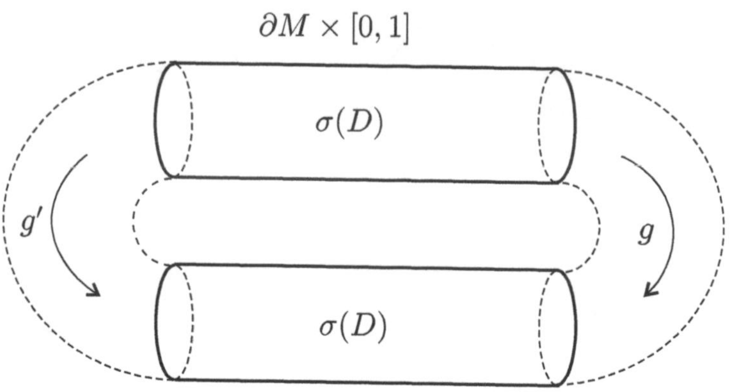


Figure 5.5. The operators $T_{g^{-1}g'}$ and $T_{g(g')^{-1}}$ are isomorphic.

From the last relation, we readily obtain

$$\operatorname{ind} T_{g(g')^{-1}} = \operatorname{ind} T_{(g')^{-1}g} = -\operatorname{ind} T_{g^{-1}g'}$$

(the second equality is the well-known property $\operatorname{ind} T_{f^{-1}} = -\operatorname{ind} T_f$ of the index on twisted tori).

Hence $\operatorname{ind} T_{g^{-1}g'} = 0$. □

*Remark* 5.20. In the general case, the invariant of the interior symbol depends on the choice of symmetry. For example, the symbol of the Cauchy–Riemann operator $D$ on the circle $\mathbb{S}^1 = \partial\mathbb{D}^2$ viewed as the boundary of a disk has the form $p + i\xi$. One can readily verify that the symbol is invariant with respect to the symmetries

$$G_0 = (\varphi \mapsto -\varphi, 1, -1), \qquad G_1 = (\varphi \mapsto -\varphi, e^{i\varphi}, -e^{i\varphi}),$$

which are involutions. We have $G_0G_1 = (1, e^{i\varphi}, e^{i\varphi})$. In other words, the operator $T_{G_0^{-1}G_1}$ on the torus $\mathbb{T}^2$ is just the Cauchy–Riemann operator with coefficients in a bundle with nonzero Chern class. Its index is equal to one. Thus $\operatorname{ind}(2D_{G_0}) - \operatorname{ind}(2D_{G_1}) \neq 0$.

## 5.3. Invariants of the Edge Symbol

In this section, we obtain homotopy invariants of the edge symbol for the case in which the boundary symbol (which is simultaneously the interior symbol of the edge symbol by virtue of the compatibility condition) satisfies the equivariance condition (5.13) or the anti-equivariance condition (5.14). The situation here is somewhat different from the one that takes place for the interior symbol.

The main difficulty in defining the invariants of the interior symbol is the fact that the symbol is not defined on the cotangent bundle of a closed manifold and hence one cannot directly assign an elliptic operator to it.

Edge symbols, on the contrary, are defined on the cotangent bundle of $X$, which is a closed compact manifold, but the difficulty is different: edge symbols fail to have compact fiber variation (we have considered this property in Section 3.3), and hence one cannot assign an elliptic operator to an edge symbol directly. Thus our task in this section is to use the equivariance conditions so as to construct an elliptic symbol of compact fiber variation from the edge symbol. Then the index of the corresponding elliptic operator can be taken as the desired invariant of the edge symbol.

### 5.3.1. Transformations of the edge covariables

We shall discuss invariants of the edge symbol only for operators whose boundary symbols are symmetric with respect to the edge covariables, i.e., the variables dual to directions along the edge $X$. Let us describe this class of symmetries.

Let

$$\widetilde{\alpha} : T^*X \longrightarrow T^*X \tag{5.30}$$

be an automorphism of the cotangent bundle of the edge (for example, the antipodal involution $\xi \longmapsto -\xi$). We use the expansion (1.58)

$$\partial T^*\mathcal{M} \simeq \pi^*(T^*X) \oplus \partial T^*K$$

and lift the mapping (5.30) to the boundary $\partial T^*\mathcal{M}$ by defining it to be the identity mapping on the second component:

$$\alpha = \pi^*(\widetilde{\alpha}) \oplus 1 : \partial T^*\mathcal{M} \longrightarrow \partial T^*\mathcal{M}. \tag{5.31}$$

Thus the transformation $\alpha$ affects only the edge coordinates; it is orientation-reversing (orientation-preserving) if and only if so is $\widetilde{\alpha}$.

*Remark* 5.21. One can also consider the case in which $\widetilde{\alpha}$ is a bundle isomorphism over a diffeomorphism $\widetilde{g} : X \longrightarrow X$. To lift $\alpha$ to $\partial T^*\mathcal{M}$, one needs a lift $g : \partial T^*\mathcal{M} \longrightarrow \partial T^*\mathcal{M}$ of the diffeomorphism $\widetilde{g}$; then one can set $\alpha = \pi^*\widetilde{\alpha} \oplus \{(g_x^*)^{-1}\}$, where

$$g_x = g|_{\pi^{-1}(x)} \oplus 1 : K_x \longrightarrow K_x.$$

Then the subsequent results in this section remain valid with appropriate modifications. We consider only the simplest case $\widetilde{g} = \mathrm{id}$, $g = \mathrm{id}$ to avoid complicated notation.

### 5.3.2. The construction of invariants

Suppose that the automorphism (5.31) is orientation-reversing (orientation-preserving), and let $\sigma_\wedge \in \mathfrak{S}_\wedge$ be an elliptic edge symbol whose interior symbol $\sigma(\sigma_\wedge)$ satisfies condition (5.13) (respectively, condition (5.14)). How can we assign a homotopy invariant to this symbol? The elliptic edge symbol $\sigma_\wedge(x,\xi)$ is defined on $T_0^*X$. Hence the first idea is to define the invariant of the edge symbol as the index

$$\operatorname{ind}\sigma_\wedge\left(x,-i\frac{\partial}{\partial x}\right)$$

of the corresponding operator with operator-valued symbol. However, this approach fails. We know from Section 3.3 that to assign a Fredholm operator to an elliptic symbol uniquely modulo compact operators, we need the symbol to satisfy the compact fiber variation condition. This is usually not the case.

EXERCISE 5.22. Show that the edge symbol has a compact fiber variation if and only if its interior symbol is independent of the covariables.

Thus the edge symbol $\sigma_\wedge(x,\xi)$ itself does not usually define a Fredholm operator. (Nor does it define an element in $K$-theory with compact supports on $T^*X$.)

However, using the $\alpha$-(anti-)equivariance of the boundary symbol, we still can construct a new edge symbol from $\sigma_\wedge(x,\xi)$. This new symbol has a unit interior symbol and hence already enjoys the compact fiber variation condition. The index of the corresponding operator is the desired homotopy invariant.

The construction goes as follows. Since $\alpha$ affects only the edge covariables and does not touch the covariables in $T^*K$, we can define the counterparts of the transformations (5.16) and (5.17) (which act on boundary symbols) for edge symbols themselves:

$$G^*\sigma_\wedge = f^{-1}(\widetilde{\alpha}^*\sigma_\wedge)e^{-1} \tag{5.32}$$

in the equivariant case ($\alpha$ is orientation-reversing), and

$$G^*\sigma_\wedge = e(\widetilde{\alpha}^*\sigma_\wedge)^{-1}f \tag{5.33}$$

in the anti-equivariant case ($\alpha$ is orientation-preserving). Now it follows from the (anti-)equivariance condition for $\sigma(\sigma_\wedge)$ that the edge symbols $G^*\sigma_\wedge$ and $\sigma_\wedge$ have the same interior symbol. Hence the interior symbol of the ratio $(G^*\sigma_\wedge)^{-1}\sigma_\wedge$ is equal to unity. By Exercise 5.22, this ratio has a compact fiber variation, and the index of the elliptic operator

$$\left[(G^*\sigma_\wedge)^{-1}\sigma_\wedge\right]^\wedge = \left[(G^*\sigma_\wedge)^{-1}\sigma_\wedge\right]\left(x,-i\frac{\partial}{\partial x}\right) \tag{5.34}$$

on the edge is well defined.

DEFINITION 5.23. We set

$$f_2(\sigma_\wedge) = \frac{1}{2}\operatorname{ind}\left[(G^*\sigma_\wedge)^{-1}\sigma_\wedge\right]\left(x,-i\frac{\partial}{\partial x}\right). \tag{5.35}$$

By construction, $f_2(\sigma_\wedge)$ is a homotopy invariant of an (anti-)equivariant elliptic edge symbol $\sigma_\wedge$.

### 5.3.3. The invariant of the edge symbol and the index

Let us obtain an important property of the invariant (5.35). Namely, let us show that it is the contribution of the edge symbol to the index, that is, the difference $\operatorname{ind} D - f_2(\sigma_\wedge(D))$ is determined by the interior symbol of $D$.

First, we state an auxiliary lemma. Let $X$ be a closed smooth manifold, and let $a$ be an elliptic symbol of compact fiber variation on $T_0^*X$. By

$$\widehat{a} = a\left(x, -i\frac{\partial}{\partial x}\right)$$

we denote the corresponding Fredholm operator.

LEMMA 5.24. *Let $g : T^*X \to T^*X$ be a fiberwise linear orientation-reversing diffeomorphism. Then*

$$\operatorname{ind} \widehat{g^*a} = -\operatorname{ind} \widehat{a}.$$

*(In other words, the induced mapping $g^*$ inverses the sign of the index of elliptic operators.) For orientation-preserving diffeomorphisms, the index is preserved.*

*Proof.* 1. By Luke's index Theorem 3.89, it suffices to prove the assertion for finite-dimensional symbols.

2. If we have an operator with a finite-dimensional symbol, then we can pass to the signature operator with coefficients on the Atiyah–Bott–Patodi space (see Section 5.2.2) preserving the index. Now the desired property of the index follows from the (anti-)invariance of the signature operator under orientation-preserving (orientation-reversing) transformations. □

PROPOSITION 5.25. *Let $D \in \mathrm{Ell}_\alpha$. Then the number*

$$\operatorname{ind} D - f_2(\sigma_\wedge(D)) \tag{5.36}$$

*depends only on the interior symbol $\sigma(D)$ and is a homotopy invariant of $\sigma(D)$.*

*Proof.* Let $D_1, D_2 \in \mathrm{Ell}_\alpha$, and suppose that

$$\sigma(D_1) = \sigma(D_2).$$

We claim that the difference (5.36) is the same for both operators. Indeed, let

$$\sigma_\wedge(D_1) = a_\wedge, \quad \sigma_\wedge(D_2) = c_\wedge.$$

On the one hand, we have

$$\operatorname{ind} D_1 - \operatorname{ind} D_2 = \operatorname{ind}[a_\wedge c_\wedge^{-1}]\left(x, -i\frac{\partial}{\partial x}\right), \tag{5.37}$$

which follows from the logarithmic property $\operatorname{ind} A - \operatorname{ind} B = \operatorname{ind} AB^{-1}$ of the index and Proposition 3.94.

On the other hand,

$$f_2(a_\wedge) - f_2(c_\wedge) = \frac{1}{2} \operatorname{ind}\big[(G^* a_\wedge)^{-1} a_\wedge c_\wedge^{-1} (G^* c_\wedge)\big] \left(x, -i\frac{\partial}{\partial x}\right). \tag{5.38}$$

The index on the right-hand side in (5.38) can be computed by Luke's theorem, and we obtain

$$f_2(a_\wedge) - f_2(c_\wedge) = \frac{1}{2} p_! \operatorname{ind}\big[(G^* a_\wedge)^{-1} a_\wedge c_\wedge^{-1} (G^* c_\wedge)\big], \tag{5.39}$$

where $p_! : K_c(T^*X) \longrightarrow K_c(pt)$. Now we make use of the following lemma.

LEMMA 5.26. *If $F$ and $G$ are invertible operator-valued symbols on $S^*X$ such that the symbol $FG - 1$ is compact-valued, then*

$$\operatorname{ind} FG = \operatorname{ind} GF \in K_c(T^*X).$$

◀ Indeed, the symbols $FG$ and $GF$ are homotopic in the class of symbols of the form $1 + K$, where $K$ is compact-valued; the homotopy has the form

$$H_t = U_t F G U_t^{-1},$$

where $U_t$ is a homotopy of $G$ to the unit symbol through invertible symbols on $S^*X$, which exists by Kuiper's theorem (see Atiyah 1989). ▶

Hence we can make a cyclic permutation of factors on the right-hand side in (5.39) and, applying Luke's theorem once more, obtain

$$\begin{aligned} f_2(a_\wedge) - f_2(c_\wedge) &= \frac{1}{2} p_! \operatorname{ind}\big[a_\wedge c_\wedge^{-1} G^*(c_\wedge a_\wedge^{-1})\big] \\ &= \frac{1}{2} \operatorname{ind}[a_\wedge c_\wedge^{-1}] \left(x, -i\frac{\partial}{\partial x}\right) + \frac{1}{2} \operatorname{ind}\big[G^*(c_\wedge a_\wedge^{-1})\big] \left(x, -i\frac{\partial}{\partial x}\right). \end{aligned} \tag{5.40}$$

(The symbols $a_\wedge c_\wedge^{-1}$ and $c_\wedge a_\wedge^{-1}$ have compact fiber variation.)

By applying Lemma 5.24 to the second term on the right-hand side of the last equation and by using (5.37), we obtain the desired relation

$$f_2(a_\wedge) - f_2(c_\wedge) = \operatorname{ind} D_1 - \operatorname{ind} D_2.$$

This completes the proof of Proposition 5.25. □

## 5.4. Index Theorems

### 5.4.1. The index under symmetry with respect to the conormal variable

First, we consider elliptic operators whose boundary symbols $\sigma_\partial(D)$ satisfy the symmetry condition (5.13) in which the mapping $\alpha$ is the inversion $p \longmapsto -p$ of the conormal variable.

Recall that under this condition the symbol can be extended naturally to the double of the manifold.

Examining the doubling construction more closely, we see that we deal with surgery, whereby from the disjoint sum $\mathcal{M} \sqcup \mathcal{M}$ we cut out two copies of the neighborhood $U$ of the singularity and then paste the resulting two smooth manifolds with boundary together along the boundary. (This surgery is shown in the left column in Fig. 5.6.) Note that the deleted parts can also be pasted together, forming a manifold $2U$ with two edges (the upper row in the figure). The manifold $2U$ is the bundle of suspensions (see Example 3.93) corresponding to the smooth bundle $\partial M \to X$ over the edge.

For an operator $D$ with symmetric symbol, this surgery results in an operator $2D$ on the double of the manifold (see Section 5.2.1) and the operator on the suspension bundle $2U$. Let us denote the latter operator by $\widehat{2\sigma_\wedge(D)}$. The operators $2D$ and $\widehat{2\sigma_\wedge(D)}$ are determined modulo compact operators by the interior symbol and the edge symbol, respectively, of the original operator.

Let us state the index theorem.

THEOREM 5.27. *Suppose that the interior symbol of an elliptic operator $D$ satisfies the symmetry condition*

$$\sigma_\partial(D)(-p) = f\sigma_\partial(D)(p)e \tag{5.41}$$

*with respect to the conormal variable. Then the index formula*

$$\operatorname{ind} D = \frac{1}{2}\left\{\operatorname{ind}(2D) + \operatorname{ind}(\widehat{2\sigma_\wedge(D)})\right\} \tag{5.42}$$

*is valid, where the first term is determined by the interior symbol and the second term by the edge symbol of $D$.*

*Proof.* The formula follows by an application of the relative index theorem in Section 4.3 to the surgery diagram shown in Fig. 5.6. The right bottom corner of the diagram contains an operator with zero index on the torus $\mathbb{S}^1 \times \partial M$. (Its symbol is independent of $t$.) □

Formula (5.42) gives a partition of the index of the edge-degenerate operator: the index of $D$ is represented as the sum of two homotopy invariants, the first of which is determined by the interior symbol of the operator and the second by the

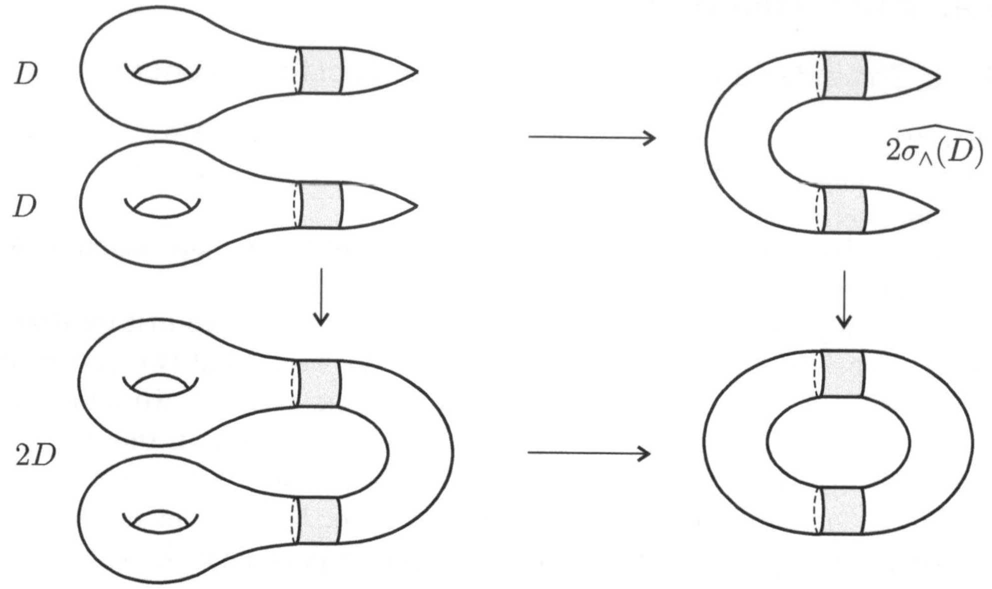


Figure 5.6. Surgery under the symmetry condition.

edge symbol. The first term is the index of an operator on a closed manifold and can be represented by the Atiyah–Singer formula (Atiyah and Singer 1968). The second term is also the index of an elliptic operator on a closed manifold, but the symbol is operator-valued. The index can be expressed via the symbol with the use of Proposition 3.94, since the suspension bundle has a smooth base and a fiber with two conical points.

In Section 5.5, we discuss specializations of the index formula (5.42) for manifolds with conical singularities.

### 5.4.2. The general index theorem

In this subsection, we state an index formula for elliptic edge-degenerate operators with boundary symbols (anti-)equivariant with respect to some symmetry in the fibers of the cotangent bundle of the edge.

The index formula will be proved under the following additional assumption.

ASSUMPTION 5.28. The mapping $\alpha$ is an involution over the identity diffeomorphism $g = \mathrm{id} : \partial M \to \partial M$:

$$\alpha^2 = \mathrm{id}\,.$$

For the equivariance conditions, the endomorphisms $e$ and $f$ are also involutions:

$$e^2 = \mathrm{id}, \quad f^2 = \mathrm{id}\,.$$

THEOREM 5.29. *Under the preceding assumptions, the index of an elliptic edge-degenerate operator $D \in \mathrm{Ell}_\alpha$ is given by the formula*

$$\operatorname{ind} D = \frac{1}{2} \operatorname{ind} \mathcal{H}_{\sigma(D)} + \frac{1}{2} \operatorname{ind}\bigl[(G^*\sigma_\wedge(D))^{-1}\sigma_\wedge(D)\bigr]^\wedge, \tag{5.43}$$

*where the operators on the right-hand side are determined by the formulas* (5.27) *and* (5.34), *respectively.*

Note a special case of this formula in which not only the interior symbol of the edge symbol (that is, the boundary symbol) but also the edge symbol itself is (anti-)equivariant. In this case, the index is expressed via the interior symbol alone.

COROLLARY 5.30. *If the edge symbol is $\alpha$-(anti-)equivariant, $G^*\sigma_\wedge(D) = \sigma_\wedge(D)$, then*

$$\operatorname{ind} D = \frac{1}{2} \operatorname{ind} \mathcal{H}_{\sigma(D)}.$$

### 5.4.3. Proof of the index theorem

Let us prove Theorem 5.29. We interpret the desired index formula as a formula that establishes the equality of two homotopy invariants of the interior symbol of $D$,

$$\operatorname{ind} D - \frac{1}{2} \operatorname{ind}\bigl[(G^*\sigma_\wedge)^{-1}\sigma_\wedge\bigr]^\wedge \quad \text{and} \quad \frac{1}{2} \operatorname{ind} \mathcal{H}_{\sigma(D)}. \tag{5.44}$$

(Recall that the first expression is a homotopy invariant of the interior symbol by virtue of Proposition 5.25.) Thus we can use homotopies of the interior symbol to reduce it to a simple form.

Now the proof consists of two steps.

**1.** We claim that an arbitrary (anti-)equivariant interior symbol is stably rationally[7] homotopic to a symbol symmetric with respect to the inversion $p \longmapsto -p$ of the conormal variable (i.e., a symbol satisfying (5.11)). Moreover, one can assume that the symmetries corresponding to the involutions $\alpha$ and $p \longmapsto -p$ satisfy the assumptions of Proposition 5.18. This topological assertion is established in Proposition 5.48 (see the Supplement 5.6).

**2.** It remains to verify the index formula for the special case in which the interior symbol is not only (anti-)equivariant but also symmetric with respect to the inversion $p \longmapsto -p$. (The general case will then follow by the homotopy invariance.)

Replacing $D$ by the operator $D \oplus D'$ if necessary, where

$$\sigma(D') = \sigma(D), \quad \sigma_\wedge(D') = G^*\sigma_\wedge(D),$$

[7]Stability refers to the direct sum with unit symbols and rationality means that we can proceed to direct sums of arbitrarily many copies of the symbol.

we can assume that the edge symbol $\sigma_\wedge(D)$ satisfies the invariance condition

$$G^*\sigma_\wedge(D) = \sigma_\wedge(D). \tag{5.45}$$

(Here we have used the fact that $G^{*2} = \mathrm{id}$.) Consequently,

$$\operatorname{ind}\left[(G^*\sigma_\wedge(D))^{-1}\sigma_\wedge(D)\right]^\wedge = 0. \tag{5.46}$$

Since $\sigma(D)$ is symmetric with respect to the conormal variable, we can compute the index of $D$ by Theorem 5.27. Note that

$$\operatorname{ind} \widehat{2\sigma_\wedge(D)} = 0.$$

◀ Indeed, in the equivariant case the vanishing of the index of the operator $\widehat{2\sigma_\wedge(D)}$ on the suspension bundle follows from the fact that its operator-valued symbol, which we denote by $2\sigma_\wedge(x,\xi)$, is invariant with respect to the involution $G^*$:

$$G^*(2\sigma_\wedge) = 2\sigma_\wedge.$$

Hence the index of the corresponding operator is zero by Lemma 5.24.

In the anti-equivariant case, we have the weaker condition

$$\left[G^*(2\sigma_\wedge)\right](x,\xi) = 2\sigma_\wedge(x,\xi) + b(x,\xi) \tag{5.47}$$

where $b(x,\xi)$ is a family of compact operators decaying as $|\xi| \to \infty$. This can be proved by writing out the expression for the operator-valued symbol on the suspension bundle (cf. Proposition 3.94). Again Lemma 5.24 shows that the index of the corresponding operator is zero. ▶

Thus the index formula (5.42) acquires the form

$$\operatorname{ind} D = \frac{1}{2}\operatorname{ind} 2D. \tag{5.48}$$

Note that the index of the operator $2D$ on the double $2M$ is equal to the index of the corresponding Hirzebruch operator on $2T^*\mathcal{M}$ by Proposition 5.16; in turn, it is equal to the index of the Hirzebruch operator $\mathcal{H}_{\sigma(D)}$ by Proposition 5.18. It remains to use (5.46).

The proof is complete. □

## 5.5. Index on Manifolds with Isolated Singularities

In this section, we freely use both conical and cylindrical representations of operators on manifolds with conical singularities.

### 5.5.1. Symmetries in the conormal variable

Formula (5.42) provides simple expressions for the index in the case of isolated singularities (where the edge degenerates into a point). In this case, the edge symbol is reduced to the conormal symbol $\sigma_c(D)$.

If the boundary symbol of $D$ is symmetric with respect to the conormal variable, then, applying Theorem 5.27, for the index of $D$ we obtain the expression

$$\operatorname{ind}_\gamma D = \frac{1}{2}\operatorname{ind} 2D + \frac{1}{2}\operatorname{ind}_\gamma \widehat{2\sigma_c(D)}. \qquad (5.49)$$

Let us describe the operator giving the index contribution of the conormal symbol in more detail. According to our construction, the operator $\widehat{2\sigma_c(D)}$ acts in a function space on the infinite cylinder $\partial M \times \mathbb{R}$ and has an interior symbol independent of $t$; its conormal symbol on one end of the cylinder coincides with that of the original operator and is obtained by the change of variables $p \to -p$ on the other end.

***Index contribution as the spectral flow.*** We know (see Proposition 4.24) that the index of the operator on the infinite cylinder with coefficients independent of $t$ is equal to the spectral flow of the linear homotopy joining the conormal symbols at $\pm\infty$. Hence the contribution of the conormal symbol to (5.49) can be rewritten in the form

$$\operatorname{ind}_\gamma \widehat{2\sigma_c(D)} = -\operatorname{sf}_{-\gamma,+\gamma}\Big\{(1-t)\sigma_c(D)(-p) + tf\sigma_c(D)(p)e\Big\}_{t\in[0,1]},$$

where the spectral flow (defined in Appendix A) of the linear homotopy is taken with respect to the weight line $\operatorname{Im} p = -\gamma$ for $t = 0$ and $\operatorname{Im} p = \gamma$ for $t = 1$. Note that changing the sign of the weight at $t = 0$ corresponds to the fact that the weight $\gamma$ in (5.49) as $t \to -\infty$ corresponds to the weight line $\operatorname{Im} p = -\gamma$.

***Index contribution as sum of multiplicities of singular points.*** There is an important special case in which the symmetry condition (5.41) holds not only for the interior symbol but also for the conormal symbol $\sigma_c(D)$:

$$\sigma_c(D)(-p) = f\sigma_c(D)(p)e.$$

Then the conormal symbols of $\widehat{2\sigma_c(D)}$ at $\pm\infty$ are isomorphic and thc index (see Proposition 4.17) can be computed as the sum of multiplicities of zeros of the conormal symbol in the strip between the weight lines $\operatorname{Im} p = \pm\gamma$ for $\gamma > 0$:

$$\operatorname{ind}_\gamma \widehat{2\sigma_c(D)} = \sum_{-\gamma<\operatorname{Im} p<\gamma} m_{\sigma_c(D)}(p). \qquad (5.50)$$

EXERCISE 5.31. Compute the contribution of the conormal symbol to (5.49) via the sum of multiplicities of zeros of the conormal symbol if the "shifted" symmetry condition with respect to some point $p_0$ of the complex plane is valid:

$$\sigma_c(D)(p_0 - p) = f\sigma_c(D)(p_0 + p)e.$$

EXAMPLE 5.32. By way of example in which the computation can be carried out up to a definitive answer in the form of a number, we compute the index of the Euler operator on a two-dimensional manifold with a single conical point. We suppose that the manifold is homeomorphic to a smooth surface of genus $g$ (see Fig. 5.7). In

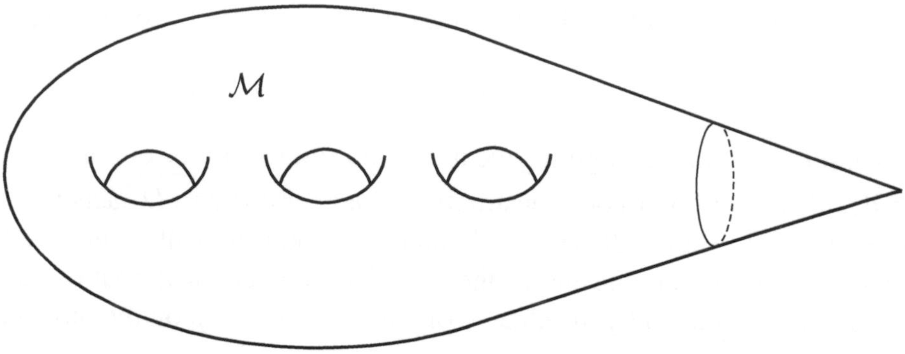


Figure 5.7. A surface of genus 3r.

a neighborhood of the conical point, we take the simplest metric $dr^2 + r^2 d\varphi^2$ with conical degeneration. Let us compute the conormal symbol of the Euler operator. A straightforward computation shows that the matrix of the Euler operator

$$d + d^* : \Lambda^0(\mathcal{M}) \oplus \Lambda^2(\mathcal{M}) \longrightarrow \Lambda^1(\mathcal{M})$$

with respect to the decomposition of forms

$$\omega = f_0\,(r,\varphi) + f_2\,(r,\varphi)\, rdr \wedge d\varphi,$$
$$\omega = f_1^1\,(r,\varphi)\, dr + f_1^2\,(r,\varphi)\, rd\varphi$$

(the functions $f$ are smooth up to the boundary of the stretched manifold $M$) has the form

$$\begin{pmatrix} \dfrac{\partial}{\partial r} & \dfrac{1}{r}\dfrac{\partial}{\partial\varphi} \\ \dfrac{1}{r}\dfrac{\partial}{\partial\varphi} & -\dfrac{\partial}{\partial r} + \dfrac{1}{r} \end{pmatrix}.$$

The conormal symbol is the following operator family on the circle:

$$\sigma_c\,(p) = \begin{pmatrix} -ip & \dfrac{\partial}{\partial\varphi} \\ \dfrac{\partial}{\partial\varphi} & ip + 1 \end{pmatrix},$$

and the inverse family has the form

$$\sigma_c^{-1}(p) = \left(-\frac{\partial^2}{\partial\varphi^2} + p^2 - ip\right)^{-1} \begin{pmatrix} ip+1 & -\dfrac{\partial}{\partial\varphi} \\ -\dfrac{\partial}{\partial\varphi} & -ip \end{pmatrix}.$$

It is clear from the last formula that the poles of the inverse family $\sigma_c^{-1}(p)$ sit at the points

$$ip = -\frac{1}{2} \pm \sqrt{n^2 + \frac{1}{4}}, \quad n \in \mathbb{Z}_+, \tag{5.51}$$

of the complex $p$-plane, the multiplicities of all poles except for $p = 0$ and $p = i$ (corresponding to the zero eigenvalue of the operator $\partial/\partial\varphi$) are equal to 2, and the multiplicity of the exceptional poles is equal to 1. By applying the index formula (5.42) with the contribution (5.50) of the conormal symbol, we obtain

$$\operatorname{ind}_\gamma(d + d^*) = \frac{1}{2}\chi(2M) - \operatorname{sgn}\gamma(1 + k_\gamma) = \chi(M) - \operatorname{sgn}\gamma(1 + k_\gamma),$$

where $k_\gamma$ is the number of poles of the family $\sigma_c^{-1}(p)$ with imaginary parts between 0 and $\gamma$. Finally, we obtain

$$\operatorname{ind}_\gamma(d + d^*) = 1 - 2g - \operatorname{sgn}\gamma(1 + k_\gamma).$$

### 5.5.2. The index under symmetry with respect to the fiber covariables

In all stratified index formulas discussed so far, the index was represented as a sum of *indices* of operators associated with the corresponding strata. In this subsection, we obtain an index formula with the contribution of the conormal symbol given by the $\eta$-invariant. (The definition of the $\eta$-invariant can be found in Appendix B.)

***Symmetries. Odd and even symbols.*** Let $\mathcal{M}$ be a manifold with conical points. Just as earlier, the covariable dual to $r$ will be denoted by $p$, and the covariables dual to the coordinates $\omega$ on the base of the cone will be denoted by $q$. We consider only two symmetries determined by the involutions

$$(q,p) \mapsto (-q,p) \quad \text{and} \quad (q,p) \mapsto (-q,-p) \tag{5.52}$$

acting in the bundle $\partial T^*\mathcal{M} = T^*\partial M \oplus \mathbb{R}$ and, accordingly, symbols of integer order $m$ invariant with respect to one of the involutions:

$$\sigma(q,p) = (-1)^m \sigma(-q,p) \quad \text{or} \quad \sigma(q,p) = (-1)^m \sigma(-q,-p). \tag{5.53}$$

A symbol is said to be *even* or *odd* depending on the parity of the order $m$. Note that for the first involution elliptic symbols exist only for even $m$.

***Contribution of the interior symbol.*** Let $D$ be an elliptic operator on $\mathcal{M}$ whose boundary symbol satisfies one of the symmetries (5.53). Since these symmetries are a special case of phase space symmetries, it follows that the contribution of the interior symbol to the index of $D$ is equal to the index of the signature operator on the twisted double of the phase space provided that the corresponding involution (5.52) is orientation-reversing. Thus all we need is to describe the contribution of the conormal symbol.

***Contribution of the conormal symbol.*** The conormal symbol $\sigma_c(D)$ of an elliptic operator $D$ is an invertible elliptic family with parameter $p$. For invertible families with a parameter, *the Melrose $\eta$-invariant* is defined in Appendix B. The fractional part of the $\eta$-invariant does not change under homotopies of the conormal symbol if the symmetry condition (5.53) holds not only for the interior symbol but also for the *conormal symbol* of the family. More precisely, a family with parameter is said to be *admissible* (see Appendix B) if the components of the asymptotic expansion of its total symbol

$$\sigma_m(q,p)+\sigma_{m-1}(q,p)+\dots$$

satisfy one of the relations

$$\sigma_k(q,p)=(-1)^k\sigma_k(-q,p) \quad \text{or} \quad \sigma_k(q,p)=(-1)^k\sigma_k(-q,-p), \tag{5.54}$$

depending on which involution is to be considered. Note that the second relation obviously holds for symbols of differential operators.

It is shown in Appendix B that under homotopies of an admissible elliptic family the $\eta$-invariant experiences only jumps corresponding to the spectral flow,

$$\eta(\sigma_c(D_1))-\eta(\sigma_c(D_0))=\operatorname{sf}\{\sigma_c(D_t)\}_{t\in[0,1]}, \tag{5.55}$$

provided that the involution (5.52) reverses the orientation. On the other hand, under homotopies the index also changes by the spectral flow of the family of conormal symbols, but with the opposite sign (Theorem 4.19). Hence the $\eta$-invariant of the conormal symbol taken with the opposite sign is the contribution of the conormal symbol to the index; that is, the sum

$$\operatorname{ind} D+\eta(\sigma_c(D))$$

is a homotopy invariant functional of the interior symbol of the operator.

***The index theorem.***

THEOREM 5.33. *Let $D$ be an elliptic operator with admissible conormal symbol on a manifold with isolated conical singularities. Then the index formula*

$$\operatorname{ind} D=\frac{1}{2}\operatorname{ind}\mathcal{H}_{\sigma(D)}-\eta(\sigma_c(D)) \tag{5.56}$$

*holds provided that the corresponding involution* (5.52) *reverses the orientation of the boundary* $\partial T^*\mathcal{M}$ *of the cotangent bundle. Here* $\mathcal{H}_{\sigma(D)}$ *is the signature operator corresponding to the involution and the interior symbol* $\sigma(D)$.

*Proof.* As was already done before in the derivation of index formulas, the proof naturally obeys the following scheme:

1. First one establishes that the sum of the index on the left-hand side in (5.56) and the $\eta$-invariant is a homotopy invariant of the interior symbol of the operator.

2. The set of elliptic symbols splits into connected components.

3. Now to prove the coincidence of the homotopy invariants

$$\operatorname{ind} D + \eta(\sigma_c(D)) \quad \text{and} \quad \frac{1}{2}\operatorname{ind}\mathcal{H}_{\sigma(D)},$$

   it suffices to verify the identity in each component for only one appropriately chosen operator.

This scheme is implemented as follows in our case.

First, the homotopy invariance of the sum of the index and the $\eta$-invariant is already known.

Second, the homotopy classification of symbols satisfying (5.53) is given in the Supplement 5.6 to this chapter. It follows from the classification that in the even case all stable homotopy classes of symbols are rationally generated by symbols constant with respect to the covariables in a neighborhood of the boundary. In other words, for an even symbol $\sigma$ there exists a positive integer $N$ such that the direct sum of $N$ copies of the symbol is homotopic to a symbol that is the symbol of an operator of multiplication in a neighborhood of the boundary. For such symbols the $\eta$-invariant is zero, and the coincidence of the indices on the left- and right-hand sides of the desired index formula (5.56) follows from the independence of the contribution of the interior symbol on the symmetry (Proposition 5.18). The proof of the index formula is complete.

One can readily reduce the odd case to the even case by observing that the formula is additive with respect to composition. Hence it suffices to verify it for only one odd operator. One can use, e.g., the Euler operator, whose $\eta$-invariant is zero. □

## 5.6. Supplement. Classification of Elliptic Symbols with Symmetry and $K$-Theory

In elliptic theory, one often faces the problem of finding the homotopy classification of elliptic symbols. The classification is possible if one supplements homotopies by the operation of *stabilization*.[8] This stable homotopy classification of symbols was given for the first time by Atiyah and Singer in terms of $K$-theory. We deal only with stable classifications throughout the following, and so the word "stable" will be omitted.

Note that finding the homotopy classification is extremely useful in applications, because the computation of $K$-groups and finding their generators can often be done by standard topological methods. Thus classification theorems permit one to use powerful topological techniques in elliptic theory.

Here we discuss homotopy classifications for *elliptic symbols with symmetries*. These classifications also naturally involve some $K$-groups. We compute these $K$-groups and discuss some applications.

The necessary facts of $K$-theory (exact sequences, the Mayer–Vietoris principle, etc.) can be found in Luke and Mishchenko (1998) and Atiyah (1989).

### 5.6.1. The homotopy classification problem

The notion of symbol in elliptic theory is naturally related to the cotangent bundle. In the supplement, we deal only with homotopy properties of symbols and so work in a wider scope: instead of the cotangent bundle, we can have an arbitrary vector bundle.

Consider a real vector bundle $V$ over a compact base $B$.

DEFINITION 5.34. A *symbol* in $V$ is a triple $(E, F, \sigma)$, where $E$ and $F$ are vector bundles over $B$ and

$$\sigma : \pi^*E \longrightarrow \pi^*F, \qquad \pi : V \setminus \{0\} \longrightarrow B,$$

is a homomorphism of the lifts of these bundles to $V$ outside the zero section. The symbol is said to be *elliptic* if $\sigma$ is an isomorphism.

On the set of elliptic symbols, one introduces an equivalence relation, referred to as *stable homotopy*, in the standard way (cf. Section 3.6.2). The set of stable homotopy classes of elliptic symbols in $V$ will be denoted by $\mathrm{Ell}(V)$. This set is a group (the inverse element is defined by the adjoint symbol).

[8]Otherwise even in the simplest cases one arrives at hard homotopy problems such as the computation of homotopy groups of spheres or unitary groups.

> *The homotopy classification problem for elliptic symbols is the problem of computing the group* $\mathrm{Ell}(V)$.

In this setting, the classification problem is easy to solve. Namely, the triple $(E, F, \sigma)$ defines the difference element (see Section 3.6.2)

$$\chi(E, F, \sigma) = [E, F, \sigma] \in K_c(V)$$

of the $K$-group with compact supports of the total space of $V$, and the *difference construction*

$$\chi : \mathrm{Ell}(V) \longrightarrow K_c(V) \tag{5.57}$$

is an isomorphism.

In what follows, we deal with finer classifications, where the symbols are either subject to additional conditions or equipped with additional structures. In this situation, the key issue is to find an appropriate analog of the $K$-group and define an analog of the difference construction (5.57).

We start from the simplest case.

### 5.6.2. Even symbols

***Even symbols.*** Consider symbols

$$\sigma : \pi^* E \longrightarrow \pi^* E$$

invariant under the antipodal involution $v \mapsto -v$ in the vector bundle $V$:

$$\sigma(v) = \sigma(-v).$$

Such symbols are said to be *even*. The group of their stable homotopy classes will be denoted by $\mathrm{Even}(V)$.

***The homotopy classification.*** Even symbols are even functions on the sphere bundle $SV \subset V$ (with respect to some metric) and can be viewed as functions over the *projectivization*

$$PV = SV/\{v \sim -v\},$$

of $SV$, which is a bundle over $B$ with fiber the real projective space $\mathbb{RP}^n$ of dimension $n = \dim V - 1$.

The map taking each even elliptic symbol to an invertible function on the projectivization defines a homomorphism

$$\mathrm{Even}(V) \longrightarrow K^1(PV) \tag{5.58}$$

into the odd $K$-group $K^1(PV)$ interpreted as the set of stable homotopy classes of invertible matrix functions.

The following homotopy classification is now an obvious consequence of the definitions.

PROPOSITION 5.35. *The mapping* (5.58) *is an isomorphism*:

$$\mathrm{Even}(V) \simeq K^1(PV).$$

***An application of the homotopy classification.*** Now we would like to use the homotopy classification and methods of $K$-theory to study the group $\mathrm{Even}(V)$ in more detail. To this end, we suppose that the involution $v \mapsto -v$ reverses the orientation of $V$, i.e., $V$ is odd-dimensional.

LEMMA 5.36 (Gilkey 1989*b*). *If* $\dim V$ *is odd, then the projection* $PV \longrightarrow B$ *induces an isomorphism*

$$K^*(B) \otimes \mathbb{Z}[1/2] \longrightarrow K^*(PV) \otimes \mathbb{Z}[1/2]$$

*modulo* 2*-torsion, where* $\mathbb{Z}[1/2]$ *is the ring of dyadic numbers.*

*Proof.* 1. We claim that the projection induces an isomorphism for a one-point space $B = pt$. Indeed, the $K$-groups

$$K^1(\mathbb{RP}^{2n}) = 0, \quad K^0(\mathbb{RP}^{2n}) = \mathbb{Z} \oplus \mathbb{Z}_{2^n}$$

of the even-dimensional projective space (e.g., Luke and Mishchenko 1998) are equal modulo 2-torsion to the $K$-groups of a point.

2. Now the isomorphism over a space $B$ with nontrivial topology follows from the Mayer–Vietoris principle (e.g., Bott and Tu 1982). □

In terms of elliptic symbols, this means that even symbols can be deformed to a very simple form.

COROLLARY 5.37. *If* $\sigma$ *is an even symbol on an odd-dimensional bundle* $V$*, then for sufficiently large* $N$ *the direct sum* $2^N\sigma$ *is homotopic in the class of even elliptic symbols to a symbol constant on the fibers of* $V$.

Let us consider a close symmetry.

***The symmetry*** $(v,p) \to (-v,p)$. Consider the bundle $V \oplus \mathbb{R}$ over $B$ and symbols even with respect to the first variables:

$$\sigma(v,p) = \sigma(-v,p).$$

The homotopy classification of elliptic symbols in this case is similar to the preceding and an analog of Corollary 5.37 holds in this case word for word.

### 5.6.3. Equivariant symbols

Here we give the homotopy classification of symbols equivariant with respect to certain involutions. We use equivariant $K$-theory.

***$\mathbb{Z}_2$-equivariant symbols.*** Let $V \in \mathrm{Vect}(B)$ be a real vector bundle with projection $\pi$ over a compact space $B$ equipped with an involution

$$\alpha : V \longrightarrow V, \quad \alpha^2 = \mathrm{id},$$

which specifies the action of the group $\mathbb{Z}_2$ on $V$. We also have the decomposition

$$V = V_+ \oplus V_-, \quad V_\pm \subset V,$$

where $V_\pm$ are the subbundles corresponding to the eigenvalues $\pm 1$ of the involution.

On $V$ we consider $\mathbb{Z}_2$-*equivariant* elliptic symbols, defined as quintuples

$$(E, F, e, f, \sigma), \tag{5.59}$$

where $E, F \in \mathrm{Vect}(B)$ are vector bundles over $B$ with involutions

$$e : E \longrightarrow E, \quad f : F \longrightarrow F$$

and

$$\sigma : \pi_0^* E \longrightarrow \pi_0^* F, \quad \pi_0 : SV \longrightarrow B,$$

is an elliptic symbol such that

$$\alpha^*(\sigma) = f \sigma e.$$

***The homotopy classification.*** By $\mathrm{Ell}_{\mathbb{Z}_2}(V)$ we denote the group of stable homotopy classes of quintuples (5.59) modulo trivial quintuples (i.e., those in which the homomorphism $\sigma$ is induced by a bundle isomorphism on the base $B$).

The quintuple $(E, F, e, f, \sigma)$ determines the *difference element*

$$[E, F, e, f, \sigma] \in K_{\mathbb{Z}_2}(V) \tag{5.60}$$

in the equivariant $K$-group with compact supports of the space $V$ with the action of $\mathbb{Z}_2$ (see Atiyah 1989). It turns out that such quintuples generate the equivariant $K$-group.

PROPOSITION 5.38. *The mapping* $(E, F, e, f, \sigma) \mapsto [E, F, e, f, \sigma]$ *defines an isomorphism*

$$\mathrm{Ell}_{\mathbb{Z}_2}(V) \xrightarrow{\simeq} K_{\mathbb{Z}_2}(V).$$

***An application of the homotopy classification.*** Now we would like to use the homotopy classification and methods of $K$-theory to study the group $\mathrm{Ell}_{\mathbb{Z}_2}(V)$ deeper. To this end, we suppose that the involution $\alpha$ reverses the orientation of $V$, i.e., the subbundle $V_- \subset V$ is odd-dimensional:

$$\dim V_- \equiv 1 \mod 2. \tag{5.61}$$

This condition will be referred to as the *parity condition*.

***Computation of the equivariant $K$-group.*** The involution $\alpha$ acts trivially on $V_+$. Hence for the $K$-group we obtain the expansion

$$K^*_{\mathbb{Z}_2}(V_+) = K^*(V_+) \oplus K^*(V_+)$$

corresponding to the trivial and nontrivial representations of $\mathbb{Z}_2$. The projections onto these summands will be denoted by $\pi_+$ and $\pi_-$.

LEMMA 5.39. *If the parity condition* (5.61) *holds, then the composition*

$$(\pi_+ - \pi_-)i^* : K^*_{\mathbb{Z}_2}(V) \longrightarrow K^*_{\mathbb{Z}_2}(V_+) \longrightarrow K^*(V_+) \tag{5.62}$$

*is an isomorphism modulo 2-torsion, where $i^*$ is induced by the natural embedding $i : V_+ \subset V$.*

*Proof.* It suffices to obtain the isomorphism for the one-point space $B = pt$ (cf. the proof of Lemma 5.36).

1. Suppose that $V_+ = 0$, i.e., $V = V_- = \mathbb{R}^k$. The pair $\mathbb{S}^{k-1} \subset \mathbb{D}^k$ induces an exact sequence in equivariant $K$-theory:

$$\cdots \longrightarrow K^*_{\mathbb{Z}_2}(\mathbb{R}^k) \xrightarrow{i^*} K^*_{\mathbb{Z}_2}(pt) \xrightarrow{p^*} K^*_{\mathbb{Z}_2}(\mathbb{S}^k) \longrightarrow K^{*+1}_{\mathbb{Z}_2}(\mathbb{R}^k) \xrightarrow{i^*} K^{*+1}_{\mathbb{Z}_2}(pt) \longrightarrow \cdots \tag{5.63}$$

Here $p^*$ is induced by the projection $p : \mathbb{S}^k \to pt$.

The antipodal $\mathbb{Z}_2$-action on the sphere is free, and hence the equivariant $K$-group is isomorphic to the $K$-group of the quotient space:

$$K^*_{\mathbb{Z}_2}(\mathbb{S}^k) \simeq K^*(\mathbb{RP}^k).$$

The latter is isomorphic modulo 2-torsion to the $K$-group of a point, and moreover, the mapping

$$p^* : K^*_{\mathbb{Z}_2}(pt) = K^*(pt) \oplus K^*(pt) \longrightarrow K^*(\mathbb{RP}^k)$$

in (5.63) takes $(x, y)$ to $x + y$ modulo 2-torsion. Thus it follows from (5.63) that $i^*$ defines an isomorphism onto the anti-diagonal subgroup of elements $(x, -x)$.

2. The case of nontrivial $V_+$ reduces to the previous situation with the use of Bott periodicity:

$$K^*_{\mathbb{Z}_2}(X \times \mathbb{R}^l) \simeq K^{*+l}_{\mathbb{Z}_2}(X),$$

where the involution is trivial on $\mathbb{R}^l$. □

***Reduction to nonequivariant symbols.*** Lemma 5.39 says that equivariant symbols in the bundle $V = V_+ \oplus V_-$ are in a one-to-one correspondence (modulo stable homotopy and 2-torsion) with the usual symbols in the subbundle $V_+$.

The following natural question arises. How can we extend an elliptic symbol

$$\sigma : \pi_0^* E \longrightarrow \pi_0^* F, \qquad \pi : V_+ \setminus 0 \longrightarrow B \tag{5.64}$$

in $V_+$ to an equivariant symbol on $V = V_+ \oplus V_-$?

This question can be answered easily if we use the notion of external product of elliptic symbols.

DEFINITION 5.40. The *external product* of symbols

$$a : \pi_+^* E \longrightarrow \pi_+^* F \qquad \text{and} \qquad b : \pi_-^* G \longrightarrow \pi_-^* H, \qquad \text{where } \pi_\pm : V_\pm \longrightarrow B,$$

on vector bundles $V_+$ and $V_-$ is the symbol on $V_+ \oplus V_-$ given by

$$a\#b := \begin{pmatrix} a \otimes 1_G & 1_F \otimes b^* \\ 1_E \otimes b & -a^* \otimes 1_H \end{pmatrix} : \pi^* \begin{pmatrix} E \otimes G \\ \oplus \\ F \otimes H \end{pmatrix} \longrightarrow \pi^* \begin{pmatrix} F \otimes G \\ \oplus \\ E \otimes H \end{pmatrix}.$$

EXERCISE 5.41. Prove that the external product is elliptic provided that $a$ is elliptic on $V_+$ and $b$ is elliptic on $V_-$. (We assume that $a$ and $b$ have equal positive orders.)

◀ To prove the triviality of the kernel, use the identity $\operatorname{Ker} x = \operatorname{Ker} x^* x$ to pass to the "Laplacian" $(a\#b)^*(a\#b)$. The "Laplacian" is a block-diagonal operator with nonnegative entries, and its kernel is easily identified. A similar reasoning works for the cokernel. ▶

Now the natural idea is to multiply the symbol $\sigma$ on $V_+$ externally by some equivariant elliptic symbol

$$c : \pi_1^* \mathbb{C}_+^{2^N} \longrightarrow \pi_1^* \mathbb{C}_-^{2^N}, \quad \pi_1 : V_- \to B, \ \ N \text{ is sufficiently large,}$$

on $V_-$. We shall assume that $c$ is Hermitian.[9]

The external product

$$c\#\sigma = \begin{pmatrix} c \otimes 1 & 1 \otimes \sigma^* \\ 1 \otimes \sigma & -c^* \otimes 1 \end{pmatrix} : \begin{matrix} \mathbb{C}_+^{2^N} \otimes E \\ \oplus \\ \mathbb{C}_-^{2^N} \otimes F \end{matrix} \longrightarrow \begin{matrix} \mathbb{C}_-^{2^N} \otimes E \\ \oplus \\ \mathbb{C}_+^{2^N} \otimes F \end{matrix} \tag{5.65}$$

[9]To define such a symbol, one can use an embedding $V_- \subset \mathbb{C}^N$ in the trivial bundle and define $c$ to be the symbol of the Euler operator:

$$c(v) = v \wedge + (v\wedge)^* : \Lambda(\mathbb{C}^N) \longrightarrow \Lambda(\mathbb{C}^N).$$

is an equivariant elliptic symbol on $V$. An easy computation shows that the symbol

$$(\pi_+ - \pi_-)i^*(c\#\sigma)$$

is the direct sum of $2^{N+1}$ copies of the symbol $\sigma$. Thus, we obtain the following corollary of Lemma 5.39.

COROLLARY 5.42. *If $\sigma$ is an equivariant symbol on $V$, then for sufficiently large $N$ the direct sum $2^N\sigma$ is stably homotopic to an external product* (5.65) *provided that the parity condition* (5.61) *is satisfied.*

### 5.6.4. Anti-equivariant symbols

In the previous subsection, we classified equivariant symbols with respect to an orientation-reversing involution and obtained normal forms for such symbols in Corollary 5.42. Here we construct a theory with similar properties for orientation-preserving involutions. It turns out that the notion of equivariance has to be replaced by *anti-equivariance.*

The main results of anti-equivariant theory are similar to those in the equivariant case. Thus, we omit the proofs and refer the reader to (Nazaikinskii, Savin, Schulze and Sternin 2003) for details.

***Anti-equivariant symbols.*** Suppose that $V \in \mathrm{Vect}(B)$ is a vector bundle with involution $\alpha$, $\alpha|_{V_\pm} = \pm Id$, $V = V_+ \oplus V_-$, and $\pi : SV \to B$ is the sphere bundle of $V$.

DEFINITION 5.43. An elliptic symbol $\sigma : \pi^*E \longrightarrow \pi^*E$, where $E \in \mathrm{Vect}(B)$, is said to be *anti-equivariant* with respect to $\alpha$ if

$$\alpha^*\sigma = \sigma^{-1}. \qquad (5.66)$$

The group of stable homotopy classes of anti-equivariant symbols modulo anti-equivariant symbols lifted from the base $B$ will be denoted by $\mathrm{Ell}^{odd}_{\mathbb{Z}_2}(V)$.

*Remark* 5.44. If (5.66) is replaced by the condition $\alpha^*\sigma = \sigma^*$, then we arrive at an isomorphic group. This can be proved with the use of the polar expansion.

***The difference construction.*** The standard difference construction in terms of the $K$-group of $V$ does not apply here, since the inversion of the symbol in (5.66) corresponds to the discontinuous transformation

$$v \longrightarrow \frac{v}{|v|^2}$$

of $V$. However, the transformation becomes continuous if we supplement each fiber with the point at infinity, i.e., pass to the Atiyah–Bott–Patodi space.

We represent the Atiyah–Bott–Patodi space as the gluing

$$B_+V \bigcup_{SV} B_-V$$

of two copies $B_\pm V \xrightarrow{\pi_\pm} B$ of the unit ball bundle in $V$ along the common boundary $\partial(B_\pm V) = SV$ (see Fig. 5.3), where $S$ is the spherization of a real vector bundle. Then the Atiyah–Bott–Patodi space is just $S(V \oplus \mathbb{R})$.

To the symbol $\sigma$, we assign the vector bundle on $S(V\oplus\mathbb{R})$ obtained by clutching the lifts $\pi_\pm^* E$ of $E$ to $B_\pm V$ via the isomorphism $\sigma$; see Fig. 5.8. (Vertical arrows in the figure show identification of vectors.)

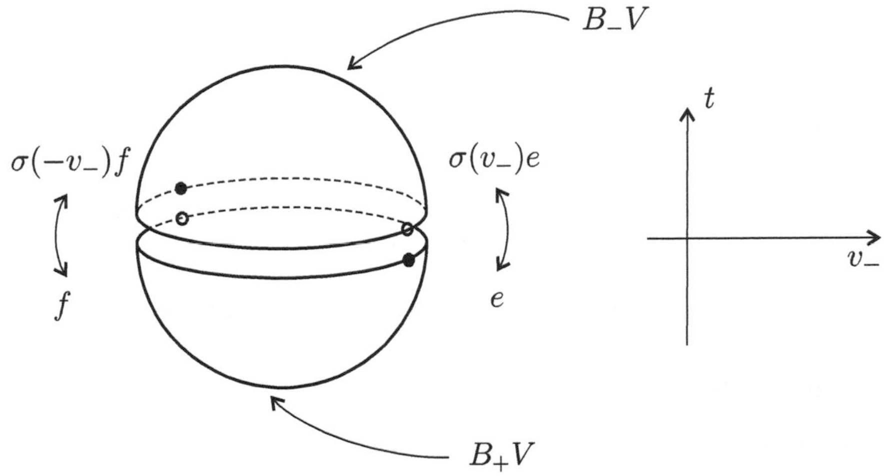


Figure 5.8. Clutching of bundles on the hemispheres $B_+V$ and $B_-V$.

The Atiyah–Bott–Patodi space $S(V \oplus \mathbb{R})$ is equipped with the involution

$$\alpha \oplus (-1) : S(V \oplus \mathbb{R}) \longrightarrow S(V \oplus \mathbb{R}),$$

which is the restriction of the corresponding involution in $V \oplus \mathbb{R}$. The involution interchanges the hemispheres $B_\pm V$. The following crucial statement is proved by an easy computation.

LEMMA 5.45. *The anti-equivariance condition* (5.66) *coincides with the consistency condition for the natural* $\mathbb{Z}_2$*-actions in the restrictions of the bundles* $\pi_\pm^* E$ *to the common boundary* $\partial(B_\pm V) = SV$ *of the balls.*

***The homotopy classification.*** We see that each anti-equivariant symbol gives rise to a $\mathbb{Z}_2$-equivariant bundle on $S(V\oplus\mathbb{R})$; the corresponding element in the $K$-group will be denoted by

$$\chi(\sigma) \in K_{\mathbb{Z}_2}(S(V \oplus \mathbb{R}_-)).$$

PROPOSITION 5.46. *The difference construction defines an isomorphism*

$$\mathrm{Ell}^{odd}_{\mathbb{Z}_2}(V) \overset{\chi}{\simeq} K_{\mathbb{Z}_2}(S(V \oplus \mathbb{R}_-))/\pi^* K_{\mathbb{Z}_2}(B), \tag{5.67}$$

*where* $\pi : S(V \oplus \mathbb{R}_-) \longrightarrow B$ *is the natural projection.*

***An application of the homotopy classification.*** By definition, the *parity condition* for anti-equivariant symbols says that the involution preserves the orientation. By analogy with the equivariant case, if this condition is satisfied, then the $K$-group (5.67) is reduced to the $K$-group of a subbundle.

LEMMA 5.47. *Let the following conditions hold*: 1) *the involution* $\alpha$ *preserves the orientation*; 2) *the bundle* $V_+$ *has at least one nonzero section. Then there is an isomorphism*

$$K_{\mathbb{Z}_2}(S(V \oplus \mathbb{R}_-))/\pi^* K_{\mathbb{Z}_2}(B) \simeq K^1(V_+) \quad \textit{modulo } 2\textit{-torsion.}$$

As elements generating the group $\mathrm{Ell}^{odd}_{\mathbb{Z}_2}(V) \otimes \mathbb{Z}[1/2]$, one can take the external products (cf. (5.65)):

$$(\sigma\#c)(v_+, v_-) = \begin{pmatrix} \sigma(v_+) \otimes 1 & 1 \otimes c(v_-) \\ -1 \otimes c(v_-) & \sigma(v_+) \otimes 1 \end{pmatrix}, \quad [\sigma\#c] \in \mathrm{Ell}^{odd}_{\mathbb{Z}_2}(V), \tag{5.68}$$

where $[\sigma(v_+)] \in K^1(V_+)$ is determined by the Hermitian isomorphism $\sigma(v_+)$.

### 5.6.5. Reduction to the simples involution

Let $V$ be a vector bundle with an involution $\alpha$. In the direct sum $V \oplus \mathbb{R}$ of vector bundles, we consider two involutions, $(v, p) \mapsto (\alpha(v), p)$ and $(v, p) \mapsto (v, -p)$. The following question is of interest:

- How many equivariant symbols are there with respect to both involutions?

The following proposition shows that such symbols rationally generate the corresponding $K$-group.

PROPOSITION 5.48. *Under the parity condition* (5.61), *for an arbitrary equivariant elliptic symbol* $a$ *there exist positive integers* $N$ *and* $N'$ *such that the direct sum*

$$\underbrace{a \oplus a \oplus \ldots \oplus a}_{2^N \textit{copies}} \oplus 1_{N'}$$

*is homotopic in the class of elliptic equivariant symbols to a symbol equivariant with respect to the inversion* $p \mapsto -p$*, where* $1_{N'}$ *is the unit symbol in the trivial bundle of dimension* $N'$ *equipped with an arbitrary involution. Moreover, the assumptions of Proposition* 5.18 *are satisfied for the two symmetries of the latter symbol.*

*A similar assertion holds for anti-equivariant symbols.*

*Proof.* To be definite, we explain the proof for equivariant symbols. (The anti-equivariant case can be treated in a similar way.)

By virtue of Corollary 5.42, to prove the proposition it suffices to realize an arbitrary class in $K^0(V_+ \times \mathbb{R})$ by a symbol $\sigma(v_+, p)$ such that the external product $c\#\sigma$ is equivariant with respect to the involution $p \mapsto -p$.

By the Bott periodicity, an arbitrary class of $K^0(V_+ \times \mathbb{R})$ is the difference element of a symbol $\sigma(v_+, p) = ip + a(v_+)$, where $a(v_+)$ is a Hermitian symbol that satisfies $a(v_+)^2 = |v_+|^2$. Then the external product $c\#\sigma$ is clearly equivariant:

$$(c\#\sigma)(-p) = U(c\#\sigma)(p)U, \quad U = \begin{pmatrix} 0 & -1 \\ 1 & 0 \end{pmatrix}. \tag{5.69}$$

It remains to verify the assumptions of Proposition 5.18 for this symbol. This is a simple exercise in linear algebra. □

## 5.7. Supplement. Proof of Proposition 5.16

The proof amounts to an application of the Atiyah–Singer formula to both sides of the desired equation (5.25) followed by the comparison of the corresponding characteristic classes. We give the main steps of the computation, since we could not find it in the literature.

1. On the one hand, the Atiyah–Singer formula on $M$ in the cohomology form due to Atiyah, Bott and Patodi (1973) reads

$$\operatorname{ind} P = \langle \operatorname{ch}[\sigma(P)]\pi^*\operatorname{Todd}(TM \otimes \mathbb{C}), [D^*M] \rangle, \quad \pi : D^*M \to M.$$

2. On the other hand, the index formula for the signature operator with coefficients on the right-hand side in (5.25) has the form

$$\operatorname{ind} \mathcal{H}_{\sigma(P)} = \Big\langle L\big(D^*M\big) \operatorname{ch}(\Lambda^*M \otimes \mathbb{C})^{-1} \operatorname{ch}[\sigma(P)], [D^*M] \Big\rangle,$$

where $L$ is the Hirzebruch $L$-class of the manifold (see Hirzebruch 1966). To compare these two cohomology expressions, we compare the characteristic classes

$$L\big(D^*M\big) \operatorname{ch}(\Lambda^*M \otimes \mathbb{C})^{-1} \quad \text{and} \quad \pi^*\operatorname{Todd}(TM \otimes \mathbb{C}). \tag{5.70}$$

The computation consists of two stages. First, we show that the $L$-class is lifted from the base $M$ in our case. Second, we establish the equality of the characteristic classes on the base.

3. One has $L(D^*M) = L(M)^2$, which follows from the vector bundle isomorphism[10]

$$TD^*M \oplus \mathbb{R} \simeq \pi^*(TM \oplus TM \oplus \mathbb{R}). \tag{5.71}$$

[10]This isomorphism is a version of the standard decomposition $T\mathbb{S}^n \oplus \mathbb{R} \simeq \mathbb{R}^{n+1}$ of vectors on the sphere into the tangential and normal components.

4. All that remains is to prove the coincidence

$$\mathrm{Todd}(TM \otimes \mathbb{C}) = \big(L(M)\big)^2 \, \mathrm{ch}(\Lambda^*(TM) \otimes \mathbb{C})^{-1} \tag{5.72}$$

of the characteristic classes on the base. In the Borel–Hirzebruch formalism (see Hirzebruch 1966), the left- and right-hand sides correspond to the functions

$$\left(\frac{x/2}{\sinh x/2}\right)^2 \quad \text{and} \quad \left(\frac{x}{\tanh x/2}\right)^2 \frac{1}{(1+e^x)(1+e^{-x})},$$

which identically coincide. The proof is complete.

# Chapter 6

# Elliptic Edge Problems

In this chapter, we study elliptic *problems* on manifolds with edges. Let us explain why such studies are important. So far, we have considered only elliptic operators, but it often happens that even though the interior symbol of an edge-degenerate operator is elliptic, the edge symbol is not invertible and the operator fails to be Fredholm. (An example to that effect can be found in the first section of this chapter.) At the same time, we always require that the conormal symbol of the edge symbol be invertible, so that the edge symbol itself is Fredholm.

This suggests that we can try to "rectify" the original problem in such a way that the new edge symbol be already invertible. There are two methods to do so.

• The first way is to add some "boundary and "coboundary" operators, that is, include our operator in some $2 \times 2$ matrix operator. Needless to say, these (co)boundary operators should be concentrated at the edge.[1]

• The second way is to perturb the operator in question by some "small" operator. In our context, a small operator is an operator with zero interior symbol. Actually, we replace our operator by a new operator within the class of operators with the same interior symbol.

In fact, these procedures are not always possible. (This is just as in the theory of boundary value problems, where posing a problem is possible only if some topological obstruction, namely, the Atiyah–Bott obstruction, is zero.)

In the present chapter, we give a procedure that implements the first method. The implementation of the second method is trivial from the viewpoint of analysis and amounts to choosing a new representative of the operator in the class of operators with the same interior symbol. We prove that the obstructions to both procedures coincide as elements of the $K$-groups. We compute (either of) these obstructions, which are represented by analytical indices, in topological terms. This computation requires some additional topological prerequisites on the part of the reader, in particular, the knowledge of basics of $K$-theory of vector bundles, including Bott periodicity.

[1] This procedure is physically rather meaningful, since it stems from boundary value problems, where, e.g., a Fredholm problem for the Laplace operator in the disk is defined by imposing some boundary conditions.

## 6.1. Morphisms

### 6.1.1. Example: The Laplace operator

We start from a simple example. We have shown in Section 2.3.1 that the Beltrami–Laplace operator $\triangle$ on a compact two-dimensional manifold $M$ with boundary is Fredholm in the weighted Sobolev spaces

$$\triangle : \mathcal{W}^{s,\gamma}(\mathcal{M}) \longrightarrow \mathcal{W}^{s-2,\gamma-2}(\mathcal{M})$$

for $\frac{1}{2} < \gamma < \frac{3}{2}$. What happens for other values of $\gamma$? The interior symbol $\sigma(\triangle)$ is always invertible (this condition is independent of $\gamma$), and hence the Fredholm property of $\triangle$ is determined by the invertibility of the edge symbol

$$\sigma_\wedge(\triangle)(\xi) = \frac{\partial^2}{\partial r^2} - \xi^2 : \mathcal{K}^{s,\gamma}(\mathbb{R}_+) \longrightarrow \mathcal{K}^{s-2,\gamma-2}(\mathbb{R}_+)$$

for nonzero $\xi$. Recall that the conormal symbol

$$\sigma_c(\sigma_\wedge(\triangle)(\xi)) = -p^2 + ip$$

of the edge symbol of $\triangle$ is invertible for $p \neq 0, i$, and so the edge symbol is Fredholm for $\gamma \neq \frac{1}{2}, \frac{3}{2}$. The kernels of the edge symbol and of its adjoint are easy to compute. Formally, they both are spanned by the function

$$u = e^{-|\xi| r}. \tag{6.1}$$

(The second linearly independent solution $e^{|\xi| r}$ grows exponentially as $r \to \infty$ and hence never belongs to $\mathcal{K}^{s,\gamma}(\mathbb{R}_+)$.) Since the weight factor $r^{-2\gamma}$ occurs in the definition of the norm in $\mathcal{K}^{s,\gamma}(\mathbb{R}_+)$, we see that the function (6.1) belongs to $\mathcal{K}^{s,\gamma}(\mathbb{R}_+)$ if and only if $\gamma < \frac{1}{2}$. We see that

1) If $\gamma < \frac{1}{2}$, then the edge symbol of $\triangle$ has the one-dimensional kernel spanned by the function (6.1) and the trivial cokernel.

2) If $\gamma > \frac{3}{2}$, then the edge symbol of $\triangle$ has the trivial kernel and the one-dimensional cokernel spanned by the function (6.1).

3) If $\frac{1}{2} < \gamma < \frac{3}{2}$, then the edge symbol of $\triangle$ is invertible (this case was already considered in Section 2.3.1).

4) If $\gamma = \frac{1}{2}$ or $\gamma = \frac{3}{2}$, then the edge symbol of $\triangle$ is not Fredholm (more precisely, it has the trivial kernel and a dense nonclosed range).

We wish to make the edge symbol invertible, or, which is the same, make the equation

$$\sigma_\wedge(\triangle)(\xi)v = F \tag{6.2}$$

uniquely solvable by imposing finitely many conditions on the solution and the right-hand side. This is clearly impossible in case 4 (the cokernel is infinite-dimensional), and so we consider only cases 1 and 2.

Case 1. Let $\gamma < \frac{1}{2}$. To make Eq. (6.2) uniquely solvable, one can equip it, say, with the condition

$$B(\xi)v \equiv \int_0^\infty \phi(|\xi|r)v(r)\,dr = g \in \mathbb{C}, \tag{6.3}$$

where the function $\phi(r) \in C_0^\infty(\mathbb{R}_+)$ is not orthogonal to $e^{-r}$ (and hence $\phi(|\xi|r)$ is not orthogonal to $e^{-|\xi|r}$). Quantizing, i.e., smoothing the symbol in a neighborhood of $\xi = 0$ and making the substitution $\xi \longmapsto -i\partial/\partial x$, we obtain the following edge problem for the Beltrami–Laplace operator for these values of $\gamma$:

$$\begin{cases} \Delta u = f, \\ Bu \equiv \displaystyle\int_0^\infty \phi\left(r\left[-i\frac{\partial}{\partial x}\right]\right)u(x,r)\,dr = g(x). \end{cases} \tag{6.4}$$

◀ Recall that $[\xi]$ is a smooth positive function such that $[\xi] = |\xi|$ for large $|\xi|$. ▶

In what spaces will this problem be Fredholm? The spaces in which $\Delta$ acts are clear, and we only need to find the natural space into which the edge boundary operator $B$ acts. This can readily be done with the help of Theorem 3.32. Namely, the symbol $B(\xi)$ of this operator satisfies the easy-to-verify estimates

$$\left||\xi|^{s+1/2}B(\xi)v\right| \le \mathrm{const}\left\||\xi|^s \varkappa_{|\xi|}^{-1} v\right\|_{\mathcal{K}^{s,\gamma}(\mathbb{R}_+)}$$

(and similar estimates for the derivatives), and hence

$$B(\xi) \in S^0_{CV}(\mathfrak{H}_1, \mathfrak{H}_2),$$

where $\mathfrak{H}_1$ is the space $\mathcal{K}^{s,\gamma}(\mathbb{R}_+)$ equipped with the family of norms

$$\|\cdot\|_\xi = \left\||\xi|^s \varkappa_{|\xi|}^{-1} \cdot\right\|_{\mathcal{K}^{s,\gamma}(\mathbb{R}_+)}$$

and $\mathfrak{H}_2$ is the one-dimensional complex space $\mathbb{C}$ equipped with the family of norms

$$\|\cdot\|_\xi = |\xi|^{s+1/2}|\cdot|.$$

By Theorem 3.32, the operator $B$ acts continuously in the spaces

$$B : \mathcal{W}^{s,\gamma}(M) \longrightarrow H^{s+1/2}(X), \quad X = \partial M.$$

Summarizing, we see that the operator $\mathbf{A}$ corresponding to our edge boundary value problem acts in the spaces

$$\mathbf{A} = \begin{pmatrix} \Delta \\ B \end{pmatrix} : \mathcal{W}^{s,\gamma}(M) \longrightarrow \begin{matrix} \mathcal{W}^{s-2,\gamma-2}(M) \\ \oplus \\ H^{s+1/2}(X) \end{matrix}. \tag{6.5}$$

This operator is Fredholm; this will follow from the general finiteness theorem stated in Section 6.1.4.

Case 2. Let $\gamma > \frac{3}{2}$. To make Eq. (6.2) uniquely solvable, one can equip it, say, with a co-condition including a numerical unknown $\mu \in \mathbb{C}$:

$$\sigma_\wedge(\triangle)(\xi)v + \mu\phi(|\xi|r) = F, \tag{6.6}$$

where $\phi$ is the same function as above. Quantizing, we obtain the problem

$$\Delta u + Cw = f, \tag{6.7}$$

where the operator $C$ is given by the formula

$$Cw = \phi\left(r\left[-i\frac{\partial}{\partial x}\right]\right)w.$$

A similar argument shows that the operator $\mathbf{A}$ corresponding to this edge coboundary value problem acts in the spaces

$$\mathbf{A} = \begin{pmatrix}\triangle & C\end{pmatrix} : \begin{matrix}\mathcal{W}^{s,\gamma}(M) \\ \oplus \\ H^{s-5/2}(X)\end{matrix} \longrightarrow \mathcal{W}^{s-2,\gamma-2}(M). \tag{6.8}$$

### 6.1.2. Edge boundary and coboundary operators

Now let us proceed to the general case. If, by a procedure similar to that given in the example, we intend to obtain a Fredholm problem for an operator $A \in \mathrm{PSD}^m_\gamma(\mathcal{M})$, then, generally speaking, we have to add edge boundary and coboundary conditions simultaneously. That is, we have to deal with matrix operators of the form[2]

$$\mathbf{A} = \begin{pmatrix}A & C\\ B & D\end{pmatrix} : \begin{matrix}\mathcal{W}^{s,\gamma}(\mathcal{M}) \\ \oplus \\ H^{s}(X)\end{matrix} \longrightarrow \begin{matrix}\mathcal{W}^{s-m,\gamma-m}(\mathcal{M}) \\ \oplus \\ H^{s-m}(X)\end{matrix}, \tag{6.9}$$

where $B$ and $C$ are "edge boundary and coboundary operators" and $D$ is a pseudodifferential operator on $X$. (The last component is usually lacking in natural statements of edge problems, but it inevitably arises in products of matrix operators of this form, and so it is expedient to include it from the very beginning.)

In this and the following subsections, we study operators of the form (6.9) regardless of whether they are elliptic or not. Issues related to ellipticity will be considered later on in this chapter. Since, however, ellipticity and smoothness of solutions are the main issues of interest to us, we shall study the matrix operators (6.9) modulo negligible operators in the sense of the following definition.

[2]For simplicity, we avoid considering operators of vector order in the sense of Douglis–Nirenberg.

DEFINITION 6.1. A matrix operator $\mathbf{A}$ acting in the spaces (6.9) for all $s \in \mathbb{R}$ is called an *operator of order* $m$ (with *weight exponent* $\gamma$). The operator $\mathbf{A}$ is said to be *negligible* if it is compact in the spaces (6.9) and continuous in the spaces

$$\mathbf{A} : \mathcal{W}^{s,\gamma}(\mathcal{M}) \oplus H^s(X) \longrightarrow \mathcal{W}^{s-m+1,\gamma-m}(\mathcal{M}) \oplus H^{s-m+1}(X)$$

for every $s \in \mathbb{R}$. The space of negligible operators of order $m$ and weight $\gamma$ will be denoted by $J_\gamma^m(\mathcal{M})$. Components of a negligible operator will also be called negligible operators.

Note that while the meaning of the objects on the main diagonal in the matrix (6.9) is completely clear, we still have to define what edge boundary and coboundary operators are. Both kinds of operators will be defined as pseudodifferential operators on $X$ with operator-valued symbols in the spirit of Section 3.3. We start from boundary operators.

***Edge boundary operators.*** Let $E$ be a vector bundle over $\mathcal{M}$; by the same letter we denote the corresponding vector bundle on the infinite wedge $W$ and by $E_x$ the restriction of this bundle to the fiber $K_x$ of the bundle $W \longrightarrow X$. Next, let $J$ be a vector bundle on $X$; by $J_x$ we denote the fiber of this bundle over a point $x \in X$.

DEFINITION 6.2. An *edge boundary symbol of order* $m$ *and weight* $\gamma$ is a smooth family of linear operators

$$B(x,\xi) : C_0^\infty(E_x) \longrightarrow J_x, \quad (x,\xi) \in T_0^*X,$$

parametrized by points of the cotangent bundle to the edge $X$ with zero section deleted such that the following conditions hold.

1. The family $B(x,\xi)$ is twisted homogeneous of degree $m$:

$$B(x,\lambda\xi) = \lambda^m B(x,\xi)\varkappa_\lambda^{-1}, \quad \lambda > 0. \tag{6.10}$$

2. The family $B(x,\xi)(|\xi|r)^l$ extends by closure to a smooth family of linear operators in the spaces

$$B(x,\xi)(|\xi|r)^l : \mathcal{K}^{s,\gamma}(K_x, E_x) \longrightarrow J_x \tag{6.11}$$

for any $s \in \mathbb{R}$ and $l \in \mathbb{Z}_+$.

Note that the symbol $B(x,\xi)$ is automatically finite rank (since $J$ is finite-dimensional) and hence compact; moreover, it follows from the twisted homogeneity and the boundedness in the spaces (6.11) that the symbols $B(x,\xi)$ belong to the spaces $S^0_{CV}(\mathfrak{H}_1,\mathfrak{H}_2)$, where $\mathfrak{H}_1$ is the space $\mathcal{K}^{s,\gamma}(K_\Omega)$ equipped with the family of norms

$$\|\cdot\|_\xi = \||\xi|^s \varkappa_{|\xi|}^{-1} \cdot \|_{\mathcal{K}^{s,\gamma}(K_\Omega)}$$

and $\mathfrak{H}_2$ is the fiber of $J$ equipped with the family of norms

$$\|\cdot\|_\xi = |\xi|^{s-m}\,\|\cdot\|,$$

where on the right-hand side we have some arbitrarily chosen Hermitian norm in the fiber of $J$. Hence the following assertion holds by Proposition 3.30.

PROPOSITION 6.3. *If $B(x,\xi)$ is an edge boundary symbol of order $m$ and weight $\gamma$, then the operator $B = B(x,-i\partial/\partial x)$ is well defined and continuous in the spaces*

$$B : \mathcal{W}^{s,\gamma}(W) \longrightarrow H^{s-m}(X) \tag{6.12}$$

*for an arbitrary $s \in \mathbb{R}$. Moreover, the operator $B$ is uniquely determined modulo negligible operators.*

The operator (6.12) acts on functions on the infinite wedge $W$ rather than on the manifold $\mathcal{M}$ itself, while we wish to have an operator on $\mathcal{M}$. However, the following proposition permits us to pass from $W$ to $\mathcal{M}$.

PROPOSITION 6.4. *Let $\psi(r)$ be an arbitrary R-function.*[3] *Then the operator*

$$B(1-\psi(r)) : \mathcal{W}^{s,\gamma}(W) \longrightarrow H^{s-m}(X) \tag{6.13}$$

*is negligible.*

◀ Indeed, we can write

$$B(x,\xi)(1-\psi(r)) = |\xi|^{-N} B(x,\xi)(|\xi| r)^N \circ r^{-N}(1-\psi(r)).$$

The multiplication by $r^{-N}(1-\psi(r))$ is compact from $\mathcal{W}^{s,\gamma}(W)$ to $\mathcal{W}^{s-1,\gamma}(W)$, and the symbol $|\xi|^{-N}B(x,\xi)(|\xi|r)^N$ gives a continuous operator from $\mathcal{W}^{s-1,\gamma}(W)$ to $H^{s-1-m+N}(X)$ by virtue of condition (6.11) and Proposition 3.30. ▶

The operator $B\psi(r)$, where $\psi(r)$ is an arbitrary $R$-function supported in the collar neighborhood $U$ of the boundary $\partial M$, is already well defined as an operator in the spaces

$$B\left(x, -i\frac{\partial}{\partial x}\right)\psi(r) : \mathcal{W}^{s,\gamma}(\mathcal{M}) \longrightarrow H^{s-m}(X). \tag{6.14}$$

DEFINITION 6.5. The operator (6.14) is called the *edge boundary operator* associated with the edge boundary symbol $B(x,\xi)$ and will usually be denoted by $B$. (The factor $\psi(r)$ will often be omitted in the notation.)

This is well defined. Indeed, it follows from Proposition 6.4 that the operator $B$ is independent of the choice of $\psi$ modulo negligible operators.

[3]Recall that an $R$-function is a smooth function of $r \in \overline{\mathbb{R}}_+$ equal to 1 for small $r$ and 0 for large $r$.

***Edge coboundary operators.*** Now we shall introduce the notion of edge coboundary operators. This notion is in fact dual to that of edge boundary operators. Let $E$ be again a vector bundle over $\mathcal{M}$, and let $J$ be a vector bundle over $X$.

DEFINITION 6.6. An *edge coboundary symbol of order $m$ and weight $\gamma$* is a smooth family of linear operators

$$C(x,\xi) : J_x \longrightarrow \mathcal{D}'(E_x), \quad (x,\xi) \in T_0^*X,$$

parametrized by points of the cotangent bundle to the edge $X$ with zero section deleted such that the following conditions hold.

1. The family $C(x,\xi)$ is twisted homogeneous of degree $m$:

$$C(x,\lambda\xi) = \lambda^m \varkappa_\lambda C(x,\xi), \quad \lambda > 0. \tag{6.15}$$

2. The family $(|\xi|r)^l C(x,\xi)$ extends by closure to a smooth family of linear operators in the spaces

$$(|\xi|r)^l C(x,\xi) : J_x \longrightarrow \mathcal{K}^{s,\gamma-m}(E_x) \tag{6.16}$$

for any $s \in \mathbb{R}$ and $l \in \mathbb{Z}_+$.

Again, the symbol $C(x,\xi)$ is finite rank and hence compact; the twisted homogeneity and the boundedness in the spaces (6.16) imply that the symbols $C(x,\xi)$ belong to the spaces $S^0_{CV}(\mathfrak{H}_2,\mathfrak{H}_1)$, where $\mathfrak{H}_1$ and $\mathfrak{H}_2$ are the same as for edge boundary symbols (except that the weight exponent in $\mathfrak{H}_1$ is $\gamma-m$). Hence the following assertion holds.

PROPOSITION 6.7. *Suppose that $C(x,\xi)$ is an edge coboundary symbol of order $m$ and weight $\gamma$. Then the operator $C = C(x,-i\partial/\partial x)$ is well defined and continuous in the spaces*

$$C : H^s(X) \longrightarrow \mathcal{W}^{s-m,\gamma-m}(W) \tag{6.17}$$

*for an arbitrary $s \in \mathbb{R}$. Moreover, the operator $C$ is uniquely determined modulo negligible operators.*

The operator (6.17) acts into a space of functions on the infinite wedge $W$ rather than on the manifold $\mathcal{M}$ itself, and the following proposition permits us to pass from $W$ to $\mathcal{M}$.

PROPOSITION 6.8. *Let $\psi(r)$ be an arbitrary $R$-function. Then the operator*

$$(1-\psi(r))C : H^s(X) \longrightarrow \mathcal{W}^{s-m,\gamma-m}(W) \tag{6.18}$$

*is negligible.*

The operator $\psi(r)C$, where $\psi(r)$ is an arbitrary $R$-function supported in the collar neighborhood $U$ of the boundary $\partial M$, is already well defined as an operator in the spaces

$$\psi(r)C\left(x, -i\frac{\partial}{\partial x}\right) : H^s(X) \longrightarrow \mathcal{W}^{s-m,\gamma-m}(\mathcal{M}). \tag{6.19}$$

DEFINITION 6.9. The operator (6.19) is called the *edge coboundary operator* associated with the edge coboundary symbol $C(x,\xi)$ and will usually be denoted by $C$ (the factor $\psi(r)$ will be omitted).

This is well defined. Indeed, it follows from Proposition 6.8 that the operator $C$ is independent of the choice of $\psi$ modulo negligible operators.

*Remark* 6.10. The continuity conditions imposed on the families (6.11) and (6.16) essentially indicate that our edge boundary and coboundary operators are concentrated on the edge. Outside an arbitrarily small neighborhood of the edge, they are infinitely smoothing. This justifies the names "boundary and coboundary operators" which we use here. Recall that even in the classical theory of boundary value problems, along with differential boundary operators that are exactly concentrated on the boundary one also has pseudodifferential boundary operators and of course coboundary operators, which are concentrated on the boundary only in the sense of singular support.

### 6.1.3. The calculus of edge morphisms

Now that we have defined edge boundary and coboundary operators and proved boundedness theorems for these operators, we can introduce a class of operators (6.9) that is of interest to us.

DEFINITION 6.11. A *morphism of order* $m$ *and weight* $\gamma$ is an operator $\mathbf{A}$ representable modulo negligible operators of order $m$ and weight $\gamma$ in the form

$$\mathbf{A} = \begin{pmatrix} A & C \\ B & D \end{pmatrix} : \begin{matrix} \mathcal{W}^{s,\gamma}(\mathcal{M}, E) \\ \oplus \\ H^s(X, J_1) \end{matrix} \longrightarrow \begin{matrix} \mathcal{W}^{s-m,\gamma-m}(\mathcal{M}, F) \\ \oplus \\ H^{s-m}(X, J_2) \end{matrix}, \tag{6.20}$$

where $A$ is a pseudodifferential operator of order $m$ and weight $\gamma$ on $\mathcal{M}$, $B$ and $C$ are edge boundary and coboundary operators, respectively, of order $m$ and weight $\gamma$, and $D$ is a (classical) pseudodifferential operator of order $m$ on the edge $X$. The set of morphisms of order $m$ and weight $\gamma$ is denoted by $\mathrm{Mor}^m_\gamma(\mathcal{M})$.

Next, we introduce the notions of interior symbol and edge symbol for morphisms.

DEFINITION 6.12. The *interior symbol* of the morphism $\mathbf{A}$ (6.20) is the interior symbol of the pseudodifferential operator $A$:

$$\sigma(\mathbf{A}) = \sigma(A).$$

The *edge symbol* of the morphism $\mathbf{A}$ (6.20) is the operator family

$$\sigma_\wedge(\mathbf{A}) = \begin{pmatrix} \sigma_\wedge(A)(x,\xi) & C(x,\xi) \\ B(x,\xi) & D(x,\xi) \end{pmatrix} : \begin{matrix} \mathcal{K}^{s,\gamma}(K_x, E) \\ \oplus \\ J_{1x} \end{matrix} \longrightarrow \begin{matrix} \mathcal{K}^{s-m,\gamma-m}(K_x, F) \\ \oplus \\ J_{2x} \end{matrix},$$
$$(x,\xi) \in T_0^*X, \quad (6.21)$$

where $B(x,\xi)$, $C(x,\xi)$, and $D(x,\xi)$ are the symbols of the operators $B$, $C$, and $D$, respectively.

The *principal symbol* of the edge morphism $\mathbf{A}$ is the pair

$$\mathfrak{s}(\mathbf{A}) = \big(\sigma(\mathbf{A}), \sigma_\wedge(\mathbf{A})\big). \quad (6.22)$$

In the following, we write

$$B(x,\xi) = \sigma_\wedge(B), \quad C(x,\xi) = \sigma_\wedge(C), \quad D(x,\xi) = \sigma_\wedge(D)$$

(even though the notation $\sigma$ instead of $\sigma_\wedge$ might be more appropriate), since all these symbols are entries of the edge symbol of $\mathbf{A}$. One should however have in mind that these symbols are just the (operator-valued) principal symbols of $B$ and $C$ and the principal symbol of the pseudodifferential operator $D$ in the ordinary sense.

PROPOSITION 6.13. *An elliptic morphism $\mathbf{A} \in \mathrm{Mor}_\gamma^m(\mathcal{M})$ is uniquely determined by its principal symbol $\mathfrak{s}(\mathbf{A})$ modulo negligible operators of order $m$ and weight $\gamma$.*

◀ This follows from Propositions 6.3 and 6.7 and the corresponding results for pseudodifferential operators. ▶

Further, we define the boundary symbol of an edge morphism $\mathbf{A}$ as the restriction of the interior symbol of $\mathbf{A}$ to the boundary $\partial T^*\mathcal{M}$ and the interior symbol of the edge symbol of an edge morphism as the interior symbol of the edge symbol of $A$:

$$\sigma_\partial(\mathbf{A}) = \sigma(A)\,\big|_{\partial T^*\mathcal{M}} = \sigma_\partial(A), \quad \sigma(\sigma_\wedge(\mathbf{A})) = \sigma(\sigma_\wedge(A)).$$

With these definitions, the following assertion is tautological.

PROPOSITION 6.14. *The components of the principal symbol of an edge morphism $\mathbf{A}$ satisfy the compatibility condition*

$$\sigma_\partial(\mathbf{A}) = \sigma(\sigma_\wedge(\mathbf{A})). \quad (6.23)$$

*This condition is necessary and sufficient for a pair (interior symbol, edge symbol) to be the principal symbol of some morphism $\mathbf{A}$.*

Let us state the main theorem of the calculus of edge morphisms.

THEOREM 6.15 (composition theorem for edge morphisms). (1) *Suppose that the product* $\mathbf{A}_1\mathbf{A}_2$ *of two elliptic morphisms* $\mathbf{A}_1$ *and* $\mathbf{A}_2$ *is well defined (that is, the bundles and the indices of Sobolev spaces fit). Then the product is also an edge morphism, and*

$$\mathfrak{s}(\mathbf{A}_1\mathbf{A}_2) = \mathfrak{s}(\mathbf{A}_1)\mathfrak{s}(\mathbf{A}_2), \tag{6.24}$$

*where the composition of principal symbols is defined componentwise: one separately multiplies the interior symbols and the matrix edge symbols of the morphisms.*

(2) *The adjoint of an edge morphism* $\mathbf{A}$ *is again an edge morphism, and*

$$\mathfrak{s}(\mathbf{A}^*) = \mathfrak{s}(\mathbf{A})^*. \tag{6.25}$$

*Remark* 6.16. Here we should possibly mention that adjoints in weighted Sobolev spaces are taken with respect to the inner product in $\mathcal{W}^{0,0}$ (or $\mathcal{K}^{0,0}$ for the case of symbols).

*Proof of the composition theorem.* Assertion (2) is trivial; it pertains to separate entries and follows from the corresponding assertions for edge-degenerate pseudodifferential operators and operators in sections in Hilbert bundles.

Let us prove (1). To this end, we should show that, modulo negligible operators,

1) the product $A_1A_2$ of two edge-degenerate pseudodifferential operators on $\mathcal{M}$ is again an edge-degenerate pseudodifferential operator on $\mathcal{M}$, and moreover, $\mathfrak{s}(A_1A_2) = \mathfrak{s}(A_1)\mathfrak{s}(A_2)$

2) the product $D_1D_2$ of two pseudodifferential operators on $X$ is again a pseudodifferential operator on $X$ with $\sigma_\wedge(D_1D_2) = \sigma_\wedge(D_1)\sigma_\wedge(D_2)$

3) a product of the form $BC$, where $B$ is an edge boundary operator and $C$ is an edge coboundary operator, is a pseudodifferential operator on $X$ with $\sigma_\wedge(BC) = \sigma_\wedge(B)\sigma_\wedge(C)$

4) a product of the form $CB$ is a pseudodifferential operator on $\mathcal{M}$ with symbols $\sigma(CB) = 0$ and $\sigma_\wedge(CB) = \sigma_\wedge(C)\sigma_\wedge(B)$

5) a product of the form $BA$, where $A$ is an edge-degenerate pseudodifferential operator on $\mathcal{M}$ and $B$ is an edge boundary operator, is again an edge boundary operator, and $\sigma_\wedge(BA) = \sigma_\wedge(B)\sigma_\wedge(A)$

6) a product of the form $AC$, where $C$ is an edge coboundary operator, is again an edge coboundary operator, and $\sigma_\wedge(AC) = \sigma_\wedge(A)\sigma_\wedge(C)$

7) products of the form $DB$ and $CD$, where $D$ is a pseudodifferential operator on $X$, are again edge boundary and coboundary operators, respectively, and moreover, $\sigma_\wedge(DB) = \sigma_\wedge(D)\sigma_\wedge(B)$ and $\sigma_\wedge(CD) = \sigma_\wedge(C)\sigma_\wedge(D)$

Assertions 1 and 2 are well known; these are just the composition theorems for edge-degenerate and classical pseudodifferential operators. Assertions 3 , 4 , and 7 follow from the composition theorem for pseudodifferential operators in sections of Hilbert bundles. (To observe that the interior symbol of $CB$ is zero in assertion 4, it suffices to note that the operator $CB$ is infinitely smoothing outside the edge.)

It remains to prove assertions 5 and 6. Let us prove 5; the proof of 6 is similar. We have

$$BA = \sigma_\wedge(B)\left(x, -i\frac{\partial}{\partial x}\right)\psi(r)A,$$

where $\psi$ is an arbitrary $R$-function. Let us use the representation (3.58) for the edge-degenerate pseudodifferential operator $A$; then we obtain

$$BA = \sigma_\wedge(B)\left(x, -i\frac{\partial}{\partial x}\right)\left[\psi(r)\sigma_\wedge(A)\left(x, -i\frac{\partial}{\partial x}\right) + r\psi(r)Q\left(x, -i\frac{\partial}{\partial x}\right)\right],$$

where $Q(x,\xi) \in S^0_{CV}(\mathcal{K}^{s,\gamma}(K_\Omega), \mathcal{K}^{s-m,\gamma-m}(K_\Omega))$. Since $R$-functions commute with edge symbols modulo lower-order symbols, we see that the symbol of $BA$ is equal to the product $\sigma_\wedge(B)\sigma_\wedge(A)$ modulo terms whose quantization produces negligible operators. (The second term in the sum gives a zero contribution to the edge symbol, since it is of lower order of twisted homogeneity owing to the presence of the extra factor $r$, which is homogeneous of order $-1$.) □

*Remark* 6.17. Note that the assertion concerning the products $BC$ and $CB$ is essentially trivial in the theory of edge morphisms. This is in contrast with the theory of Sobolev problems (see Chapter 8), where the pseudodifferential nature of the product $BC$ is one of the main results of the analytical part of the theory and the product $CB$ is not pseudodifferential at all; it is known as a *Green operator* in that theory. In the theory of edge-degenerate differential operators, the notion of a Green operator is trivial if we do not consider operators in weighted spaces with asymptotics; (e.g., Egorov and Schulze 1997): a Green operator is an edge-degenerate pseudodifferential operator whose interior symbol is zero and whose edge symbol is compact. Green operators play no role in our exposition, and we do not dwell on the topic.

### 6.1.4. Ellipticity, finiteness, and index theorems

The composition theorem readily implies the finiteness theorem.

DEFINITION 6.18. A morphism $\mathbf{A} \in \mathrm{Mor}^m_\gamma(\mathcal{M})$ is said to be *interior elliptic* if its interior symbol $\sigma(\mathbf{A})$ is invertible up to the boundary of the cotangent bundle

$T_0^*\mathcal{M}$ without the zero section. The morphism $\mathbf{A}$ is said to be *elliptic* if it is interior elliptic and its edge symbol is invertible on $T_0^*X$.

THEOREM 6.19. *If* $\mathbf{A} \in \mathrm{Mor}_\gamma^m(\mathcal{M})$ *is an elliptic morphism, then* $\mathbf{A}$ *is Fredholm in the spaces* (6.9) *for all* $s \in \mathbb{R}$. *The kernel, cokernel, and index of* $\mathbf{A}$ *are independent of* $s$.

The proof is the same as for pseudodifferential operators.

It turns out that the index Theorem 5.29 remains valid for elliptic edge morphisms under an appropriate modification of the definitions. By $\mathbf{Ell}_\alpha$ we denote the set of elliptic edge morphisms $\mathbf{A}$ whose interior symbol $\sigma(\mathbf{A})$ satisfies the symmetry condition associated with the involution $\alpha$. We define invariants $f_1$ and $f_2$ of the interior and edge symbols of morphisms $\mathbf{A} \in \mathbf{Ell}_\alpha$ by the same formulas as in Chapter 5. Then the following theorem holds.

THEOREM 6.20. *Let Assumption* 5.28 *hold. Then the index of an elliptic edge morphism* $\mathbf{A} \in \mathbf{Ell}_\alpha$ *is given by the formula*

$$\operatorname{ind}\mathbf{A} = f_1(\sigma(\mathbf{A})) + f_2(\sigma_\wedge(\mathbf{A})). \tag{6.26}$$

## 6.2. The Obstruction to Ellipticity

Let

$$A : \mathcal{W}^{s,\gamma}(\mathcal{M}, E) \longrightarrow \mathcal{W}^{s,\gamma}(\mathcal{M}, F)$$

be an interior elliptic edge-degenerate operator on a manifold $\mathcal{M}$ with edge $X$. We assume that $\operatorname{ord} A = 0$.

◀ The condition that the order of $A$ is zero is not restrictive, since the general case can be reduced to this by the order reduction procedure (see Section 3.6.4). ▶

We shall consider the following two problems for $A$.

***Problem 1.*** *Equip the operator* $A$ *with edge boundary and coboundary operators* $B$ *and* $C$ *and a pseudodifferential operator* $D$ *on the edge such that the resulting edge problem*

$$\mathbf{A}u = v, \tag{6.27}$$

*where* $\mathbf{A}$ *is the edge morphism* (6.20), *is Fredholm.*

Clearly, this can be possible only if the edge symbol $\sigma_\wedge(A)$ is Fredholm (or, which is the same, the conormal symbol $\sigma_c(\sigma_\wedge(A))$ of the edge symbol is invertible). If this is not the case, then there is no Fredholm edge problem for the operator $A$, and so one has to weaken Problem 1. Namely, we consider also the following problem.

***Problem 2.*** *Find a Fredholm edge-degenerate operator $A'$ with $\sigma(A') = \sigma(A)$.*

Note that we could ask for a Fredholm edge *problem* for the modified operator $A$. However, this is superfluous: it turns out that if we are allowed to modify the operator $A$ anyway, conditions under which there exists a Fredholm edge problem for a modified operator coincide with the conditions under which there exists a Fredholm modified operator.

There are obstructions to both problems. We find these obstructions in the two subsections below; we shall see that they coincide. The obstruction will be computed in topological terms in the next section.

### 6.2.1. Obstruction to the existence of Fredholm problems for a given operator

Here we shall prove the following theorem concerning the solvability of Problem 1.

THEOREM 6.21. *Suppose that $A$ is an interior elliptic operator with Fredholm edge symbol on a manifold $\mathcal{M}$ with edge $X$.*

1. *A necessary and sufficient condition for the existence of an elliptic edge problem for the operator $A$ is that*

$$\operatorname{ind} \sigma_\wedge(A) = 0 \in K(S^*X)/\pi_0^* K(X), \tag{6.28}$$

*where $\pi_0 : S^*X \to X$ is the natural projection and $\pi_0^*K(X) \subset K(S^*X)$ is the subgroup generated by bundles lifted from $X$.*

2. *The $K$-theory element* (6.28) *is determined by the interior symbol of $A$.*

*Proof.* Let us prove 1.

***Necessity.*** Suppose that there exists a Fredholm edge problem (6.20) for the operator $A$. By the ellipticity of $\mathbf{A}$, the edge symbol $\sigma_\wedge(\mathbf{A})$ is invertible. Then

$$\begin{aligned} 0 = \operatorname{ind} \sigma_\wedge(\mathbf{A}) &= \operatorname{ind} \sigma_\wedge \begin{pmatrix} A & 0 \\ 0 & 0 \end{pmatrix} \\ &= \operatorname{ind} \sigma_\wedge(A) + \operatorname{ind}(0 : \pi_0^* J_1 \to \pi_0^* J_2) \\ &= \operatorname{ind} \sigma_\wedge(A) + [\pi_0^* J_1] - [\pi_0^* J_2]. \end{aligned}$$

Here we have replaced by zeros the operators whose edge symbols are compact and hence do not affect the index. The difference $[\pi_0^* J_1] - [\pi_0^* J_2]$ belongs to $\pi_0^* K(X)$, and so we arrive at (6.28).

***Sufficiency.*** We carry out our considerations for $(x, \xi) \in S^*X$. Since $\sigma_\wedge(A)(x, \xi)$ is Fredholm, it follows that for sufficiently large $N$ there exists a bundle homomorphism

$$C(x, \xi) : \mathbb{C}^N \longrightarrow \mathcal{K}^{s,\gamma}(K_x, F_x)$$

such that the mapping

$$\begin{pmatrix}\sigma_\wedge(A) & C\end{pmatrix} : \mathcal{K}^{s,\gamma}(K_x, E_x) \oplus \mathbb{C}^N \longrightarrow \mathcal{K}^{s,\gamma}(K_x, F_x) \tag{6.29}$$

is an epimorphism.

◀ The existence of such a $C$ in a neighborhood of each point is trivial, and then one can globalize the assertion by using a partition of unity. ▶

Then the kernel of (6.29) is a vector bundle over $S^*X$. Suppose that (6.28) holds, that is,

$$\operatorname{ind} \sigma_\wedge(A) \in \pi_0^* K(X).$$

But

$$\operatorname{ind} \sigma_\wedge(A) = [\operatorname{Ker} \begin{pmatrix}\sigma_\wedge(A) & C(x,\xi)\end{pmatrix}] - [\mathbb{C}^N],$$

and hence the kernel of (6.29) is stably isomorphic to a bundle lifted from $X$. Augmenting $C$ by sufficiently many zero columns, we can assume that the kernel of (6.29) is isomorphic to a bundle $J$ lifted from $X$; let the isomorphism be

$${}^t(B(x,\xi), D(x,\xi)) : J_x \longrightarrow \mathcal{K}^{s,\gamma}(K_x, E_x) \oplus \mathbb{C}^N.$$

We extend $(B(x,\xi), D(x,\xi))$ by zero to the orthogonal complement of the kernel and then continue $B(x,\xi)$, $C(x,\xi)$, and $D(x,\xi)$ to the entire $T_0^*X$ by twisted homogeneity. This gives the invertible edge symbol

$$\begin{pmatrix}\sigma_\wedge(A)(x,\xi) & C(x,\xi) \\ B(x,\xi) & D(x,\xi)\end{pmatrix}$$

of the desired Fredholm edge problem.

Now let us prove 2. Let $A$ and $A'$ be two operators with the same interior symbols and with Fredholm edge symbols. Then

$$\operatorname{ind} \sigma_\wedge(A) - \operatorname{ind} \sigma_\wedge(A') = \operatorname{ind} \sigma_\wedge(A)[\sigma_\wedge(A')]^{-1} = 0 \in K(S^*X)/\pi_0^* K(X), \tag{6.30}$$

where we have used the fact that the family $\sigma_\wedge(A)[\sigma_\wedge(A')]^{-1}$ (parametrized by $S^*X$) of cone-degenerate operators has a unit interior symbol and a conormal symbol independent of the point in the fiber of the projection $\pi_0 : S^*X \to X$. Hence the index of this family is the pullback of some element of the group $K(X)$. □

### 6.2.2. Obstruction to the existence of Fredholm operators with a given interior symbol

THEOREM 6.22. *Suppose that $A$ is an interior elliptic operator on a manifold $\mathcal{M}$ with edge $X$.*

1. *There always exists a Fredholm edge symbol compatible with $\sigma(A)$.*

2. *Condition (6.28) (for the edge symbol mentioned in item 1) is necessary and sufficient for the existence of an elliptic operator $A'$ with $\sigma(A) = \sigma(A')$.*

*Remark* 6.23. Thus the obstruction coincides with the one obtained in the preceding subsection.

*Proof.* First, we shall prove that we can always make the edge symbol of $A$ Fredholm. It is necessary and sufficient to make the conormal symbol $\sigma_c(\sigma_\wedge(A))$ invertible.

By interior ellipticity, we find that the conormal symbol is a family of Fredholm operators parametrized by $X \times \mathbb{R}$ and invertible outside a compact set. It follows that its index lies in the $K$-group with compact supports of the parameter space (see Section 3.6.2). It turns out that this index is trivial.

LEMMA 6.24. *The index of the conormal symbol is zero*:

$$\operatorname{ind} \sigma_c(\sigma_\wedge(A)) = 0 \in K_c(X \times \mathbb{R}).$$

*Proof.* The proof is based on functorial properties of the index of families with parameters (see the end of Section 3.6.2). □

Now we shall use the following quite standard lemma (cf. Booß-Bavnbek and Bleecker 1985).

LEMMA 6.25. *A Fredholm operator can be made invertible by the addition of a compact operator if and only if its index is zero. Similar assertions hold for a) families of Fredholm operators; b) families of elliptic operators with parameters.*

Thus there exists an operator $M_0$ with zero interior symbol such that the conormal symbol $\sigma_c(\sigma_\wedge(A+M_0))$ is invertible,[4] and in solving Problem 2 we can assume from the very beginning that the edge symbol of $A$ is Fredholm.

Now we can proceed to the proof of the second part of the theorem.

***Necessity*** readily follows from (6.30).

***Sufficiency.*** Let

$$\operatorname{ind} \sigma_\wedge(A) = \pi_0^* x \quad \text{for some } x \in K(X).$$

One can show that an arbitrary element of $K(X)$ can be realized as the index of a family of elliptic cone-degenerate operators with unit interior symbol.

Indeed, the index of the operator with unit interior symbol and conormal symbol

$$\sigma_c = \frac{p+i}{p-i} Q + (1-Q),$$

where $Q$ is a finite-dimensional projection in the function space on the base $\Omega$ of the cone is equal to the rank of the projection $Q$. If we consider a continuous family of operators

[4] Additional smoothing provides the desired smoothness and estimates.

of this type, then the index is equal to the vector bundle $\operatorname{Im} Q$. Now the desired assertion follows from the fact that every bundle can be realized as the range of some projection.

Then consider the family with index $-x$, and let $1+G_0$ be the edge-degenerate operator with unit interior symbol and with edge symbol equal to the given family. By construction, the index of the edge symbol of the product $A(1+G_0)$ is zero; by applying Lemma (6.25), we obtain an invertible edge symbol and hence an elliptic edge-degenerate operator. □

## 6.3. A Formula for the Obstruction in Topological Terms

In this section, we compute the index (6.28) of a Fredholm edge symbol $\sigma_\wedge(D)$ in the quotient group $K(S^*X)/\pi_0^*K(X)$. The definitive index formula is given in Theorem 6.30, and the reader who is not interested in details can skip the other material.

### 6.3.1. Reduction to an operator stabilizing at infinity

Consider a Fredholm edge symbol $\sigma_\wedge$ of zero order.[5] The computation of its index is complicated by the fact that the behavior of the edge symbol as an operator on the infinite cone $K_\Omega$ as $r \to 0$ and as $r \to \infty$ is different: if the operator near zero corresponds to conical theory and its ellipticity is determined by the conormal symbol, the ellipticity at infinity is determined by the interior symbol, and the conormal symbol at infinity merely does not exist.

In this subsection, we explain how, without affecting the index, one can replace the edge symbol in a neighborhood of infinity in $K_\Omega$ by an operator that stabilizes at infinity, that is, has a conormal symbol at infinity.

We deal with the representation of the edge symbol as a function of the operators

$$\sigma_\wedge(x,\xi) = D\left(\overset{2}{e^{-t}\xi}, \overset{1}{-i\frac{\partial}{\partial t}}\right). \tag{6.31}$$

Let us explain the notation. We work in the cylindrical representation, where $r = e^{-t}$. By $D(\eta, p)$ we denote a family of zero-order pseudodifferential operators acting on the base of the cone $K_\Omega$ and elliptic with parameters $(\eta, p) \in T^*X \times \mathbb{R}$ (see Section 3.2.5).

EXERCISE 6.26. Check that every edge symbol can be represented in the form (6.31) modulo compact operators.

◀ Indeed, it suffices to construct an edge symbol with given compatible interior and conormal symbols, since these symbols determine the edge symbol modulo compact operators.

---

[5]This does not restrict generality, since the order reduction is always possible.

It suffices to take a family $D(\eta, p)$ with $D(0, p)$ being equal to the given conormal symbol and $\sigma(D(\eta, p))$ being equal to the given interior symbol (in the sense of Agranovich–Vishik). This is possible owing to the compatibility condition between the two symbols. Now the family (6.31) is an edge symbol, which follows from the estimates carried out in Section 3.5.5 (for $h = 1$). ▶

Let $\chi_1(t)$ be a smooth positive function on $\mathbb{R}$ with the following behavior:

$$\chi_1(t) = \begin{cases} 1 & \text{for } t < -1, \\ e^{-t} & \text{for } t > -1/2. \end{cases}$$

LEMMA 6.27. *One has the relation*

$$\operatorname{ind} D\left(\overset{2}{e^{-t}\xi}, -i\overset{1}{\frac{\partial}{\partial t}}\right) = \operatorname{ind} D\left(\overset{2}{\chi_1(t)\xi}, -i\overset{1}{\frac{\partial}{\partial t}}\right) \in K(S^*X) \tag{6.32}$$

*for the indices of Fredholm families on the cosphere bundle $S^*X$ of sufficiently large radius $R = |\xi|$.*

*Remark* 6.28. The index of the edge symbol is independent of the radius $R$ of the spheres where the operator is considered. This follows from twisted homogeneity.

*Proof.* 1. The family on the right-hand side of the desired relation is a family of stabilizing operators on the infinite cylinder and is indeed Fredholm: the interior symbol is elliptic, and so is the conormal symbol as $t \to +\infty$ (it is equal to the conormal symbol $D(0, p)$ of the edge symbol). As $t \to -\infty$, the conormal symbol is equal to $D(\xi, p)$ and is invertible for all $p$ provided that $|\xi| > R$. (The constant $R$ is determined by the symbol and is independent of the variable $p$). The last assertion is a standard property of elliptic parameter-dependent families.

2. To prove that the indices are the same, we use the surgery technique. Consider the following surgery diagram of elliptic operators:

$$\begin{array}{ccc} D\left(\overset{2}{e^{-t}\xi}, -i\overset{1}{\frac{\partial}{\partial t}}\right) & \longrightarrow & D\left(\overset{2}{\chi_2(t)\xi}, -i\overset{1}{\frac{\partial}{\partial t}}\right) \\ \downarrow & & \downarrow \\ D\left(\overset{2}{\chi_1(t)\xi}, -i\overset{1}{\frac{\partial}{\partial t}}\right) & \longrightarrow & D\left(\overset{2}{\chi_3(t)\xi}, -i\overset{1}{\frac{\partial}{\partial t}}\right). \end{array} \tag{6.33}$$

The graphs of the functions

$$\chi_2(t) = \begin{cases} e^{-t} & \text{for } t < 1/2, \\ 1 & \text{for } t > 1 \end{cases} \quad \text{and} \quad \chi_3(t) = \begin{cases} \chi_1 & \text{for } t < 1/2, \\ \chi_2 & \text{for } t > -1/2 \end{cases}$$

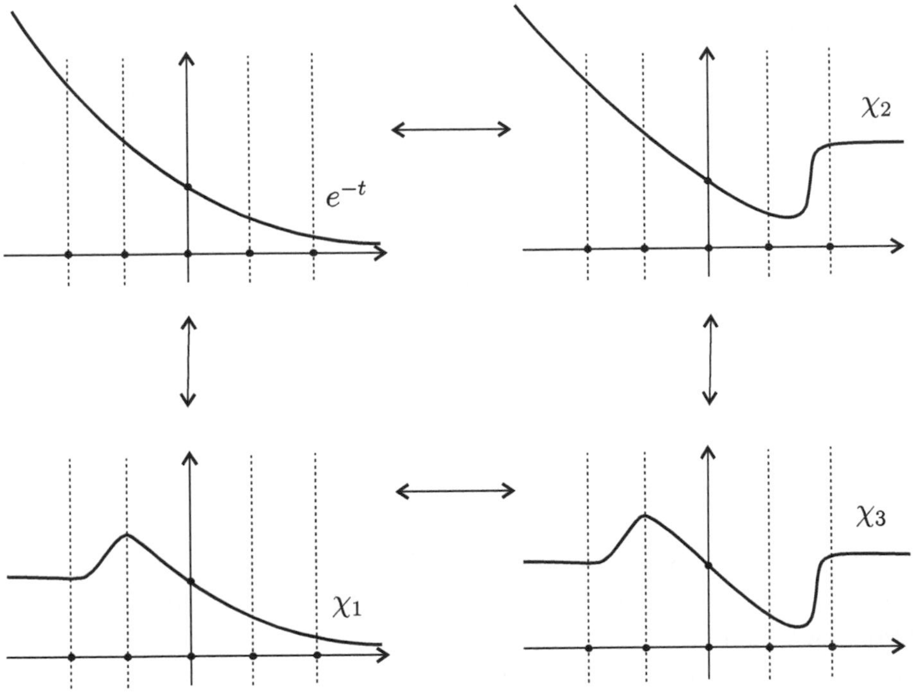


Figure 6.1. The functions $\chi_2(t)$ and $\chi_3(t)$.

are shown in Fig. 6.1.

By the relative index theorem, the differences of the indices of the operators in the rows coincide. To obtain the desired equality of the indices, let us prove that the indices of the operators in the right column are zero.

3. The index of the operator corresponding to the function $\chi_3$ is zero. Indeed, such a function can be deformed to unity (without any deformation at infinity). Hence the index is equal to the index of a translation-invariant operator in a weight space with the same weights at $\pm\infty$, which is zero (see Section 4.4.1).

4. It turns out that for sufficiently large $\xi$ the operator $D\big(\overset{2}{\chi_2}(t)\xi, -i\overset{1}{\frac{\partial}{\partial t}}\big)$ is invertible. We claim that an almost inverse (modulo an operator with small norm) is $D^{-1}\big(\overset{2}{\chi_2}(t)\xi, -i\overset{1}{\frac{\partial}{\partial t}}\big)$. Indeed, consider the composition formula (3.10) for these operators.

EXERCISE 6.29. One has the estimates

$$\left\| \frac{\partial^\alpha D(\chi_2(t)\xi, p)}{\partial t^\alpha} \right\| \leq C_\alpha,$$

$$\left\| \frac{\partial^{\alpha+\beta} D(\chi_2(t)\xi, p)}{\partial t^\alpha \partial p^\beta} \right\| \leq C_{\alpha\beta}(1+|\xi|)^{-1/2}(1+|p|)^{-|\beta|+1/2},$$

$|\alpha| = 0, 1, 2, \ldots, |\beta| = 1, 2, \ldots,$ and similar estimates for the inverse family.

◀ *Hint.* The estimates are based on the fact that the function $\chi_2$ is strictly separated from zero. ▶

It follows that in the composition formula (3.10) for pseudodifferential operators with symbols $D(\chi_2(t)\xi, p)$ and $D^{-1}(\chi_2(t)\xi, p)$ the remainder is of the order of $(1+|\xi|)^{-1/2}$ for large $\xi$. (The proof of this remainder estimate is similar to that in the chapter on elliptic operators with parameters in Shubin (1985).) Therefore, the composition of the corresponding operators is equal to unity plus a small norm operator, whence the invertibility for large $\xi$ follows.

The proof of formula (6.32) is complete. □

### 6.3.2. The index of the stabilizing operator family and the main theorem

The index of a stabilizing operator family on an infinite cylinder in (6.32) is computed by standard methods.

***Reduction to the cross-section of the cylinder.*** The index of the operator on the cylinder is equal to the spectral flow of the corresponding homotopy of conormal symbols (see Section 4.4.4). On the other hand, the spectral flow is expressed in terms of the index via Bott periodicity (Atiyah, Patodi and Singer 1976). This gives the following formula for the index of a family of operators on the cylinder:

$$\operatorname{ind} D\left(\overset{2}{\chi_1}(t)\xi, -i\overset{1}{\frac{\partial}{\partial t}}\right) = \beta^{-1} \operatorname{ind} D(\chi_1(t)\xi, p) \in K(S^*X), \tag{6.34}$$

where $\beta : K(S^*X) \simeq K_c(S^*X \times \mathbb{R}_t \times \mathbb{R}_\rho)$ is the Bott periodicity isomorphism in $K$-theory.

***Index in the quotient $K$-group.*** Let us now compute the index (6.34) in the quotient $K(S^*X)/\pi_0^*K(X)$. Quite remarkably, the projection

$$K(S^*X) \longrightarrow K(S^*X)/\pi_0^*K(X)$$

admits an alternative description. Namely, we realize $S^*X \times \mathbb{R}_+$ as the complement

$$S^*X \times \mathbb{R}_+ \simeq T_0^*X \overset{i}{\subset} T^*X$$

of the zero section in $T^*X$. Here $\mathbb{R}_+$ corresponds to rays issuing from the origin. This embedding permits us to define the diagram

$$\begin{array}{ccc} K(S^*X) & \xrightarrow{\beta} & K_c(S^*X \times \mathbb{R}_+ \times \mathbb{R}) \\ \downarrow & & \downarrow{\scriptstyle i_!} \\ K(S^*X)/\pi_0^*K(X) & \xrightarrow[\alpha]{} & K_c(T^*X \times \mathbb{R}). \end{array} \tag{6.35}$$

Here thc direct image mapping $i_!$ is induced by the embedding $i$, and the lower horizontal arrow is the *Atiyah–Patodi–Singer isomorphism* (Atiyah, Patodi and Singer 1976).

By definition, this isomorphism is induced by the coboundary mapping $\partial : K(S^*X) \to K_c^1(T^*X)$ of the pair $S^*X \subset B^*X$. Here $B^*M$ is the bundle of unit balls in $T^*X$ and the sphere bundle $S^*X$ is its boundary.

The Atiyah–Patodi–Singer isomorphism is induced by the composition $i_!\beta$; i.e., the diagram (6.35) is commutative.

It can be shown that the direct image $i_!$ in this case takes the index of the family $D(\chi_1(t)\xi, p)$ over $S^*X \times \mathbb{R}^2$ to the index of the family $D(\xi, p)$ over $T^*X \times \mathbb{R}$. By taking the projection of (6.34) onto the quotient and by using (6.35), we obtain

$$\begin{aligned}[\operatorname{ind}\sigma_\wedge] &= \left[\beta^{-1}\operatorname{ind} D(\chi_1(t)\xi, p)\right] = \alpha^{-1} i_! \operatorname{ind} D(\chi_1(t)\xi, p) \\ &= \alpha^{-1}\operatorname{ind} D(\xi, p) \in K(S^*X)/\pi_0^* K(X).\end{aligned} \tag{6.36}$$

**Final formula for the obstruction.** Thus we have expressed the obstruction in terms of the family $D(\xi, p)$ of parameter-dependent elliptic operators. An index formula for such Fredholm families was found by Victor Nistor (we discuss it in Appendix C). For our family $D = D(\xi, p)$ with parameters $(\xi, p) \in T^*X \times \mathbb{R}$, this formula expresses the index in terms of the principal symbol of the family $\sigma(D)$. (The principal symbol is defined in $T^*\partial M \times \mathbb{R}$ and is equal to the interior symbol of $\sigma_\wedge$.) The formula reads

$$\operatorname{ind} D = \pi_![\sigma(D)] \in K_c(T^*X \times \mathbb{R}), \tag{6.37}$$

where $\pi_!$ is the direct image map induced by the projection

$$\pi : T^*\partial M \times \mathbb{R} \longrightarrow T^*X \times \mathbb{R}.$$

Using this formula in (6.36), we finally obtain:

$$[\operatorname{ind}\sigma_\wedge] = \alpha^{-1}[\operatorname{ind} D] = \alpha^{-1}\pi_![\sigma(D)].$$

Thus we arrive at a formula for the index of Fredholm edge symbols.

THEOREM 6.30. *Let $\sigma_\wedge$ be a Fredholm edge symbol. Then its index as an element of the quotient $K(S^*X)/\pi_0^*K(X)$ (here $\pi_0 : S^*X \to X$) is expressed by*

$$[\operatorname{ind}\sigma_\wedge] = \alpha^{-1}\pi_![\sigma]. \tag{6.38}$$

*Here $\sigma$ is the interior symbol of $\sigma_\wedge$.*

*Remark* 6.31. In the following section, we show that the obstruction (6.38) may or may not be zero for interior elliptic operators; i.e., elliptic edge problems may or may not exist.

The obstruction has a topological nature and is an analog of the famous Atiyah–Bott obstruction (Atiyah and Bott 1964) in the theory of classical boundary value problems, which also singles out operators for which there exists a well-posed boundary value problem. In fact, our obstruction contains the Atiyah–Bott obstruction, which is given by the formula

$$[\sigma|_{\partial T^*M}] \in K_c(\partial T^*M) \simeq K_c^1(T^*\partial M),$$

as the special case in which $\partial M = X$, $\pi = \mathrm{id}$, and hence $\mathcal{M}$ is a manifold with boundary.

***The obstruction to the existence of Fredholm edge symbols is trivial.*** Finally, we note that invertible conormal symbols (or, equivalently, Fredholm edge symbols) always exist for an elliptic interior symbol. Indeed, the following proposition holds.

PROPOSITION 6.32. *The index of the family of conormal symbols is always zero; i.e., for each elliptic interior symbol on a manifold $\mathcal{M}$ with edge $X$ there exists a compatible Fredholm edge symbol.*

*Proof.* Consider the cotangent bundle ${}^vT^*\partial M$ of the fibers of the projection

$$\pi : \partial M \to X.$$

Choosing a metric on $\partial M$, we obtain an embedding

$$i : {}^vT^*\partial M \to T^*\partial M,$$

and so the following diagram can be defined:

$$\begin{array}{ccc} K_c(T^*\partial M \times \mathbb{R}) & \xrightarrow{i^*} & K_c({}^vT^*\partial M \times \mathbb{R}) \\ {\scriptstyle \mathrm{ind}}\downarrow & & \downarrow{\scriptstyle \mathrm{ind}} \\ K_c(T^*X \times \mathbb{R}) & \xrightarrow[j^*]{} & K_c(X \times \mathbb{R}). \end{array}$$

The horizontal maps are induced by $i$ and the embedding $j : X \to T^*X$ of the zero section. The vertical index maps are defined by taking the index of parameter-elliptic families on $\Omega$ and parametrized by $T^*X \times \mathbb{R}$ and $X \times \mathbb{R}$, respectively. Atiyah, Patodi and Singer (1976) proved that the restriction $j^*$ is zero. (In fact, $j^* = 0$ is equivalent to the bijectivity of $\alpha$ in (6.35).) Since the diagram commutes, it follows that the index is zero: for $[\sigma(a_\wedge)] \in K_c(T^*\partial M \times \mathbb{R})$, we have

$$\mathrm{ind}\, \sigma_c(a_\wedge) = \mathrm{ind}\, i^*[\sigma(a_\wedge)] = j^*\, \mathrm{ind}[\sigma(a_\wedge)] = 0 \in K_c(X \times \mathbb{R}).$$

The proof of the proposition is complete. □

## 6.4. Examples. Obstructions for Geometric Operators

In this section, we analyze the vanishing of the obstruction to the existence of elliptic edge problems for the classical operators of index theory, which are most important in applications.

### 6.4.1. The Euler operator

Let $\mathcal{M}$ be an oriented even-dimensional Riemannian manifold with edge $X$. Let $M$ be the stretched manifold of $\mathcal{M}$ equipped with the edge-degenerate metric

$$\pi^*(dx^2) + dr^2 + r^2 d\omega^2(x)$$

near the boundary, where $dx^2$ is a smooth metric on the edge $X$ and $d\omega^2(x)$ is a smooth family of metrics on the fibers of $\pi$. One can show that the Euler operator

$$d + d^* : \Lambda^{ev}(M^\circ) \longrightarrow \Lambda^{odd}(M^\circ)$$

acting from the space of even forms to the space of odd forms is an edge-degenerate operator. On the other hand, even the Atiyah–Bott obstruction

$$[\sigma(d + d^*)|_{\partial T^*\mathcal{M}}] \in K^1_c(T^*\partial M)$$

is zero, since the Dirichlet problem for the Euler operator on a manifold with boundary is Fredholm.

EXERCISE 6.33. Prove the vanishing of the Atiyah–Bott obstruction for the Euler operator independently by an explicit computation (see Section 5.1.4).

The obstruction (6.38) for the Euler operator is all the more trivial.

### 6.4.2. The Hirzebruch (signature) operator

In addition to the assumptions of the previous example, we require that $X$ be oriented. The bundle $\pi$ must also be oriented, i.e., each fiber should be equipped with an orientation continuously depending on the point of the base. Moreover, the orientations must be compatible in the sense that the orientation of the boundary must be consistent with the orientations of the base and the fiber. In this case, the signature operator

$$\mathcal{H}_M = d + d^* : \Lambda^+(M^\circ) \longrightarrow \Lambda^-(M^\circ),$$

acting from self-dual to antiself-dual forms is also edge degenerate.[6] The Atiyah–Bott obstruction for the signature operator

$$[\sigma(\mathcal{H}_M)|_{\partial T^*\mathcal{M}}] \in K^1_c(T^*\partial M)$$

is *never* zero (see Section 5.1.4). However, the obstruction (6.38) sometimes vanishes. Let us study this possibility.

[6] To show this, one has first to continue the bundles $\Lambda^\pm(M^\circ)$ to vector bundles on $M$.

***The product structure near the boundary.*** Since the obstruction is independent of the metric chosen, we consider the product metric on the cylinder $[0,1) \times \partial M$, where the nondegenerate metric on $\partial M$ is given by

$$\pi^*(dx^2) + d\omega^2(x). \tag{6.39}$$

It is well known (see Dai 1991) that the signature operator on the fibered product $[0,1) \times \partial M$ inherits the product structure. We shall explain the product structure of the signature operator at the level of principal symbols. (This will be important for the computation of the obstruction (6.38).) More precisely, the cotangent bundle at the boundary has the decomposition

$$\partial T^*M \simeq \pi^*T^*X \oplus ({}^vT^*\partial M \oplus \mathbb{R}) \tag{6.40}$$

into components horizontal, vertical, and normal to the boundary.

***Even-dimensional edges.*** To proceed further, we shall assume for definiteness that the edge is even-dimensional. Hence, there is a signature operator $\mathcal{H}_X$ on the edge and a family of signature operators $\mathcal{H}_{\Omega\times[0,1)}$ on the fibers of $\partial M \times [0,1) \to X$. It turns out that at the symbolic level the total signature operator is isomorphic to the external product (see Definition 5.40) of the signature operators on the factors:

$$\sigma(\mathcal{H}_M)|_{T^*(\partial M\times[0,1))} \simeq \sigma(\mathcal{H}_X)\#\sigma(\mathcal{H}_{\Omega\times[0,1)}). \tag{6.41}$$

EXERCISE 6.34. Prove (6.41).

***Computation of the obstruction.*** The external product is important in that it gives an explicit realization of the standard product

$$K_c(G) \times K_c(H) \longrightarrow K_c(G \times H)$$

in $K$-theory (Atiyah 1989, Luke and Mishchenko 1998). Namely, the product of $[a] \in K_c(G)$ by $[b] \in K_c(H)$ is equal to the equivalence class of the external product of the symbols:

$$[a][b] = [a\#b].$$

By taking $a = \sigma(\mathcal{H}_X)$ and $b = \sigma(\mathcal{H}_{\Omega\times[0,1)})$, we obtain

$$[\sigma(\mathcal{H}_M)|_{r=0}] = [\sigma(\mathcal{H}_X)][\sigma(\mathcal{H}_{\Omega\times[0,1)})|_{r=0}]$$

Moreover, the direct image map

$$\pi_! : K_c(T^*\partial M \times \mathbb{R}) \to K_c(T^*X \times \mathbb{R})$$

is multiplicative:

$$\pi_!\big([\sigma(\mathcal{H}_X)][\sigma(\mathcal{H}_{\Omega\times[0,1)})|_{r=0}]\big) = [\sigma(\mathcal{H}_X)]\pi'_![\sigma(\mathcal{H}_{\Omega\times[0,1)})|_{r=0}],$$

where

$$\pi'_! : K_c({}^vT^*\partial M \times \mathbb{R}) \to K_c(X \times \mathbb{R})$$

is the direct image map induced by the family of projections

$$\pi' = \{\pi_x : \Omega_X \to \{x\}\}_{x \in X}.$$

(Analytically, it corresponds to taking the index of the family of elliptic operators on the fibers.) Finally, the last term can be rewritten as the index $\operatorname{ind} \mathcal{H}_\Omega \in K^1(X)$ of the odd signature operators $\mathcal{H}_\Omega$ (see Section 5.1.4) on the fibers, and

$$\pi_![\sigma(\mathcal{H}_M)|_{\partial M}] = [\sigma(\mathcal{H}_X)]\pi'_![\sigma(\mathcal{H}_\Omega)]. \tag{6.42}$$

PROPOSITION 6.35. *For the signature operator, the obstruction* (6.38) *vanishes if* $\dim X$ *is even.*

*Proof.* Let us apply (6.42). The vanishing of the obstruction essentially follows from Hodge theory: the kernel of the self-adjoint signature operator $\mathcal{H}_\Omega$ on an odd-dimensional manifold has a constant dimension (since it coincides with the space of harmonic forms), so that the index of this family of elliptic self-adjoint operators is trivial as an element of the group $K^1(X)$ corresponding to the parameter space (see Appendix A or Dai and Zhang 1998). □

***Odd-dimensional edges.*** A similar computation can be carried out in this case. For odd-dimensional edges, we can consider the different product

$$[\sigma(\mathcal{H}_{X\times[0,1)})|_{r=0}][\sigma(\mathcal{H}_\Omega)],$$
$$[\sigma(\mathcal{H}_{X\times[0,1)})|_{r=0}] \in K_c^{\dim X+1}(T^*X), \quad [\sigma(\mathcal{H}_\Omega)] \in K_c^{\dim \Omega}({}^vT^*\partial M)$$

of the signature operators $\mathcal{H}_{X\times[0,1)}$ on the cylinder $X \times [0,1]$ by the family of signature operators $\mathcal{H}_\Omega$ on the fibers $\Omega_x$.

The free $K^*(X)\otimes\mathbb{Q}$-module $K^{\dim X+*}(T^*X)\otimes\mathbb{Q}$ is generated by the signature operator $\mathcal{H}_X$ (see Palais 1965). This leads to the following proposition.

PROPOSITION 6.36. *For the signature operator on a manifold with an odd-dimensional edge, the obstruction* (6.38) *rationally vanishes if and only if the index*

$$\operatorname{ind} \mathcal{H}_\Omega \in K^{\dim \Omega}(X) \otimes \mathbb{Q} \tag{6.43}$$

*of the family of signature operators on the fibers is zero. This condition is satisfied, for example, under the additional assumption that the cohomology of the fiber in dimension $k$ is trivial. (We assume that the dimension of the fiber is $2k$.)*

*Proof.* It suffices to apply (6.42). Indeed, by the Atiyah–Singer index theorem for families, $\pi'_![\sigma]$ is equal to the index of the corresponding operator family

$$\pi'_![\sigma(\mathcal{H}_\Omega)] = \operatorname{ind} \mathcal{H}_\Omega.$$

Let us prove the last part of the statement dealing with the vanishing of the index for families of signature operators.

Connes, Sullivan and Teleman (1994) showed that the signature operator, modulo invertible operators, is equivalent to the elliptic operator

$$dd^* - d^*d : \Lambda^+(\Omega) \cap \Lambda^k(\Omega) \longrightarrow \Lambda^-(\Omega) \cap \Lambda^k(\Omega).$$

The kernel and cokernel of the latter operator are trivial as subspaces of the space of harmonic forms of degree $k$, since by the assumption the fiber has no dimension $k$ cohomology. Thus we arrive at the desired relation $\operatorname{ind} \mathcal{H}_\Omega = 0$. □

### 6.4.3. The Dirac operator

In addition to the assumptions of the previous example, let as assume that $M$ and $X$ are equipped with *spin* structures (Lawson and Michelsohn 1989). We also assume that the fibers of $\pi$ are equipped with a *spin* structure. The induced *spin* structure on the boundary must coincide with the structure induced by the spin structures of the base and the fibers. In this case, the Dirac operator

$$D : \mathbb{S}^+(M^\circ) \longrightarrow \mathbb{S}^-(M^\circ)$$

on $M$ (e.g. Lawson and Michelsohn 1989) acting from even spinors to odd spinors is edge degenerate, and its difference element in $K$-theory is decomposed for the metric (6.39) as follows:

$$[\sigma(D_M)|_{\partial M}] = [\sigma(D_X)][\sigma(D_{\Omega\times[0,1)})|_{r=0}],$$

where

$$[\sigma(D_X)] \in K_c^{\dim X}(T^*X), \quad [\sigma(D_{\Omega\times[0,1)})|_{r=0}] \in K_c^{\dim \Omega+1}({}^vT^*\partial M)$$

are the difference elements of the Dirac operator on the base and the family of Dirac operators on the fibers of $\pi$. Therefore, the obstruction is also the product

$$\pi_![\sigma(D_M)|_{\partial M}] = [\sigma(D_X)]\pi'_![\sigma(D_{\Omega\times[0,1)})|_{r=0}].$$

Since $D_X$ is a generator of the free $K^*(X)$-module $K_c^{\dim X+*}(T^*X)$, we arrive at the following proposition.

PROPOSITION 6.37. *For the Dirac operator, the obstruction* (6.38) *is trivial if and only if the index*

$$\operatorname{ind} D_\Omega \in K^{\dim \Omega}(X)$$

*of the Dirac operators on the fibers is zero. For example, this condition is satisfied if the fibers of $\pi$ are equipped with a metric of positive scalar curvature continuously depending on the fiber.*

*Proof.* The latter assertion follows from the fact that the Dirac operator on a manifold of positive scalar curvature metric is invertible (e.g., Lawson and Michelsohn 1989). Therefore, the index is trivial independent of the dimension. □

### 6.4.4. Operators on a trivial bundle

Finally, consider the Cartesian product

$$\pi : \partial M = X \times \Omega \longrightarrow X.$$

The Künneth formula in $K$-theory gives

$$K_c^1(T^*\partial M) \simeq \left[K_c^1(T^*X) \otimes K_c(T^*\Omega)\right] \oplus \left[K_c(T^*X) \otimes K_c^1(T^*\Omega)\right]$$

modulo torsion. Let us represent $[\sigma(D)|_{\partial M}]$ as a sum $ab + cd$ in accordance with the above decomposition. Then the obstruction is

$$\pi_![\sigma(D)|_{\partial M}] = a\pi_!'(b) + c\pi_!'(d) = (\operatorname{ind} b)a,$$

modulo torsion, where $\operatorname{ind} b \in \mathbb{Z}$ is the index of an elliptic operator with difference element $b \in K_c(T^*\Omega)$. Thus the obstruction depends only on the index contribution of the even "vertical" component and the odd "horizontal" component $a$.

# Part IV

# Applications and Related Topics

# Chapter 7

# Fourier Integral Operators on Singular Manifolds

This chapter deals with Fourier integral operators on the simplest class of manifolds with singularities, namely, on manifolds with conical singular points. We give the definition of these operators, study their basic properties, and establish an index theorem resembling that for elliptic (pseudo)differential operators and representing the index of a Fourier integral operator satisfying certain symmetry conditions as the sum of contributions of its interior symbol and conormal symbol. Further, we present an application of this theorem to the index problem for quantized canonical transformations in the singular manifold setting. An example of a Fourier integral operator satisfying the assumptions of the theorem concludes the chapter.

## 7.1. Homogeneous Canonical (Contact) Transformations

Let $\mathcal{M}$ be a compact manifold with conical singularities. For simplicity, we assume that $\mathcal{M}$ has only one conical point $\alpha$ and denote the base of the corresponding model cone by $\Omega$. We shall study homogeneous canonical transformations of the compressed cotangent bundle $\widetilde{T}^*\mathcal{M}$ (Melrose 1981).[1]

◀ Recall that $\widetilde{T}^*\mathcal{M}$ is a vector bundle over the stretched manifold $M$ corresponding to $\mathcal{M}$ and that it is equipped with the symplectic form $\omega^2$ defined over the interior $M^\circ$ of $M$ and given in the conical coordinates near the boundary by the formula

$$\omega^2 = -\frac{dp \wedge dr}{r} + \omega^2_{T^*\Omega}, \tag{7.1}$$

where $r$ is the radial variable, $p$ is the conormal variable, and $\omega^2_{T^*\Omega}$ is the standard symplectic form on the cotangent bundle $T^*\Omega$. The form (7.1) has a singularity on the boundary $\partial\widetilde{T}^*\mathcal{M}$. ▶

### 7.1.1. Definition of canonical transformations

Let

$$\widetilde{T}^*_0\mathcal{M} = \widetilde{T}^*\mathcal{M} \setminus \{0\}$$

[1]A theory related to homogeneous canonical transformations of the *stretched* cotangent bundle can also be developed and would in fact lead to similar results.

be the compressed cotangent bundle of $\mathcal{M}$ with the zero section deleted. It is equipped with the natural action of the group $\mathbb{R}_+$ of positive numbers (multiplication in the fibers).

DEFINITION 7.1. A *homogeneous canonical transformation* of the space $\widetilde{T_0^*}\mathcal{M}$ is a diffeomorphism

$$g : \widetilde{T_0^*}\mathcal{M} \longrightarrow \widetilde{T_0^*}\mathcal{M}$$

of manifolds with boundary such that

1) $g$ commutes with the action of $\mathbb{R}_+$

2) $g$ preserves the form $\omega^2$:

$$g^*\omega^2 = \omega^2 \quad \text{in } (\widetilde{T_0^*}\mathcal{M})^\circ$$

Thus the only difference with the classical case is that the transformation is defined (and smooth) on a manifold with boundary and preserves a symplectic form defined only in the interior; on the boundary, the symplectic form has a singularity.

*Remark* 7.2. Owing to homogeneity, the transformation $g$ descends to a diffeomorphism

$$\widetilde{g} : \widetilde{T_0^*}\mathcal{M}/\mathbb{R}_+ \longrightarrow \widetilde{T_0^*}\mathcal{M}/\mathbb{R}_+$$

of the quotient space $\widetilde{S}^*\mathcal{M} = \widetilde{T_0^*}\mathcal{M}/\mathbb{R}_+$, which is known as the *compressed cosphere bundle* of $\mathcal{M}$ and whose interior bears a well-defined contact structure associated with the symplectic structure on $(\widetilde{T_0^*}\mathcal{M})^\circ$ in the standard way (e.g., Arnold 1989). The mapping $\widetilde{g}$ preserves the former, i.e., is a *contact transformation* of $\widetilde{S}^*\mathcal{M}$. One could also give all statements in terms of contact transformations, but we prefer the language of homogeneous canonical transformations.

### 7.1.2. Canonical coordinates and generating functions

To define Fourier integral operators associated with a given homogeneous canonical transformation $g$, we need the representations of $g$ in special coordinates, known as *canonical coordinates*, via *generating functions*.

***Smooth manifolds.*** Recall that a canonical transformation

$$g : T_0^*X \longrightarrow T_0^*X$$

of the cotangent bundle of a smooth $n$-dimensional manifold $X$ always admits the following coordinate representation (Arnold 1989). Let $(y, q)$ be canonical coordinates on $T^*X$ in a neighborhood of some point $m = (y_0, q_0)$, and let $(x, p)$ be canonical coordinates on $T^*X$ in a neighborhood of the point $g(m) = (x_0, p_0)$. Then there exists a subset $I \subset \{1, \dots, n\}$ and a smooth real-valued function $S_I(x_I, p_{\bar{I}}, q)$ defined in a neighborhood of the point $(x_{0I}, p_{0\bar{I}}, q_0)$, where $x_I = \{x_j\}_{j\in I}$ and $p_{\bar{I}} = \{p_j\}_{j\notin I}$, such that

1) The Hessian

$$\operatorname{Hess} S_I = \det \frac{\partial^2 S_I}{\partial(x_I, p_{\bar{I}})\partial q}$$

of the function $S_I$ at the point $(x_{0I}, p_{0\bar{I}}, q_0)$ is nonzero.

2) The transformation $g$ is implicitly defined in a neighborhood of the point $m$ by the formulas

$$p_I = \frac{\partial S_I}{\partial x_I}, \quad x_{\bar{I}} = -\frac{\partial S_I}{\partial p_{\bar{I}}}, \quad y = \frac{\partial S_I}{\partial q}. \tag{7.2}$$

(Note that the implicit function theorem can be applied to (7.2) in view of the first condition.)

The functions $(x_I, p_{\bar{I}}, q)$ form a local coordinate system on the graph

$$\operatorname{graph} g \subset T_0^* X \times T_0^* X$$

of the transformation $g$ in a neighborhood of $(x_{0I}, p_{0\bar{I}}, q_0)$. They are called *canonical local coordinates*. One can readily see that if $X$ is compact and $g$ is homogeneous, then $T_0^* X$ can be covered by finitely many $\mathbb{R}_+$-invariant canonical charts (i.e., charts with canonical local coordinates).

The function $S_I(x_I, p_{\bar{I}}, q)$ is called a *generating function* of $g$ in the coordinates $(x_I, p_{\bar{I}}, q)$. In the $\mathbb{R}_+$-homogeneous case, the function $S_I$ can be chosen to be homogeneous, and then it is unique.

***Manifolds with conical singularities.*** Our aim is to obtain a similar representation of homogeneous canonical transformations in the case of manifolds $\mathcal{M}$ with conical singularities. In the interior $(\widetilde{T_0^*\mathcal{M}})^\circ$, the situation does not differ in any way from the classical case, and the above-mentioned results apply. Thus the problem is to obtain such a representation in a neighborhood of the conical point.

First, we reduce the symplectic form (7.1) to the standard expression. To this end, we use the cylindrical representation (see Definition 1.30) near the conical point. Recall that this means that we make the change of variables $r = e^{-t}$ in a collar neighborhood of the boundary in the interior $M^\circ$ of the stretched manifold $M$ corresponding to $\mathcal{M}$, so that this collar neighborhood is represented by the semi-infinite cylinder $(0, \infty) \times \Omega$ (see Fig. 7.1). Then the symplectic form (7.1) in the new coordinates becomes

$$\omega^2 = dp \wedge dt + \omega^2_{T^*\Omega}. \tag{7.3}$$

The canonical transformation near the conical point is represented by a mapping (which we denote by the same letter)

$$g : T_0^*[(0, \infty) \times \Omega] \longrightarrow T_0^*[(0, \infty) \times \Omega], \tag{7.4}$$

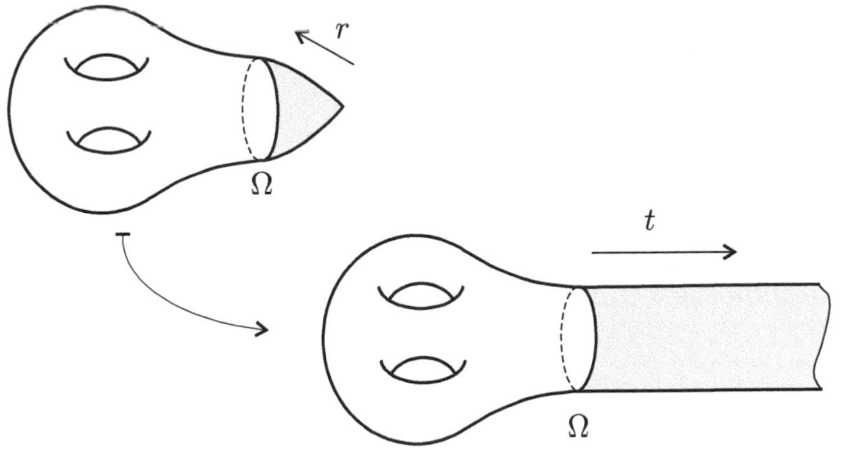


Figure 7.1. Passage to the cylindrical representation.

defined for sufficiently large $t$. We shall use coordinates $(t, \omega)$ and $(\tau, \psi)$ on the source and target spaces, respectively, in Eq. (7.4). Here $\omega = (\omega_1, \dots, \omega_{n-1})$, $n = \dim M$, and similarly for $\psi$. The dual covariables in the fibers of the cotangent bundle will be denoted by $(p, q)$ and $(\xi, \eta)$, respectively. Thus in local coordinates

$$g : (t, \omega, p, q) \mapsto (\tau, \psi, \xi, \eta).$$

Since the mapping (7.4) represents the reduction to $(\widetilde{T_0^*}\mathcal{M})^\circ$ of a smooth mapping of $\widetilde{T_0^*}\mathcal{M}$, it follows that it stabilizes as $t \to \infty$ at an exponential rate. Therefore, it is no surprise that the following assertion is true.

PROPOSITION 7.3. *The graph of $g$ can be covered by finitely many $\mathbb{R}_+$-invariant canonical charts with coordinates $(\tau, \psi_I, \eta_{\bar{I}}, q)$, $I \subset \{1, \dots, n-1\}$, in which the transformation is determined by generating functions of the form*

$$S_I(\tau, \psi_I, \eta_{\bar{I}}, p, q) = p\tau + S_{1I}(e^{-\tau}, \psi_I, \eta_{\bar{I}}, p, q). \tag{7.5}$$

*The functions $S_{1I}(r, \psi_I, \eta_{\bar{I}}, p, q)$ are smooth up to $r = 0$ and $\mathbb{R}_+$-homogeneous:*

$$S_{1I}(e^{-\tau}, \psi_I, \lambda\eta_{\bar{I}}, \lambda p, \lambda q) = \lambda S_{1I}(e^{-\tau}, \psi_I, \eta_{\bar{I}}, p, q), \quad \lambda > 0.$$

*The equations determining the transformation via the generating functions have the usual form*

$$t = \frac{\partial S_I}{\partial p}, \quad \xi = \frac{\partial S_I}{\partial \tau}, \quad \omega = \frac{\partial S_I}{\partial q}, \quad \eta_I = \frac{\partial S_I}{\partial \psi_I}, \quad \psi_{\bar{I}} = -\frac{\partial S_I}{\partial \eta_{\bar{I}}}.$$

◀ Let us explain why this proposition is true (see Nazaikinskii, Schulze and Sternin 2001). First, since the symplectic form is given by the standard expression (7.3), the same reasoning as in the smooth case shows the existence of a canonical chart and an $\mathbb{R}_+$-homogeneous generating function in a neighborhood of each point of graph $g$. Next, it follows from the smoothness of the transformation up to the boundary that the $r$-component of $g$ has the form

$$r' = rF(r, \omega, p, q)$$

with $F > 0$ for small $r \geq 0$, or, in the cylindrical coordinates,

$$\tau = t - \ln F(e^{-t}, \omega, p, q).$$

The $t$-derivatives of the second term decay exponentially as $t \to \infty$, and hence $\partial\tau/\partial t \neq 0$ for large $\tau$. Hence it follows that for large $t$ one can always take canonical coordinates including the variable $\tau$ rather than the dual momentum $\psi$. From this, one can also conclude that the generating functions have the form (7.5). Finally, the fact that finitely many canonical charts suffice follows from the stabilization of the transformation as $t \to \infty$. ▶

*Remark* 7.4. The generating functions do not stabilize as $t \to \infty$. (This could not be expected anyway, since even the generating function $S = p\tau + \psi q$ of the identity transformation does not stabilize.) However, the nonstabilizing term $p\tau$ has the same form for any canonical transformation.

## 7.2. Definition of Fourier Integral Operators

Now we are in a position to define Fourier integral operators. We use the standard method of defining Fourier integral operators via the Maslov canonical operator (Maslov 1972) as described in (Mishchenko, Shatalov and Sternin 1990, Nazaikinskii, Oshmyan, Sternin and Shatalov 1981). Let us briefly describe how this scheme applies indicating where novelties related to the presence of singularities (or, in the cylindrical representation, the noncompactness of the manifold) occur.

To a homogeneous canonical transformation

$$g : \widetilde{T}_0^*\mathcal{M} \to \widetilde{T}_0^*\mathcal{M},$$

we assign a class of operators acting in weighted Sobolev spaces $H^{s,\gamma}(\mathcal{M})$. First, we consider the case $\gamma = 0$. (Then in cylindrical coordinates we deal with the usual Sobolev spaces without an exponential weight depending on $t$). The passage to nonzero $\gamma$ will be carried out at the end of the next section.

***The Lagrangian manifold.*** The first ingredient in the construction of the canonical operator is a Lagrangian manifold. The graph

$$L_g = \{(\beta, \overline{\beta}) \mid \overline{\beta} = g(\beta)\} \subset \widetilde{T_0^*}\mathcal{M} \times \widetilde{T_0^*}\mathcal{M} \tag{7.6}$$

is a homogeneous Lagrangian manifold in $T_0^*\mathcal{M} \times T_0^*\mathcal{M}$ with respect to the symplectic structure

$$\widetilde{\omega}^2 = \pi_2^*\omega^2 - \pi_1^*\omega^2,$$

where

$$\pi_j : \widetilde{T_0^*}\mathcal{M} \times \widetilde{T_0^*}\mathcal{M} \longrightarrow \widetilde{T_0^*}\mathcal{M}, \quad j = 1, 2,$$

is the natural projection onto the $j$th factor and $\omega^2$ is the symplectic structure (7.1) on $(\widetilde{T_0^*}\mathcal{M})^\circ$. As explained in the preceding subsection, this manifold is equipped with a finite cover by canonical charts, where the corresponding canonical coordinates and generating functions are defined. Following the established tradition, we denote the charts by $\mathcal{U}_I$ (even though actually the same indexing set may be shared by more than one chart).

***The measure.*** The next ingredient in the construction of the canonical operator is a measure. We take some smooth measure $\mu$ on $L_g$ which satisfies the following conditions:

1) $\mu$ is *quantized* (e.g., Mishchenko, Shatalov and Sternin 1990)

2) $\mu$ is $n$th-order homogeneous with respect to the action of $\mathbb{R}_+$, that is,

$$\lambda^*\mu = \lambda^n \mu,$$

where $\lambda^*$ is the mapping induced on differential forms on $L_g$ by the mapping $\lambda : L_g \longrightarrow L_g$ (multiplication by $\lambda$ in the momentum variables)

Such a measure $\mu$ always exists (Mishchenko, Shatalov and Sternin 1990). To simplify some formulas, we shall assume that $\mu = \pi_1^*\mu_0$, where

$$\mu_0 = \frac{1}{n!}\left(\omega^2\right)^{\wedge n}$$

is the standard volume form on $(\widetilde{T_0^*}\mathcal{M})^\circ$. If this measure is not quantized, then one should take some other, quantized measure $\mu$ (such measures always exist), and then some of the subsequent formulas become more complicated owing to occurrences of the ratio $\mu/\pi_1^*\mu_0$.

***The canonical operator.*** Once we have the Lagrangian manifold $L_g$ equipped with the quantized measure $\mu$, we can consider Maslov's canonical operator (Maslov 1972) associated with the pair $(L_g, \mu)$ in the homogeneous setting (Nazaikinskii, Oshmyan, Sternin and Shatalov 1981, Mishchenko, Shatalov and Sternin 1990). The canonical operator is the operator

$$\mathcal{K}_g : \mathcal{O}(L_g) \to \mathcal{D}'(M^\circ \times M^\circ)$$

acting from the space

$$\mathcal{O}(L_g) = \bigcup_m \mathcal{O}^m(L_g)$$

of asymptotically $\mathbb{R}_+$-homogeneous functions on $L_g$ (here $\mathcal{O}^m(L_g)$ is the space of asymptotically homogeneous functions of order $\leq m$) into the space $\mathcal{D}'(M^\circ \times M^\circ)$ of distributions on $M^\circ \times M^\circ$ and given by the formula

$$\mathcal{K}_g a = \sum_I \chi_I \mathcal{K}_I(e_I a), \tag{7.7}$$

$$\begin{aligned} \mathcal{K}_I\varphi(t, \tau, \omega, \psi) = \left(\frac{i}{2\pi}\right)^{|\bar{I}|/2} \iiint e^{i[\eta_{\bar{I}}\psi_{\bar{I}} + \omega q + pt + S_{1I}(\tau, \psi_I, \eta_{\bar{I}}, p, q)]} \\ \times \left[\sqrt{\mu_I}\varphi\right](\tau, \psi_I, \eta_{\bar{I}}, p, q)\, d\eta_{\bar{I}}\, dq\, dp, \end{aligned} \tag{7.8}$$

where the integral is treated as an oscillatory integral in the standard sense. Here $\{e_I\}$ is an $\mathbb{R}_+$-homogeneous partition of unity on $L_g$ subordinate to the cover $\{\mathcal{U}_I\}$, the $\chi_I$ are appropriate cutoff functions on $M^\circ \times M^\circ$ equal to unity in neighborhoods of the projections of the respective charts $\mathcal{U}_I$ onto $M^\circ \times M^\circ$, and $\mu_I$ is the density of the measure $\mu$ in the canonical coordinates $(\tau, \psi_I, \eta_{\bar{I}}, p, q)$ of the chart $\mathcal{U}_I$.

***Fourier integral operators.*** We define Fourier integral operators as operators with Schwartz kernels obtained by an application of the canonical operator to a function $a \in \mathcal{O}(\widetilde{T_0^*\mathcal{M}})$. We assume that $M^\circ$ is equipped with some volume form $d\,\mathrm{vol}$ that has the expression $dt \wedge d\omega$ in cylindrical coordinates in a neighborhood of the conical point, where $d\omega$ is a volume form independent of $t$ on $\Omega$; this permits us to treat Schwartz kernels as functions rather than densities.

DEFINITION 7.5. Let $a \in \mathcal{O}(\widetilde{T_0^*\mathcal{M}})$. The integral operator $T(g, a)$ on $M^\circ$ with Schwartz kernel

$$\mathcal{K}_g[\pi_1^* a](x, y), \quad x, y \in M^\circ, \tag{7.9}$$

is called the *Fourier integral operator with amplitude $a$ associated with the canonical transformation $g$.*

*Remark* 7.6. Note that the canonical operator $\mathcal{K}_g$ and hence the operator $T(g, a)$ depend on the choice of a quantized measure $\mu$. This is not important if we take the standard measure $\mu = \pi_1^*\mu_0$.

The canonical operator thus defined acts *a priori* from the space $C_0^\infty(M^\circ)$ of compactly supported smooth functions on $M^\circ$ into the distribution space $\mathcal{D}'(M^\circ)$. Actually, it has much better mapping properties, as we shall see in the following section.

## 7.3. Properties of Fourier Integral Operators

***Notation.*** By $\mathrm{Op}_\gamma^m$ we denote the space of operators $A$ continuous in the spaces

$$A : H^{s,\gamma}(\mathcal{M}) \to H^{s-m,\gamma}(\mathcal{M}), \quad s \in \mathbb{R}. \tag{7.10}$$

Next, by $I_\gamma^m$ we denote the set of operators $A$ that belong to $\mathrm{Op}_\gamma^{m-1}$ and are compact in the spaces (7.10).

***Boundedness in the spaces*** $H^{s,0}(\mathcal{M})$**.** Fourier integral operators defined in the preceding section act continuously in the scale of weighted Sobolev spaces on $\mathcal{M}$. (For now, we consider only the case of the zero weight exponent $\gamma$. The case of nonzero $\gamma$ will be treated later on in this section.)

THEOREM 7.7. *Let $a \in \mathcal{O}^m(\widetilde{T}_0^*\mathcal{M})$. Then the operator $T(g,a)$ belongs to* $\mathrm{Op}_0^m$, *i.e., is continuous in the spaces*

$$T(g,a) : H^{s,0}(\mathcal{M}) \to H^{s-m,0}(\mathcal{M}), \quad s \in \mathbb{R}. \tag{7.11}$$

This theorem is an immediate consequence of the following lemma with regard to properties of cone-degenerate pseudodifferential operators on $\mathcal{M}$.

LEMMA 7.8. *Let $a \in \mathcal{O}^m(\widetilde{T}_0^*\mathcal{M})$. Then the operator $T^*(g,a)T(g,a)$ is a $2m$th-order cone-degenerate pseudodifferential operator with interior symbol $|a|^2$ in the scale $H^{s,0}(\mathcal{M})$.*

Here $T^*(g,a)$ is the adjoint of $T(g,a)$ with respect to the inner product in $L^2(M^\circ, d\,\mathrm{vol})$. (The volume form near the conical point is $d\,\mathrm{vol} = dt\,d\omega$ in the cylindrical coordinates.)

*The proof* of Lemma 7.8 is rather lengthy. It can be found, e.g., in Nazaikinskii, Schulze and Sternin (2001). □

***The conormal symbol.*** The canonical transformation and amplitude alone do not define the Fourier integral operator uniquely modulo compact operators in the spaces (7.11). Just as in the theory of pseudodifferential operators, one needs to introduce the notion of a *conormal symbol.*

The conormal symbol $\sigma_c(T)(p)$ of the operator $T = T(g,a)$ is defined as follows. (Cf. the definition of conormal symbols of (pseudo)differential operators in Chapters 1 and 3.)

DEFINITION AND THEOREM 7.9. If $T = T(g, a)$ is a Fourier integral operator on a manifold $\mathcal{M}$ with conical singularities, then for each $\varphi \in C_0^\infty(M^\circ)$ supported in a small neighborhood of the boundary $\partial M$ of the stretched manifold there exists a strong limit

$$\sigma_c(T)\varphi = \lim_{\lambda\to\infty} (g_\lambda^*)^{-1} T g_\lambda^* \varphi, \tag{7.12}$$

where $[g_\lambda^*\varphi](r, \omega) = \varphi(r\lambda, \omega)$. In cylindrical coordinates, this limit is a translation-invariant operator on the infinite cylinder $(-\infty, \infty) \times \Omega$ and hence is equivalent via the Fourier transform to an operator-valued function

$$\sigma_c(T)(p) : H^s(\Omega) \longrightarrow H^{s-m}(\Omega), \quad s \in \mathbb{R}, \tag{7.13}$$

in Sobolev spaces on the base $\Omega$ of the cone. This function is called the *conormal symbol* of the Fourier integral operator $T$.

*Remark* 7.10. The conormal symbol corresponds to the homogeneous (with parameter $p$) family $g_0(p)$ of canonical transformations of $T_0^*\Omega$ obtained from $g$ by restriction to the boundary $\partial\widetilde{T}^*\mathcal{M}$.

PROPOSITION 7.11. *The operator $T(g, a)$ is uniquely determined modulo $I_0^m$ by the transformation $g$, the amplitude $a \in \mathcal{O}^m(\widetilde{T}_0^*\mathcal{M})$, and the conormal symbol $\sigma_c(T)(p)$.*

*Remark* 7.12. Here we do not discuss the compatibility condition between the amplitude and the conormal symbol of a Fourier integral operator as well as the construction of a Fourier integral operator with given amplitude and conormal symbol. Nevertheless, we write

$$T = T(g, a, c)$$

to denote a Fourier integral operator, associated with a transformation $g$, with amplitude $a$ and conormal symbol $c$. This operator is determined uniquely modulo $I_0^m$, $m = \operatorname{ord} a$ (if it exists at all).

***Calculus.*** Now we can describe the behavior of Fourier integral operators under products and taking adjoints.

THEOREM 7.13. (1) *The composition formula*

$$T(g, a, c)T(g_1, a_1, c_1) = T(gg_1, a * a_1, cc_1) \mod I_0^{\operatorname{ord} a + \operatorname{ord} a_1 - 1} \tag{7.14}$$

*is valid for Fourier integral operators, where the amplitude $a * a_1$ is expressed by the formula*

$$a * a_1 = g_1^*(a)a_1. \tag{7.15}$$

(2) *The adjoint of a Fourier integral operator is given by the formula*

$$T^*(g, a, c) = T(g^{-1}, (g^{-1})^*\overline{a}, c^*) \mod I_0^{\operatorname{ord} a - 1}$$

*where the bar stands for complex conjugation.*

*The proof* of this theorem is rather complicated. We refer the interested reader to the paper by Nazaikinskii, Schulze and Sternin (2001). □

***Definition and properties of Fourier integral operators for nonzero weight exponents.*** Let us now consider the case $\gamma \neq 0$. To define the operator $T(g,a)$ in the scale $H^{s,\gamma}(\mathcal{M})$ with an arbitrary $\gamma$, we should require all functions occurring in the definition of the canonical operator in a neighborhood of the singular point to be analytic in $p$ in some neighborhood of the weight line

$$\mathcal{L}_\gamma = \{\operatorname{Im} p = \gamma\}.$$

To ensure this, we have to modify the definition of the canonical operator. Namely, all homogeneous functions occurring in that definition (the generating function, the amplitude, the density of the measure, and the elements of the partition of unity) are replaced by functions asymptotically homogeneous in $(p, q, \eta_{\bar{I}})$ with the same asymptotics at infinity and with the following property: all new functions are smooth everywhere, including the point $(p, q, \eta_{\bar{I}}) = (0, 0, 0)$, and admit analytic continuation with respect to $p$ into some strip $|\operatorname{Im} p| < K$ with $K > |\gamma|$; moreover, the analytic continuation should still be asymptotically homogeneous in this strip.

This so-called *analytic smoothing* can be done in numerous ways. Note, however, that various modifications will produce operators $T(g,a)$ differing by noncompact terms. Indeed, the dependence of the conormal symbol $\sigma_c(T)(p)$ on $p$ in the complex domain is determined by the choice of the analytic smoothing.

The integral operators $T(g,a)$ thus obtained will be well defined in weighted Sobolev spaces with nonzero weight exponent $\gamma$. The following result is a straightforward generalization of Theorems 7.7 and 7.13 and can readily be derived from them (Nazaikinskii, Schulze and Sternin 2001).

THEOREM 7.14. *Let*

$$g : \widetilde{T_0^*}\mathcal{M} \longrightarrow \widetilde{T_0^*}\mathcal{M}$$

*be a homogeneous canonical transformation, let $a \in \mathcal{O}^m(T_0^*\mathcal{M})$, and let all elements in the construction of the canonical operator $\mathcal{K}_g$ be analytic with respect to $p$ in the strip $|\operatorname{Im} p| < K$ and satisfy the above-mentioned conditions. Then the Fourier integral operator $T = T(g,a)$ belongs to $\operatorname{Op}_\gamma^m$, that is, is continuous in the weighted Sobolev spaces*

$$T : H^{s,\gamma}(\mathcal{M}) \to H^{s-m,\gamma}(\mathcal{M}), \quad s \in \mathbb{R}, \tag{7.16}$$

*for $|\gamma| < K$.*

*Moreover, $T^*T$, where $T^*$ stands for the adjoint of $T$ with respect to the inner product in $L^{2,\gamma}(\mathcal{M}) \equiv H^{0,\gamma}(\mathcal{M})$, is a pseudodifferential operator of order $2m$ in the scale $\{H^{s,\gamma}(\mathcal{M})\}$ for $|\gamma| < K$. The interior symbol of $T^*T$ is equal to $|a|^2$.*

*Finally, assertions* (1) *and* (2) *of Theorem* 7.13 *remain valid for operators $T(g,a)$ acting in weighted Sobolev spaces with nonzero $\gamma$ if we replace $I_0^m$ by $I_\gamma^m$.*

*Remark* 7.15. The reason for using the inner product of $L^{2,\gamma}(M)$ in the definition of the adjoint operator is that for this definition the conormal symbol of $T^*T$ is an analytic function of $p = \operatorname{Re} p + i\gamma$.

***Ellipticity and the Fredholm property.*** Let us find conditions under which the operator (7.16) is Fredholm in the scale $\{H^{s,\gamma}(\mathcal{M})\}$ with given $\gamma$.

DEFINITION 7.16. The Fourier integral operator $T(g, a)$ is said to be *interior elliptic* if the amplitude $a$ is nonzero for large $|p| + |q|$ (or, in other words, the *symbol* of $T(g, a)$, which is the leading term of the amplitude, is nonzero on $\widetilde{T}_0^*\mathcal{M}$).

Let $T(g, a)$ be interior elliptic. By Theorem 7.14, the operator $T^*T$ is a pseudodifferential operator with interior symbol $|a|^2$; the conormal symbol of $T^*T$ is $\sigma_c(T)^*(p)\sigma_c(T)(p)$. A similar assertion is valid for $TT^*$. Now we can apply elliptic theory for pseudodifferential operators on manifolds with conical singularities (see Chapter 3). It follows that $\sigma_c(T)^*(p)\sigma_c(T)(p)$ and hence $\sigma_c(T)(p)$ are finite-meromorphically invertible in the strip $|\operatorname{Im} p| < K$ and have finitely many poles in each strip $|\operatorname{Im} p| < K - \varepsilon$, $\varepsilon > 0$. We conclude that the operator $T$ is Fredholm in the scale $\{H^{s,\gamma}(\mathcal{M})\}$ provided that the weight line $\operatorname{Im} p = \gamma$ does not contain poles of the family $\sigma_c(T)(p)$. Regularizers of $T$ can be taken in the form $ST^*$ or $T^*S_1$, where $S$ and $S_1$ are regularizers of $T^*T$ and $TT^*$, respectively.

Thus the following theorem holds.

THEOREM 7.17. *The operator $T = T(g, a)$ is Fredholm in the scale $\{H^{s,\gamma}(\mathcal{M})\}$ provided that the following conditions are satisfied*:

1) *$T(g, a)$ is interior elliptic*
2) *the conormal symbol $\sigma_c(T)(p)$ is invertible on the weight line $\{\operatorname{Im} p = \gamma\}$*

## 7.4. The Index of Elliptic Fourier Integral Operators

This section deals with Weinstein's (1997) index problem for Fourier integral operators in the singular manifold setting. We give the index formula due to Nazaikinskii, Schulze and Sternin (2001). This formula is similar to the formula (5.49), (5.50) for the index of elliptic (pseudo)differential operators on manifolds with conical singularities. Just as with $\psi$DO, we begin the analysis of the index problem by stating the *relative index formula* for Fourier integral operators. Later, in conjunction with symmetry conditions and surgery, this will give the desired index formula.

The exposition in this section is rather brief, and we refer the interested reader to the cited paper for details.

### 7.4.1. The relative index formula

Let $T = T(g, a)$ be an interior elliptic Fourier integral operator on $\mathcal{M}$ with conormal symbol $\sigma_c(T) = \sigma_c(T)(p)$, and let $\gamma_1 < \gamma_2$ be weight exponents such that the weight lines $\mathcal{L}_{\gamma_1}$ and $\mathcal{L}_{\gamma_2}$ do not contain poles of the family $\sigma_c(T)(p)^{-1}$. Then the operators

$$T_{\gamma_i} : H^{s,\gamma_i}(\mathcal{M}) \to H^{s-m,\gamma_i}(\mathcal{M})$$

induced by $T$ in the corresponding scales of weighted Sobolev spaces are elliptic and hence Fredholm. Under these conditions, the following theorem holds.

THEOREM 7.18 (the relative index formula).

$$\operatorname{ind} T_{\gamma_2} - \operatorname{ind} T_{\gamma_1} = - \sum_{\gamma_1 < \operatorname{Im} p_j < \gamma_2} m_j, \tag{7.17}$$

*where the* $m_j$ *are the multiplicities* (*see Appendix* A) *of the poles of the family* $\sigma_c(T)(p)^{-1}$.

*Remark* 7.19. Formula (7.17) exactly coincides in form with the corresponding formula for $\psi$DO. This is by no means accidental. We obtain the latter as a special case if $g$ is the identity transformation.

### 7.4.2. Symmetry conditions and the index formula

In this subsection, we apply the surgery technique to obtain an index formula for Fourier integral operators.

***Fourier integral operators as operators in collar spaces.*** First of all, we shall show that Fourier integral operators on a manifold $\mathcal{M}$ with conical singularities can always be viewed as elliptic operators in collar spaces of some special form.

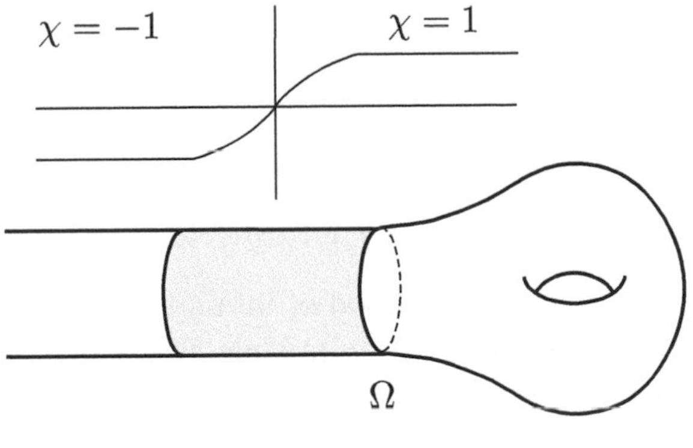


Figure 7.2. Manifold with a collar.

Namely, we treat $\mathcal{M}^\circ$ as a manifold with a cylindrical end and equip $H^{s,\gamma}(\mathcal{M})$ with the structure of collar space by the standard construction: the action of a function $\varphi \in C^\infty([-1,1])$ on an element $u(x) \in H^{s,\gamma}(\mathcal{M})$ is given by the multiplication by the function $\varphi(\chi(x))$:

$$[\varphi u](x) = \varphi(\chi(x))\, u(x),$$

where $\chi(x)$ is a smooth function growing from $-1$ to $1$ in the collar and constant outside the collar (see Fig. 7.2, where the collar is dashed).

Let $T = T(g, a)$ be an elliptic Fourier integral operator on $\mathcal{M}$. We claim that it can be included in a continuous family $T(\delta)$ of operators such that

(i) $\{T(\delta)\}$ is a proper operator (see Definition 4.5) in $H^{s,\gamma}(\mathcal{M})$ for each $\gamma$

(ii) $\{T_\gamma(\delta)\}$ is elliptic for all $\gamma$ such that $T_\gamma$ is Fredholm

Indeed, we can ensure condition (ii) by using only homotopies that preserve the conormal symbol:

$$\sigma_c(T(\delta)) \equiv \sigma_c(T),$$

whereas the amplitude ranges in the class of elliptic symbols.

As to condition (i), the possibility to ensure it can be derived from the following lemma.

LEMMA 7.20. *There is a homotopy*

$$g_t : \widetilde{T_0^*}\mathcal{M} \to \widetilde{T_0^*}\mathcal{M}$$

*in the class of homogeneous canonical transformations such that* $g_1 = g$ *and* $g_0$ *commutes with translations along the* $t$*-axis for sufficiently large* $t$ *on the cylindrical end. Moreover, this homotopy can be chosen in such a way that the conormal family of* $g_t$ *is independent of* $t$.

***Index formula.*** Let us introduce the simplest *symmetry condition*

$$\sigma_c(T)(p) = \sigma_c(T)(-p). \tag{7.18}$$

Then the conormal family

$$g_p : T^*\Omega \to T^*\Omega, \quad \forall p \in \mathbb{R},$$

of the canonical transformation $g$ satisfies the condition

$$g_p = g_{-p} \quad \forall p \in \mathbb{R}. \tag{7.19}$$

Now if we cut away a neighborhood of infinity in the cylindrical end of $T^*M^\circ$ and paste the resulting two manifolds with boundary with an overlap, we obtain the canonical transformation

$$2g : T_0^*(2M) \longrightarrow T_0^*(2M).$$

(See Fig. 7.3, where the pasting is shown; the collar is dashed as usual.) One can also prove that the symmetry condition permits the smooth pasting of the amplitudes.

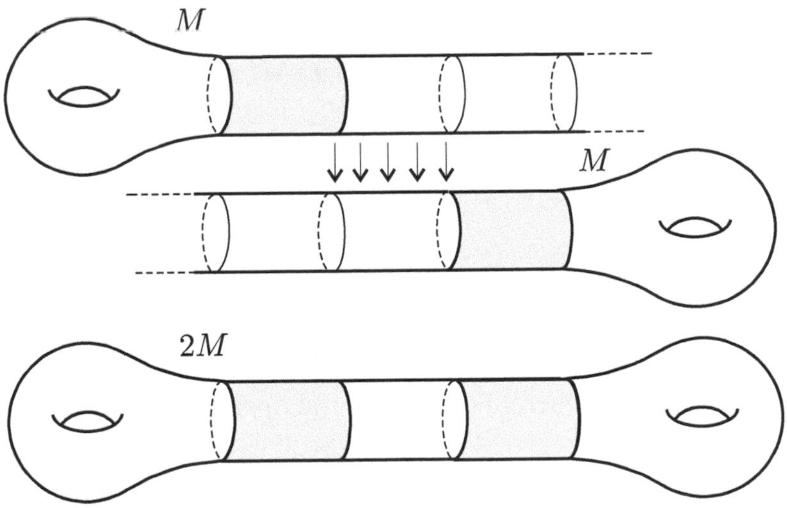


Figure 7.3. The pasting with overlap.

THEOREM 7.21. *Let $T$ be an interior elliptic Fourier integral operator on a manifold $\mathcal{M}$ with conical singularities satisfying the symmetry condition* (7.18). *Then*

$$\operatorname{ind} T_\gamma = \frac{1}{2}\Big(\operatorname{ind} 2T - \sum_{-\gamma<\operatorname{Im} p_j<\gamma} m_j\Big), \tag{7.20}$$

*for any $\gamma \geq 0$ such that the conormal symbol $\sigma_c(T)(p)$ is invertible on $\mathcal{L}_\gamma$. Here the $p_j$ are the poles of the family $\sigma_c(T)^{-1}(p)$, and the $m_j$ are their multiplicities. For $\gamma < 0$, one has a formula similar to* (7.20)*, with "+" instead of "−" on the sum of multiplicities.*

*Proof.* The proof coincides with that of the index theorem for pseudodifferential operators on manifolds with conical singularities. □

*Remark* 7.22. 1. The index of the Fourier integral operator on the closed manifold $2M$ can be expressed by a formula due to Leichtnam, Nest and Tsygan (2001).

2. The formula (7.20) remains valid if the operator $T$ acts between function spaces on different manifolds $\mathcal{M}_1$ and $\mathcal{M}_2$; note that the Leichtnam–Nest–Tsygan formula still applies to $\operatorname{ind} 2T$ in this case.

## 7.5. Application to Quantized Contact Transformations

***Quantized contact transformations and the index problem.*** There is a special case of Fourier integral operators known as *quantized contact transformations* (e.g., see Epstein and Melrose (1998), where they are defined for the smooth case via Szegö projections and the Guillemin transform, and references therein). In this case, the amplitude $a$ is equal to 1 and the measure on the Lagrangian manifold $L_g$ (recall that the measure is involved in the definition of the canonical operator) corresponds to some *complex polarization* on $T^*M$.

◀ Thus for different polarizations we obtain different quantized contact transformations $T(g,1)$. However, the index of $T(g,1)$ is independent of the choice of polarization (Epstein and Melrose 1998). ▶

Consider the problem of computing the index of quantized contact transformations. For smooth manifolds, this problem was solved by Epstein and Melrose (1998) in a special case prior to the solution by Leichtnam, Nest and Tsygan (2001) of the index problem for general Fourier integral operators on smooth closed manifolds.

We are interested in the same problem for manifolds with conical singularities. In this case, the polarization on $T^*\mathcal{M}$ near the conical point must be chosen so that the symbol will have an exponential rate of stabilization as $t \to \infty$. (Here $t$ is the cylindrical coordinate.) For example, a polarization independent of $t$ in the cylindrical coordinates near the conical point is always admissible.

We denote the quantized contact transformation corresponding to $g$ by $T(g)$. (We do not explicitly indicate the polarization.) In the case of manifolds with conical singularities, we keep in mind that there is additional ambiguity in the definition of $T(g)$ modulo compact operators, since the conormal symbol depends on the choice of analytic smoothing in the construction of the canonical operator.

***The index formula.*** Now we can present the result for quantized contact transformations. Consider a quantized contact transformation $T = T(g)$ with conormal symbol $\sigma_c(T)(p)$. Suppose that $\gamma \geq 0$. (For $\gamma < 0$, the statement is quite similar.)

THEOREM 7.23 (the index of a quantized contact transformation). *Suppose that the weight line $\mathcal{L}_\gamma$ does not contain any poles of $\sigma_c(T)^{-1}(p)$. If the symmetry condition* (7.18) *is valid, then*

$$\operatorname{ind}(T_\gamma) = \frac{1}{2}\, c\text{-}deg\,(2g) - \frac{1}{2} \sum_{-\gamma<\operatorname{Im} p_j<\gamma} \operatorname*{Res}_{p=p_j}\left(\sigma_c(T)^{-1}(p)\frac{\partial\sigma_c(T(p))}{\partial p}\right), \qquad (7.21)$$

*where* c*-deg is the contact degree* (*Epstein and Melrose* 1998) *of a contact transformation and the $p_j$ are the poles of $\sigma_c(T)^{-1}(p)$.*

◀ The index formula (7.21) can be derived as follows from the formula (7.20). To the first term, we apply the theorem due to Epstein and Melrose (1998) saying that the index of a quantized contact transformation on a closed manifold is equal to the contact degree of the transformation. In the second term, we use the expression of the multiplicities of poles in terms of residues; see Appendix A. ▶

## 7.6. Example

In conclusion, we give an example of a canonical transformation $g$ for which there exists a Fourier integral operator $T$ with conormal symbol satisfying the symmetry condition $\sigma_c(T)(p) = \sigma_c(T)(-p)$. More precisely, we describe the transformation $g$ near the conical point. (Only this is essential to the behavior of the conormal symbol.) Let the base $\Omega$ of the cone be the circle $\mathbb{S}^1$ with $\bmod 2\pi$-coordinate $\omega$. We must specify the canonical transformation for large $t$ on the infinite cylinder $C = (-\infty, \infty) \times \mathbb{S}^1$ with coordinates $(t, \omega)$. Let $p$ and $q$ be the coordinates dual to $t$ and $\omega$, respectively, and let

$$g : (t, \omega, p, q) \mapsto (t', \omega', p', q')$$

be the transformation determined by the generating function

$$S(p, q, t', \omega') = pt' + q\omega' + \sqrt{p^2 + q^2} \tag{7.22}$$

according to the formulas

$$\begin{aligned} t &= \frac{\partial S}{\partial p} = t' + \frac{p}{\sqrt{p^2+q^2}}, \quad & p' &= \frac{\partial S}{\partial t'} = p, \\ \omega &= \frac{\partial S}{\partial q} = \omega' + \frac{q}{\sqrt{p^2+q^2}}, \quad & q' &= \frac{\partial S}{\partial \omega'} = q. \end{aligned}$$

Expressing the primed variables via nonprimed, we see that the transformation is given by the formulas

$$t' = t - \frac{p}{\sqrt{p^2+q^2}}, \quad p' = p, \quad \omega' = \omega - \frac{q}{\sqrt{p^2+q^2}}, \quad q' = q. \tag{7.23}$$

The last two equations in (7.23) specify the conormal family

$$g_p : T^*\mathbb{S}^1 \to T^*\mathbb{S}^1$$

of $g$. One can readily see that the symmetry condition $g_p = g_{-p}$ holds. To construct some operator $T$ corresponding to $g$ in a local chart near the singular point, we must smooth the function (7.22). Let $\chi(\tau)$ be a smooth compactly supported function such that $\chi(0) = 1$ and $\chi(-\tau) = \chi(\tau)$, and let $\widetilde{\chi}(p)$ be its Fourier transform. (Note that $\widetilde{\chi}(p)$ is real-valued.) The convolution

$$\dot{S}(p, q) = \sqrt{p^2+q^2} * \widetilde{\chi}(p) \equiv \int_{-\infty}^{\infty} \widetilde{\chi}(p - \eta)\sqrt{\eta^2 + q^2}\, d\eta$$

has the same asymptotics for large $|p| + |q|$, $p \in \mathbb{R}$, as the function (7.22) but is analytic in the entire $p$-plane. Moreover,

$$\dot{S}(-p, q) = \dot{S}(p, q). \tag{7.24}$$

Now we define the Fourier integral operator $T$ on $(-\infty, \infty) \times \mathbb{S}^1$ by setting

$$T = \exp\left\{ i\dot{S}\left( -i\frac{\partial}{\partial t}, -i\frac{\partial}{\partial \omega} \right) \right\}.$$

The conormal symbol $\sigma_c(T)(p)$ of this operator has thc form

$$\sigma_c(T)(p) = \exp\left\{ i\dot{S}\left( p, -i\frac{\partial}{\partial \omega} \right) \right\}$$

and obviously satisfies the symmetry condition (7.18).

◀ More explicitly, if

$$\psi = \sum_{k=-\infty}^{\infty} e^{i\omega k}\psi_k,$$

then

$$\sigma_c(T)(p)\psi = \sum_{k=-\infty}^{\infty} e^{i[\dot{S}(p,k)+\omega k]}\psi_k;$$

in other words, $\sigma_c(T)(p)$ multiplies the Fourier coefficients of $\psi$ in the system $\{e^{i\omega k}\}$ by the factors $e^{i\dot{S}(p,k)}$. ▶

Obviously, $\sigma_c(T)(p)$ satisfies the symmetry condition (7.18). Note that the inverse $\sigma_c(T)(p)^{-1}$ has no poles, and so only the topological term survives in the index formula for a Fourier integral operator with this conormal symbol.

# Chapter 8

# Relative Elliptic Theory

Relative elliptic theory is an elliptic theory associated with a pair $(M, X)$, where $M$ is a smooth compact manifold and $i : X \hookrightarrow M$ is a smooth compact submanifold. In this chapter, we shall describe a natural class of "operator morphisms" associated with the pair $(M, X)$. This class generalizes the class of pseudodifferential operators (which arises in this context as a special case with $X = \varnothing$).

These morphisms have a wide range of applications, the most important of which is Sobolev problems, which will be discussed in some detail. We also touch upon the topological aspects of the theory. We present only the main results. Further details and proofs can be found in the survey by Nazaikinskii and Sternin (2004) and references therein.

## 8.1. Analytic Aspects of Relative Elliptic Theory

Relative elliptic theory deals with *operator morphisms*. Let us describe what such morphisms are.

Let $X$ and $M$, where $X$ is embedded in $M$, be smooth compact manifolds. We consider matrix operators $A$ of the form

$$A = \begin{pmatrix} D_{MM} & D_{MX} \\ D_{XM} & D_{XX} \end{pmatrix} : \begin{pmatrix} H^s(M) \\ H^\sigma(X) \end{pmatrix} \longrightarrow \begin{pmatrix} H^{s'}(M) \\ H^{\sigma'}(X) \end{pmatrix},$$

where the entries belong to specific classes described below. The simplest case of such matrices are so-called *elementary operators*, where

- $D_{MM}$ and $D_{XX}$ are pseudodifferential operators on $M$ and $X$
- $D_{XM} = i^*$ is the *boundary* (restriction) *operator*, and $D_{MX} = i_*$ is the adjoint of $i^*$ and is called the *coboundary operator*.[1] It acts by the formula

  $$D_{MX} = i_* : g \mapsto g \otimes \delta_X,$$

  where $\delta_X$ is the delta function on $X$.

[1]Note that the operators $D_{XM}$ and $D_{MX}$ cannot be pseudodifferential operators, because they act between function spaces on different manifolds. Thus they necessarily are of a somewhat more general nature.

We point out that elementary operators alone are not sufficient, because the *product of operators* gives new kinds of entries. Indeed, if we multiply two operators $A$ and $A'$ of the form shown above, then in general we obtain a matrix like

$$B = \begin{pmatrix} D_{MM} + G & \widetilde{D}_{MX} \\ \widetilde{D}_{XM} & D_{XX} \end{pmatrix}$$

where

- $\widetilde{D}_{XM} = i^* D_{1MM}$ and $\widetilde{D}_{MX} = D_{2MM} i_*$ are general (co)boundary operators (obtained by the multiplication of elementary boundary and coboundary operators by pseudodifferential operators)
- in the right bottom entry we still have a pseudodifferential operator on $X$ (a composition of the form $i^* D_{MM} i_*$ is a pseudodifferential operator on $X$, so we do not get a new kind of operator there)
- in the left top entry, along with pseudodifferential operators, we obtain new elements of the form $G = D'_{MM} i_* D_{XX} i^* D''_{MM}$ (or linear combinations of such operators), which are called *Green operators*

All in all, we see that there are five distinct types of operators occurring in the entries of a general morphism:

- pseudodifferential operators on $M$
- pseudodifferential operators on $X$
- boundary operators
- coboundary operators
- Green operators

The last three types are not pseudodifferential operators, but nevertheless all five types admit a very simple common description. Namely, they are Fourier integral operators (e.g., Mishchenko, Shatalov and Sternin 1990, Nazaikinskii, Schulze and Sternin 2002) associated with certain Lagrangian manifolds. These manifolds are as follows:

$$\begin{aligned} L_M &= N^*(\Delta_M) \subset T^*M \times T^*M \\ L_X &= N^*(\Delta_X) \subset T^*X \times T^*X \\ L_B &= N^*(\operatorname{graph} i) \subset T^*X \times T^*M \\ L_C &= N^*({}^T\operatorname{graph} i) \subset T^*M \times T^*X \\ L_G &= N^*W \subset T^*M \times T^*M \end{aligned}$$

Here the subscripts on $L$ stand for the type of the corresponding operator ($B$, $C$, and $G$ stand for "boundary," "coboundary," and "Green," respectively), $\Delta_M$ is the diagonal in $M \times M$, $N^*$ is the conormal bundle, and $W = (X \times X) \cap \Delta_M$ is the subset of the diagonal in $M \times M$ generated by points that belong to $X$. The superscript $^T$ stands for transposition, i.e., interchanging the factors.

*Remark* 8.1. This description of entries of morphisms as Fourier integral operators provides a clear method for obtaining composition formulas for morphisms.

Note, however, that the last three manifolds $L_B$, $L_C$, and $L_G$ do not satisfy Hörmander's nondegeneracy condition, which is always assumed in the standard theory of Fourier integral operators.

◀ Recall that this condition requires the manifolds to be contained in $T_0^*X \times T_0^*M$ or $T_0^*M \times T_0^*M$ and closed in $(T^*X \times T^*M)_0$ or, respectively, $(T^*M \times T^*M)_0$. ▶

Hence the corresponding three types of operators are not classical Fourier integral operators, and that is why they have the following specific features, distinguishing them from classical pseudodifferential and Fourier integral operators.

- **These operators are not continuous in the *entire* Sobolev scale.** For example, the elementary boundary operator is continuous in the spaces

$$i^* : H^s(M) \longrightarrow H^{s-\nu/2}(X)$$

*only* for $s > \nu/2$. Here $\nu = \operatorname{codim} X$ is the codimension of $X$.

- **Compositions are not defined for arbitrary operators.** For example, the composition $i^* D i_*$ is well defined only if $D$ has a sufficiently *large negative* order. In particular, $D$ can never be a differential operator.

- And finally, to obtain a meaningful, consistent theory **one has to use symbols that are separately homogeneous in several groups of variables.**

### 8.1.1. Ellipticity and index in a special case

Now that we have introduced operator morphisms, we will discuss how one can check the Fredholm property and compute the index for these morphisms. It turns out that there are two substantially different cases for operator morphisms

$$A = \begin{pmatrix} D_M + G & C \\ B & D_X \end{pmatrix}.$$

- The case $G = 0$ (there is no Green operator in the morphism)

- The much more difficult case $G \neq 0$ (where there is a nontrivial Green operator)

Let us start from the first case, where it is possible to state a criterion for the Fredholm property in *finite* terms.

DEFINITION 8.2. A morphism $A$ (with $G = 0$) is said to be *elliptic* if the pseudodifferential operator $D_M$ is elliptic and the *noncommutative determinant*

$$\det A = D_X - BD_M^{-1}C$$

(which is a pseudodifferential operator on $X$) is also elliptic. (Here $D_M^{-1}$ is an almost inverse of $D_M$.)

The symbol of the noncommutative determinant can be effectively computed as an integral over the fibers of the conormal bundle of $X$.

With this definition, the following theorem takes place:

THEOREM 8.3. *If the morphism $A$ is elliptic, then it is Fredholm and*

$$\operatorname{ind} A = \operatorname{ind} D_M + \operatorname{ind} \det A.$$

### 8.1.2. Ellipticity in the general case

The second case, in which there is a nontrivial Green operator, is much more complicated. To analyze it, we first study operators of the form $1 + G$.

In other words, we consider *Green equations*, that is, equations of the form $(1 + G)u = f$. How to deal with this equation? First, note that outside a tubular neighborhood of $X$ the operator $G$ is smoothing and hence can be neglected as long as we are interested only in the Fredholm property and the index.

Next, in the tubular neighborhood of $X$ the operator $G$ can be represented as a pseudodifferential operator on $X$ with operator-valued symbol acting in the $L^2$ spaces on the fibers of the normal bundle $NX$:

$$G = H\left(\overset{2}{x}, -i\frac{\overset{1}{\partial}}{\partial x}\right),$$

$$H(x,p)v(t) = \iint e^{it\tau} a(x,p,\tau,\tau')\widetilde{v}(\tau')\,d\tau d\tau'.$$

Here $a$ is the symbol of $G$ viewed as a Fourier integral operator and $t$ is the coordinate in the fibers of $NX$.

The operator-valued symbol $H(x,p)$ is itself a Green operator in the space of functions of $t$. It depends on the parameters $x$ and $p$.

Moreover, the following assertion holds.

PROPOSITION 8.4. *The equation $(1 + H(x,p))v = f$ is a family of Fredholm integral equations of the second kind parametrized by points of the cotangent bundle of $X$ without the zero section.*

Thus the solvability of this equation is equivalent to the triviality of the kernel. From this, we obtain the Fredholm property theorem.

THEOREM 8.5. *The operator* $1+G$ *is Fredholm if and only if it is elliptic (i.e.,* $1+H(x,p)$ *is invertible on the cosphere bundle of* $X$*).*

Now let us consider a morphism

$$A=\begin{pmatrix} D_M+G & C \\ B & D_X \end{pmatrix}$$

containing a nontrivial Green operator. We have

$$A=\begin{pmatrix} D_M & 0 \\ 0 & 1 \end{pmatrix}\begin{pmatrix} 1+D_M^{-1}G & D_M^{-1}C \\ B & D_X \end{pmatrix}.$$

Note that $D_M^{-1}G$ is again a Green operator and $D_M^{-1}C$ is a coboundary operator. The second factor can be represented as an operator on $X$

$$\begin{pmatrix} 1+D_M^{-1}G & D_M^{-1}C \\ B & D_X \end{pmatrix}=H\left(\overset{2}{x},-i\overset{1}{\frac{\partial}{\partial x}}\right)\equiv H_X$$

with operator-valued symbol $H(x,\xi)$ of compact fiber variation (see Section 3.3).

DEFINITION 8.6. The morphism $A$ is said to be *elliptic* if $\sigma(D_M)$ is elliptic and $H$ is elliptic.

Then the following theorem holds.

THEOREM 8.7. *The morphism* $A$ *is Fredholm if and only if* $A$ *is elliptic. In this case, we have*

$$\operatorname{ind} A=\operatorname{ind} D_M+\operatorname{ind} H_X.$$

### 8.1.3. Sobolev problems

Relative elliptic theory allows one to solve *Sobolev problems*, that is, problems for elliptic operators where the equation is assumed to be satisfied outside a submanifold (possibly, of higher codimension) and conditions on the solution are posed on the same submanifold.

Let us first discuss the *physical background* of Sobolev problems. Suppose that we push a soap film spanning a one-dimensional contour with a thin needle. In the case of perfect nonwetting, the film collapses immediately unless we actually only touch it not trying to change its shape. In the case of perfect wetting, the needle passes through the film, whose shape remains unchanged. However, if we push a plate with the same needle, then the plate shape changes. In other words, *one*

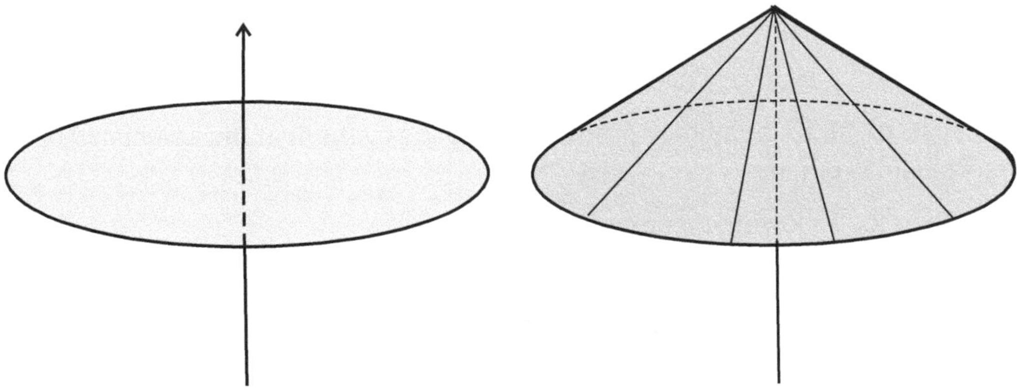

Figure 8.1. A needle passes through a thin film. The needle supports a thick plate.

*cannot pose "boundary conditions" for the thin film equation at a single point, but such conditions are possible for the plate equations* (see Fig. 8.1).

What does this mean *mathematically*? The interpretation is obvious. One cannot pose conditions at a point in $\mathbb{R}^2$ for the second-order equation $\Delta u = 0$ (removable singularity), but this is possible for the fourth-order equation $\Delta\Delta u = 0$.

Now let us describe the *general Sobolev problem*. It has the form

$$\begin{cases} Du \equiv f \mod H^{s-m}(M, X), \\ i^* Bu = g \in H^{s-b-\nu/2}(X) \end{cases} \tag{8.1}$$

for an unknown function $u \in H^s(M)$. Here the first equation in (8.1) is regarded as an equation $Du = f$ for an unknown function $u$, which should be satisfied modulo the subspace $H^{s-m}(M, X) \subset H^{s-m}(M)$ of distributions supported on $X$, and accordingly one poses boundary conditions $i^* Bu = g$ on $X$. Here $b$ is the order of $B$ and $\nu$ is the codimension of $X$.

It turns out that the Sobolev problem can be equivalently represented by an *operator morphism*; namely, it has the special form

$$A = \begin{pmatrix} D & i_{l*} \\ i^* B & 0 \end{pmatrix}, \qquad l = l(s, m, \nu), \tag{8.2}$$

where the range of the coboundary operator $i_{l*}$ consists of linear combinations of the delta function on $X$ and its derivatives up to order $l$; the coefficients of these linear combinations are of course functions on $X$. An explicit formula for the number $l(s, m, \nu)$ can be found in Nazaikinskii and Sternin (2004).

### 8.1.4. Fredholm property and index

By applying the general theory to the morphism (8.2) equivalent to a Sobolev problem, one obtains a criterion for the Fredholm property and an index formula for this problem.

First, we give a very natural definition.

DEFINITION 8.8. A Sobolev problem is said to be *elliptic* if the associated operator morphism is elliptic.

Then the following theorem holds.

THEOREM 8.9. *If a Sobolev problem is elliptic, then it is Fredholm. Moreover, the index of an elliptic Sobolev problem is given by the formula*

$$\operatorname{ind}(D, B) = \operatorname{ind} D + \operatorname{ind} i^* B D^{-1} i_{l*}.$$

The second term is of course the index of the corresponding noncommutative determinant.

Finally, let us discuss the following question: what is the *adjoint problem* of a given Sobolev problem? The answer is readily obtained if we pass to the equivalent morphisms. Namely, the adjoint problem of the Sobolev problem has the form

$$\begin{cases} Du + Bi_* v = f, \\ i_l^* u = 0, \end{cases}$$

where $i_l^*$ is the operator taking each function $u$ to its normal jet of order $l$ on $X$.

## 8.2. Topological Aspects of Relative Elliptic Theory

Now let us discuss in some detail certain topological aspects of relative elliptic theory. We shall work in the *category of smooth embeddings*. Its objects are smooth closed manifolds, and morphisms are smooth embeddings of such manifolds.

Given an embedding $i : X \hookrightarrow M$, there is a well-defined restriction map in $K$-theory

$$i^!_{top} : K_c(T^*M) \longrightarrow K_c(T^*X).$$

Indeed, one can define isomorphisms between tangent and cotangent spaces with the help of a Riemannian metric and use the natural embedding of tangent spaces to obtain an embedding of the cotangent spaces. The map $i^!_{top}$ is independent of the choice of the metric.

Now we have the following task:

- Give an analytical construction of a mapping associating (elliptic) pseudodifferential operators on $X$ to (elliptic) pseudodifferential operators on $M$
- Compare this mapping with the functor $i^!_{top}$

### 8.2.1. The metric trace of an elliptic operator

Such a mapping can indeed be constructed. Consider an embedding $i : X \hookrightarrow M$, and suppose that both manifolds are equipped with Riemannian metrics $\rho_M$ and $\rho_X$, $i^*\rho_M = \rho_X$; then we also have the natural embedding

$$i_m : T^*X \hookrightarrow T^*M$$

of the cotangent spaces. Now we can define the metric trace as follows.

DEFINITION 8.10. The *metric trace* on the submanifold $X$ of a pseudodifferential operator $D$ defined on $M$ is a pseudodifferential operator on $X$ whose principal symbol is the restriction to $T^*X$ of the principal symbol of $D$:

$$i_m^*(D) = \widehat{i_m^*(\sigma(D))}.$$

This is obviously well defined modulo lower-order terms, and if $D$ is elliptic, then its metric trace is also elliptic.

Let $\mathrm{Ell}(M)$ be the group of classes of elliptic operators on $M$ modulo stable homotopies (see Section 3.6). The metric trace induces a well-defined mapping

$$\begin{aligned} i_m^* : \mathrm{Ell}(M) &\longrightarrow \mathrm{Ell}(X), \\ [D] &\longmapsto [i_m^* D], \end{aligned}$$

(we denote the map of Ell-groups by the same symbol) of stable homotopy classes of elliptic operators on $M$ and $X$. This mapping is independent of the metric and functorial:

$$\text{if} \quad Y \overset{j}{\hookrightarrow} X \overset{i}{\hookrightarrow} M, \quad \text{then} \quad (i \circ j)_m^* = j_m^* \circ i_m^*.$$

In other words, $(\mathrm{Ell}, i_m^*)$ is a functor from the category of smooth embeddings into the category of Abelian groups. Obviously, the diagram

$$\begin{array}{ccc} \mathrm{Ell}(M) & \xrightarrow{i_m^*} & \mathrm{Ell}(X) \\ \chi \downarrow & & \downarrow \chi \\ K_c(T^*M) & \xrightarrow[i_{top}^!]{} & K_c(T^*X) \end{array}$$

commutes, where the vertical arrows are given by the Atiyah–Singer difference construction (see Section 3.6).

### 8.2.2. The pullback of an elliptic operator

Now it is natural to ask the following question. The metric trace is defined in terms of *symbols*. Can one possibly define a mapping directly in terms of *operators*?

The theory of operator morphisms suggests an answer. Suppose that we have an elliptic pseudodifferential operator $D$ on $M$. Then we can give the following definition.

DEFINITION 8.11. The *pullback* $i_a^!(D)$ of $D$ is the noncommutative determinant of the Dirichlet morphism

$$\begin{pmatrix} -D^{-1} & i_* \\ i^* & 0 \end{pmatrix}.$$

Of course, we also have the explicit formula

$$i_a^!(D) = i^* \circ D \circ i_*$$

This "naive" definition has however some *disadvantages*:

- the pullback is defined only if $\operatorname{ord} D < -\nu/2$, $\nu = \operatorname{codim} X$
- still worse, the pullback is not necessarily elliptic if $D$ is elliptic

### 8.2.3. Regularization

The remedy is the same for both disadvantages. We use regularization. Consider the Laplacians $\Delta_M$ and $\Delta_X$ on $M$ and $X$. Then the formula

$$i_a^!(D)(k) = (1 - \Delta_X)^{k-\nu/2} i_a^!\big((1 - \Delta_M)^{-k} D\big) \tag{8.3}$$

where $k$ is a parameter, permits us to regularize the pullback.

LEMMA 8.12. *The following assertions hold.*

- *The operator* (8.3) *is well defined for sufficiently large $k$.*
- *If $D$ is elliptic, then* (8.3) *is elliptic for large $k$.*
- *The class of* (8.3) *in* $\operatorname{Ell}(X)$ *is independent of $k$ for large $k$.*

Thus we define the *regularized pullback* by the formula (8.3) as $k \to \infty$. More precisely, we define a class in $\operatorname{Ell}(X)$ by the formula

$$i_{reg}^!(D) \stackrel{\text{def}}{=} \big[i_a^!(D)(k)\big] \in \operatorname{Ell}(X), \quad k \to \infty.$$

Then the following theorem holds.

THEOREM 8.13. *This class coincides with the class defined by the metric trace*:

$$i_{reg}^!(D) = i_m^*(D).$$

COROLLARY 8.14. *The Riemann–Roch diagram*

$$\begin{array}{ccc} \operatorname{Ell}(M) & \xrightarrow{\;i_{reg}^!\;} & \operatorname{Ell}(X) \\ {\scriptstyle\chi}\big\downarrow & & \big\downarrow{\scriptstyle\chi} \\ K_c(T^*M) & \xrightarrow[\;i_{top}^!\;]{} & K_c(T^*X) \end{array}$$

*commutes.*

### 8.2.4. Finitely regularized pullback

The role of regularization in the construction of the pullback is twofold.

- First, it makes the order of the operator $(1-\Delta_M)^{-k}D$ sufficiently large negative so as to ensure that the pullback $i_a^!(D)(k)$ is well defined ("*finite regularization*").
- Next, the pullback of an elliptic operator for sufficiently large $k$ (depending on the operator) will be elliptic and its class in $\mathrm{Ell}(X)$ will have an invariant meaning ("*infinite regularization*").

However, there are classes of operators for which finite regularization suffices. Let us consider two such classes.

***Operators with linear principal symbols.*** These are operators whose principal symbol coincides on the cosphere bundle with some function linear in the fibers of the cotangent bundle. For such operators, the (finitely regularized) pullback is again an operator with linear principal symbol and is homotopic through elliptic operators to the metric trace.

***External products.*** Let us identify a tubular neighborhood of $X$ in $M$ with a neighborhood of the zero section in the normal bundle $NX$ of $X$. Next, let

$$D_1 : C^\infty(X, E_1) \longrightarrow C^\infty(X, F_1)$$

be an elliptic differential operator on $X$ and

$$D_2 = \{D_{2x}\}_{x\in X} : C^\infty(N_xX, E_{2x}) \longrightarrow C^\infty(N_xX, F_{2x})$$

be a family of elliptic differential operators in the fibers of the normal bundle of $X$, parameterized by points of $X$.

DEFINITION 8.15. The *external product* of $D_1$ and $D_2$ is the operator with symbol

$$\sigma(D_1 \# D_2) = \begin{pmatrix} \sigma(D_1)\otimes 1_{E_2} & 1_{F_1}\otimes\sigma(D_2)^* \\ 1_{E_1}\otimes\sigma(D_2) & -\sigma(D_1^*)\otimes 1_{F_2} \end{pmatrix}.$$

We have the following theorem.

THEOREM 8.16. *If $D$ is an elliptic operator on $M$ equal to the external product $D_1\#D_2$ near $X$, then*

- *finite regularization suffices for the pullback to be well defined, i.e., elliptic*
- *the class of the pullback in* $\mathrm{Ell}(X)$ *is given by the formula*

$$[i_a^!(D)] = [D_1]\big([E_2|x] - [F_2|x]\big) \in \mathrm{Ell}(X)$$

Note that the operator $D_2$ itself does not affect this class, which involves only the bundles where $D_2$ acts.

### 8.2.5. Examples

Numerous examples can be obtained if we apply the preceding theorem to geometric operators on $M$.

- Consider the **Hirzebruch operator**. (We assume that $X$ and $M$ are even-dimensional and oriented.) Then the class of the pullback is given by the formula

$$[i_a^!(\mathcal{H}_M)] = [\mathcal{H}_X]\big([\Lambda_+(NX)] - [\Lambda_-(NX)]\big) \in \mathrm{Ell}(X),$$

where $\Lambda_\pm(NX)$ are the invariant subspaces of the involution $i^{j(j-1)+n/2}*$ and $*$ is the Hodge operator.

- For the pullback of the **Euler operator**, one has the similar formula

$$[i_a^!(\mathcal{E}_M)] = [\mathcal{E}_X]\big([\Lambda_{even}(NX)] - [\Lambda_{odd}(NX)]\big) \in \mathrm{Ell}(X).$$

- Consider the **Cauchy–Riemann (Todd) operator**. (Here we assume that $X$ and $M$, as well as the embedding, are complex-analytic.) Then the pullback is given by the formula

$$[i_a^!(\mathcal{T}_M)] = [\mathcal{T}_X]\big([\Lambda_{even}(NX)] - [\Lambda_{odd}(NX)]\big) \in \mathrm{Ell}(X),$$

where $NX$ is understood as a complex vector bundle.

- For the **Dirac operator**, where $X$ and $M$ are even-dimensional and equipped with compatible spin structures, we have the formula

$$[i_a^!(\not{D}_M)] = [\not{D}_X]\big([\Delta_+(NX)] - [\Delta_-(NX)]\big) \in \mathrm{Ell}(X),$$

where $\Delta_\pm$ are the half-spin representations of $Spin(\nu)$ corresponding to the $Spin(\nu)$-module structure on the normal bundle.

# Chapter 9

# Index of Geometric Operators on Manifolds with Cylindrical Ends

Chapter 5 contains a number of index formulas on manifolds with singularities. They are mainly obtained by the surgery technique. However, to compute the index we require the operators to have some symmetries. Unfortunately, many operators naturally arising in geometry do not possess such symmetries. In this chapter, we discuss an approach to index formulas for such geometric operators. To simplify the analysis, we consider operators on manifolds with infinite cylindrical ends. To obtain index formulas, we first reduce the operator to a boundary value problem with the same index on the finite part of the manifold. This is a so-called spectral Atiyah–Patodi–Singer boundary value problem. Atiyah, Patodi and Singer (1975) computed its index using the heat equation. We obtain stratified index formulas by analyzing the Atiyah–Patodi–Singer index formula.

We point out that the index formulas obtained in this chapter remain valid for operators on manifolds with conical singularities. This follows from the results of Section 1.2, where the cylindrical representation of operators on manifolds with conical singularities is discussed.

## 9.1. Operators on Manifolds with Cylindrical Ends

***Manifolds with cylindrical ends.*** Let $M$ be a smooth manifold with boundary.

We choose some collar neighborhood $U_{\partial M}$ of the boundary $\partial M$ (see Fig. 9.1) and identify it with the product $\partial M \times (-1, 0]$. The coordinate normal to the boundary will be denoted by $t$. We suppose that $t = 1$ corresponds to the boundary of our manifold.

Recall that one can construct a collar neighborhood as the set of points attainable from the boundary in time $< \varepsilon$ along the trajectories of a vector field on $M$ directed inward on the boundary. This set is an open manifold diffeomorphic to the above-mentioned product.

DEFINITION 9.1. A *manifold with an infinite cylindrical end* is the noncompact space

$$\widehat{M} = M \bigcup_{\partial M \times \{0\}} \partial M \times [0, \infty) \tag{9.1}$$

equal to the union of $M$ with the infinite cylinder $\partial M \times [0, \infty)$.

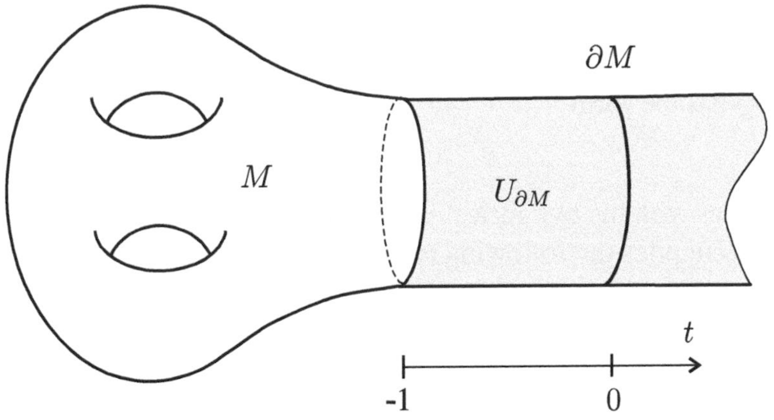


Figure 9.1. A manifold with an infinite cylinder.

***Operators.*** On a manifold $\widehat{M}$ with an infinite cylindrical end, we consider first-order interior elliptic differential operators

$$D : C^\infty(\widehat{M}, E) \longrightarrow C^\infty(\widehat{M}, F)$$

acting in spaces of smooth sections of complex vector bundles $E$ and $F$ over $\widehat{M}$.

We assume that on the cylinder $\{t \geq -1\}$ the operator $D$ can be represented in the form

$$D = \frac{\partial}{\partial t} - A, \tag{9.2}$$

where $A$ is a formally self-adjoint operator on $\partial M$ independent of $t$. It is called the *tangential operator* of $D$.

EXERCISE 9.2. Show that classical operators (Euler, signature, Dirac operators) have the form (9.2) provided that one takes the direct product metric $dt^2 + g_{\partial M}$ on the cylindrical part, where $g_{\partial M}$ is an arbitrary metric on the boundary.

***Fredholm property in weighted spaces.*** Since our operators stabilize as $t \to \infty$ (in the sense of Section 1.2.4), they can be treated as operators on a manifold with isolated singularities in the cylindrical representation. Thus it follows from Theorem 2.13 that the operator

$$D : H^{s,\gamma}(\widehat{M}, E) \longrightarrow H^{s-m,\gamma}(\widehat{M}, F) \tag{9.3}$$

is Fredholm whenever it is elliptic, i.e., satisfies the following conditions:

- the interior symbol $\sigma(D)$ is invertible on the cotangent bundle $T^*\widehat{M}$ outside the zero section
- the conormal symbol $\sigma_c(D)(p) = ip - A$ is invertible everywhere on the weight line $\{\operatorname{Im} p = \gamma\}$

***Geometric operators on infinite cylinders and cones.*** Let us compare the geometric operators corresponding to the cylindrical and conical metrics.

The standard cone-degenerate metric $dr^2 + r^2 d\omega^2$ in the cylindrical representation ($r = e^{-t}$) has the form

$$e^{-2t}(dt^2 + d\omega^2).$$

Thus it coincides with the cylindrical metric neglecting the factor $e^{-2t}$. This difference in metrics implies the following relations between the corresponding geometric operators.

- **The Euler operator** (see Brüning and Seeley 1988):

$$\mathcal{E}_{cone} \simeq e^t(\mathcal{E}_{cylinder} + E),$$

where $E = \operatorname{diag}\{e_0, \dots, e_n\}$ and $e_j = (-1)^j(j - n/2)$ on forms of degree $j$.

- **The signature operator** (see Brüning and Seeley 1988):

$$\mathcal{H}_{cone} \simeq e^t(\mathcal{H}_{cylinder} + B),$$

where $B = \operatorname{diag}\{b_0, \dots, b_n\}$ and $b_j = n/2 - j$.

- **The Dirac operator** (see Seeley 1990):

$$\mathcal{D}_{cone} = e^t \mathcal{D}_{cylinder}.$$

These equivalences imply that the index of the cone-degenerate Dirac operator is equal to the index of the Dirac operator in the cylindrical metric. However, for the Euler and signature operators there are correction terms due to the nontrivial zero-order terms[1] $B$ and $E$. These correction terms can readily be accounted for with the use of the relative index formula.

## 9.2. Index Formulas

Let us compute the index of the operator (9.3). For simplicity, suppose that $\gamma = 0$. We start by defining a special boundary value problem for our operator on the finite part of $\widehat{M}$. This boundary value problem will have the same index as the operator $D$ in (9.3).

[1] These terms do not affect the interior symbol but they occur in the conormal symbol and hence can affect the index.

### 9.2.1. The spectral boundary value problem

Let $\Pi_+(A)$ be the *nonnegative spectral projection* of $A$:

$$\Pi_+(A)e_\lambda = \begin{cases} e_\lambda, & \lambda \geq 0, \\ 0, & \lambda < 0, \end{cases}$$

where $e_\lambda$ is any eigenvector of $A$ corresponding to an eigenvalue $\lambda$. This projection is a pseudodifferential operator, since it can be represented as

$$\Pi_+(A) = \frac{A + \varepsilon + |A + \varepsilon|}{2|A + \varepsilon|} \tag{9.4}$$

for sufficiently small $\varepsilon > 0$, and one can use the following theorem (Seeley 1967): the powers of an elliptic self-adjoint positive pseudodifferential operator are pseudodifferential operators.

DEFINITION 9.3. The *Atiyah–Patodi–Singer spectral boundary value problem* for an elliptic operator $D$ is the boundary value problem

$$\begin{cases} Du = f \in C^\infty(M, F), \\ \Pi_+(A)u|_{\partial M} = g \in \operatorname{Im} \Pi_+(A) \end{cases} \tag{9.5}$$

for $u \in C^\infty(M, E)$, where $\operatorname{Im} \Pi_+(A) \subset C^\infty(\partial M, E|_{\partial M})$ is the range of the projection $\Pi_+(A)$.

*Remark* 9.4. In fact, Atiyah, Patodi and Singer (1975) considered homogeneous boundary conditions ($g = 0$). Their statement is equivalent to the one considered here: the kernels of the homogeneous and nonhomogeneous problems coincide, while a simple computation based on the surjectivity of the boundary operator

$$\Pi_+(A) : C^\infty(\partial M, E|_{\partial M}) \longrightarrow \operatorname{Im} \Pi_+(A)$$

shows that the cokernels are isomorphic.

For an elliptic operator $D$ of the form (9.2), the spectral boundary value problem defines the bounded Fredholm operator

$$(D, \Pi_+(A)) : H^s(M, E) \longrightarrow H^{s-1}(M, F) \oplus \overline{\operatorname{Im} \Pi_+(A)}, \quad s > 1/2, \tag{9.6}$$

in Sobolev spaces, where by

$$\overline{\operatorname{Im} \Pi_+(A)} \subset H^{s-1/2}(\partial M, E|_{\partial M})$$

we denote the closure of the range of $\Pi_+(A)$ in the norm of the corresponding Sobolev space. For the case of homogeneous boundary conditions, this assertion was proved by Atiyah, Patodi and Singer (1975), and the general case is equivalent to this.

Spectral problems are similar from the viewpoint of analysis to classical boundary value problems (Hörmander 1985*a*). For example, the index $\operatorname{ind}(D, \Pi_+(A))$ of the spectral problem (9.5) is independent of the exponent $s$ of the Sobolev space and can be computed in spaces of smooth functions.

On the contrary, the topological properties of spectral problems are far from those of their classical counterparts. This will be clear from the index formula, which we recall in Section 9.2.4.

But now let us show that the index of the operator $D$ on the manifold with a cylindrical end is equal to the index of the corresponding spectral boundary value problem.

### 9.2.2. Equivalence of operators

Consider the (homogeneous) spectral boundary value problem

$$\begin{cases} Du = f, \qquad u \in H^s(M, E), \quad f \in H^{s-m}(M, F), \\ \Pi_+ u|_{\partial M} = 0. \end{cases} \tag{9.7}$$

We shall prove not only that the index of this problem coincides with the index of the operator (9.3) but also that they are equivalent.

Note that, strictly speaking, we do not have an equivalence between the operator (9.3) and problem (9.7), since the function spaces where they act are different. However, if we consider the narrower class of functions $f$ supported in the submanifold $M$, then the two problems are equivalent in the following sense.

PROPOSITION 9.5. *If the operator* (9.3) *satisfies the ellipticity conditions, then the spectral problem* (9.7) *is equivalent to the operator* (9.3) *on the set of right-hand sides $f$ localized in the interior of $M$.*

*Proof.* 1) Let a function $u \in H^{s,0}(\widehat{M}, E)$ satisfy the equation

$$Du = f, \quad \operatorname{supp} f \subset M.$$

Let us show that $u$ is a solution of the homogeneous spectral problem. Indeed, the solution of the equation

$$\left(\frac{\partial}{\partial t} - A\right) u(t) = f(t)$$

has the form

$$u(t) = e^{tA} u(0) + \int_0^t e^{(t-\tau)A} f(\tau) d\tau. \tag{9.8}$$

Since $u \in H^{s,0}(\widehat{M}, E)$ and $f = 0$ for $t \geq 0$, it follows that

$$\Pi_+(A)u(0) = 0;$$

otherwise the first term in (9.8) would not lie in $H^{s,0}(\widehat{M}, E)$. Thus $u$ is indeed a solution of the spectral boundary value problem.

2) Now suppose that $u \in H^s(M, E)$ is a solution of the spectral boundary value problem. Then it follows from the condition $\Pi_+(A)u(0) = 0$ that the extension of $u$ to the entire $\widehat{M}$ by formula (9.8) on the cylindrical end is a solution of the equation $Du = f$ in the space $H^{s,0}(\widehat{M}, E)$. □

As a consequence, we readily obtain the following assertion.

THEOREM 9.6. *The operator*

$$i^* : H^{s,0}(\widehat{M}) \longrightarrow H^s(M), \quad M \overset{i}{\subset} \widehat{M},$$

*of restriction of functions to $M$ gives rise to isomorphisms between*

(a) *the kernel of Eq.* (9.3) *and the kernel of problem* (9.7)

(b) *the cokernel of Eq.* (9.3) *and the cokernel of problem* (9.7)

*In particular, the indices of the operators* (9.3) *and* (9.7) *coincide.*

*Proof.* (a) The isomorphism of the kernels follows from the preceding theorem.

(b) To establish the isomorphism of the cokernels, it suffices to verify that the cokernel of the spectral boundary value problem has a basis in which all functions vanish in a neighborhood of the boundary.

Indeed, let $f_1, \ldots, f_N \in H^{s-m}(M, F)$ be a basis of the cokernel of the spectral boundary value problem. For each element of the basis, we consider the equation

$$Du_i = \psi f_i, \quad \Pi_+(u_i|_{\partial M}) = 0,$$

where $\psi$ is a cutoff function equal to 1 in a neighborhood of the boundary of $M$. We solve this equation in a neighborhood of the boundary by using formula (9.8) and then extend the solution to the entire manifold $M'$ arbitrarily (but with smoothness preserved). We denote the corresponding function by $\widetilde{u}_i$. Then the desired basis of the cokernel of the spectral boundary value problem has the form

$$f_1 - D\widetilde{u}_1, \ldots, f_N - D\widetilde{u}_N.$$

The proof of the theorem is complete. □

This result permits one to transfer index theorems between the theory of spectral problems and the theory of equations on manifolds with cylindrical ends.

### 9.2.3. The index of the signature operator

We cite a result due to Atiyah, Patodi and Singer (1975). On a $4k$-dimensional oriented manifold $M$, consider the signature operator

$$d + d^* : \Lambda^+(M) \longrightarrow \Lambda^-(M),$$

where the $\Lambda^{\pm}(M)$ are the subspaces of forms invariant under the involution

$$\alpha : \Lambda^*(M) \longrightarrow \Lambda^*(M), \quad \alpha\big|_{\Lambda^p(M)} = (-1)^{\frac{p(p-1)}{2}+k} * .$$

(Here $*$ is the Hodge star operator.) On the boundary of $M$, we have

$$\Lambda^{\pm}(M)\big|_{\partial M} \simeq \Lambda^*(\partial M)$$

If we take a product metric in a neighborhood of the boundary, then the signature operator is equal to $\partial/\partial t + A$ modulo vector bundle isomorphisms (see Atiyah, Patodi and Singer 1975). The *tangential signature operator* $A$ acts on the boundary:

$$A : \Lambda^*(\partial M) \longrightarrow \Lambda^*(\partial M), \quad A\omega = (-1)^{k+p}(d * - \varepsilon * d)\omega,$$

where $\varepsilon\omega = (-1)^{\deg \omega}\omega$. This operator is elliptic and self-adjoint.

The index of the spectral boundary value problem is given by de Rham–Hodge theory:

$$\operatorname{ind}(d + d^*, \Pi_+(A)) = \operatorname{sign} M - \dim H^*(\partial M)/2, \tag{9.9}$$

where $\operatorname{sign} M$ is the signature of the manifold $M$ with boundary and $H^*(\partial M)$ is the cohomology of the boundary.

The formula (9.9) can be viewed as a stratified index formula: the index of the signature operator is represented as the sum of contributions of the main stratum and the boundary.

### 9.2.4. The Atiyah–Patodi–Singer index formula

To obtain the stratified index formula for the Dirac operator in Section 9.2.5 below, we need the *Atiyah–Patodi–Singer index formula* for spectral boundary value problems.

Atiyah, Patodi and Singer (1975) obtained an expression for the index of the spectral problem. Namely, an application of the heat equation method (Atiyah, Bott and Patodi 1973) results in the formula

$$\operatorname{ind}(D, \Pi_+(A)) = \int_M a(D) - \eta(A) \tag{9.10}$$

for the index of the spectral boundary value problem on a manifold $M$ for an elliptic first-order differential operator that has the expansion (9.2) into normal and tangent operators in a neighborhood of the boundary. The terms on the right-hand side in (9.10) are defined as follows.

***The interior contribution.*** Just as in the case of operators on closed manifolds, the term $a(D)$ is by definition the constant term in the local asymptotic expansion of the expression

$$\left[\operatorname{tr}\left(e^{-s\widetilde{D}^*\widetilde{D}}(x,x)\right) - \operatorname{tr}\left(e^{-s\widetilde{D}\widetilde{D}^*}(x,x)\right)\right]dx$$

as $s \to 0$. Here

- $\widetilde{D}$ is an arbitrary extension of $D$ to some closed manifold as an elliptic operator[2]
- $dx$ is the volume form and $\widetilde{D}^*$ is the adjoint with respect to the volume form
- $e^{-s\widetilde{D}^*\widetilde{D}}$ is the exponential of the operator $-s\widetilde{D}^*\widetilde{D}$
- $e^{-s\widetilde{D}^*\widetilde{D}}(x,x)$ is the restriction of Schwartz integral kernel of the operator $e^{-s\widetilde{D}^*\widetilde{D}}$ to the diagonal
- $\operatorname{tr}$ is the fiberwise trace of a vector bundle isomorphism

The value of the index density $a(D)$ at a point is an algebraic expression in the coefficients of $D$ and their derivatives at the same point. In particular, it does not depend on the choice of the extension $\widetilde{D}$.

***The boundary contribution.*** The second term is the $\eta$-*invariant* of the tangential operator $A$; see Appendix B. Unlike the first term, the boundary contribution is a nonlocal invariant, which is determined by the spectrum of the tangential operator of $D$. It cannot be expressed in terms of the principal symbol of the operator.

***Remarks on the index formula.*** Although the expression for $a(D)$ is extremely awkward in the general case, it can be described explicitly for Dirac type operators (in particular, for the Euler, signature and Dirac operators). For example, if $\mathcal{H}_E$ is the signature operator with coefficients in a bundle $E$ equipped with a connection, then

$$a(\mathcal{H}_E) = L(M)\operatorname{ch} E,$$

where $L(M) \in \Lambda^{ev}(X)$ is the Hirzebruch polynomial (Hirzebruch 1966) in the Pontryagin forms of the Riemannian manifold and $\operatorname{ch} E \in \Lambda^{ev}(M)$ is the Chern character of $E$ computed in terms of a given connection.

A similar expression for the form $a(D)$ is valid for other Dirac type operators; one should only replace the $L$-polynomial by some other polynomials.

[2]In fact, one can define an extension to the double $2M$. Namely, it follows from the identity

$$\left(\frac{\partial}{\partial t} + A\right)^* = -\frac{\partial}{\partial t} + A$$

that the operator $D$ on the first half of the double is glued smoothly with $D^*$ defined on the second half.

### 9.2.5. Index of the Dirac operator (the Kreck–Stolz invariant)

Consider the Dirac operator (e.g., Lawson and Michelsohn 1989) on a manifold $M$ of dimension $4k$ with boundary. In this subsection, we show how one can construct a stratified index formula of the form (5.2) for this operator following Kreck and Stolz (1993). Surprisingly, it turns out that this formula involves the *signature* operator on $M$. The construction will be carried out for the case in which the boundary of the manifold has zero rational Pontryagin classes.

We denote the Dirac operator by $\mathcal{D}$ and the corresponding tangential operator by $\mathcal{A}$. By the Atiyah–Patodi–Singer formula,

$$\operatorname{ind}\mathcal{D} + \eta(\mathcal{A}) = \int_M \widehat{A}(p), \tag{9.11}$$

where $\widehat{A}(p)$ is the $\widehat{A}$-polynomial in the Pontryagin forms $p_1, p_2, \dots, p_k$.

***Splitting of the integrals of decomposable terms.*** The problem of constructing a stratified index formula with homotopy invariant terms is thus reduced to the partition of the integral in (9.11) into a geometric invariant determined by the boundary and a remainder that must be a topological invariant. Such a partition can be obtained for all factorizable components of the $\widehat{A}$-polynomial except for the leading component $p_k$ by an application of the following lemma.

LEMMA 9.7. *Let $\alpha$ and $\beta$ be forms of positive degrees on $M$ whose restrictions to the boundary are exact. Then*

$$\int_M \alpha\wedge\beta = \langle j^{-1}[\alpha]\wedge j^{-1}[\beta], [M,\partial M]\rangle + \int_{\partial M} \widehat{\alpha}\wedge\beta, \tag{9.12}$$

*where $d\widehat{\alpha} = \alpha|_{\partial M}$, $j^{-1}[\alpha]$ is an arbitrary element in the preimage of the cohomology class $[\alpha]\in H^*(M)$ under the restriction mapping $j: H^*(M,\partial M)\to H^*(M)$, and $\langle\cdot,[M,\partial M]\rangle$ is the pairing with the fundamental cycle. The terms on the right-hand side are independent of the ambiguity in the choice of representatives.*

*Proof.* It suffices to apply the Stokes formula. □

Thus we can define the expressions

$$\int_{\partial M} d^{-1}(p_{i_1}\wedge p_{i_2}\wedge\ldots\wedge p_{i_s}) \stackrel{\text{def}}{=} \int_{\partial M} (\widehat{p_{i_1}}\wedge p_{i_2}\wedge\ldots\wedge p_{i_s}),$$

if we have at least two Pontryagin forms. (Note that since (9.12) is symmetric in $\alpha$ and $\beta$, we can put the hat on any of the forms $p_i$.)

***Splitting of integral of the highest Pontryagin class.*** Now it only remains to split the integral of the highest Pontryagin class. Here the signature operator $\mathcal{H}$ (with tangential operator $A$) is of help: the Atiyah–Patodi–Singer formula for this operator contains the integral of the $L$-class. In turn, the $L$-class includes the highest Pontryagin class $p_k$. One can verify that in degrees $\leq 4k$ the sum

$$\widehat{A}(p) + a_k L(p),$$

where $a_k = (2^{2k+1}(2^{2k-1} - 1))^{-1}$, contains only products of Pontryagin classes of positive degrees, i.e., does not contain the class $p_k$.

For example, for an eight-dimensional manifold we have $a_2 = 1/224$ and

$$\widehat{A}(p) = \frac{1}{5760}(7p_1^2 - 4p_2), \quad L(p) = \frac{1}{45}(7p_2 - p_1^2), \quad \widehat{A}(p) + a_2 L(p) = \frac{1}{896}p_1^2.$$

Hence, by writing out the sum of indices $\operatorname{ind}\mathcal{D} + a_k \operatorname{ind}\mathcal{H}$ by the Atiyah–Patodi–Singer formula, we obtain the relation

$$\operatorname{ind}\mathcal{D} + \eta(\mathcal{A}) - a_k\eta(A) - \int_{\partial M} d^{-1}(\widehat{A} + a_k L)(p) = t(M),$$

where $t(M)$ is the following topological invariant of manifolds with boundary:

$$t(M) = \langle(\widehat{A} + a_k L)(j^{-1}p(M)), [M, \partial M]\rangle - a_k \operatorname{ind}\mathcal{H}.$$

The expression

$$\eta(\mathcal{A}) - a_k\eta(A) - \int_{\partial M} d^{-1}(\widehat{A} + a_k L)(p),$$

which depends only on the boundary and the metric $g$ on the boundary, is called the *Kreck–Stolz invariant* of the manifold $\partial M$ with metric $g$ and is denoted by $s(\partial M, g)$. Hence we have obtained the index decomposition

$$\operatorname{ind}\mathcal{D} = t(M) - s(\partial M, g)$$

for the Dirac operator.

***Properties of the Kreck–Stolz invariant.*** Kreck and Stolz showed that the invariant $s(\partial M, g)$ of the Dirac operator is an important topological invariant of metrics of positive scalar curvature on the boundary. They proved the following theorem.

THEOREM 9.8. *The invariant $s(X, g)$ has the following properties.*

1. *It is a spin isometry invariant; that is, if $f : X \to X'$ is an isometry preserving the spin structure, then*

$$s(X, g) = s(X', g').$$

2. *It is a homotopy invariant in the class of metrics of positive scalar curvature.*

3. *It is additive with respect to connected sums of manifolds.*

# Chapter 10

# Homotopy Classification of Elliptic Operators

Informally, the classification problem for elliptic operators on smooth or singular manifolds can be understood as the following question:

| How many distinct elliptic operators are there on a given manifold? |
|---|

Prior to giving an answer, one should clarify the meaning of the words "how many" and "distinct," for the space of all elliptic operators is infinite-dimensional. Hence we start from the precise statement of the problem. Then we recall the classification on smooth manifolds and finally discuss the classification on singular manifolds.

Note that the homotopy classification is extremely useful in applications. For example, when computing the index, it suffices to consider elements that generate the set of homotopy classes of all elliptic operators. Some other applications are considered in Section 10.6.

## 10.1. The Homotopy Classification Problem

Let $M$ be a smooth or singular manifold. We shall equip the infinite-dimensional space of elliptic operators on $M$ with an equivalence relation such that the equivalence classes form a discrete set.

The order reduction procedure (e.g., see Section 3.6.4) permits us to restrict ourselves to zero-order operators in $L^2(M)$ without loss of generality.

***Stable homotopy.*** The following two equivalence relations are most important in elliptic theory.

DEFINITION 10.1. Two elliptic operators $D_0$ and $D_1$ are said to be *homotopic* if they can be joined by a norm continuous family $D_t, t \in [0,1]$, of elliptic operators. (All operators $D_t$ are assumed to act in the same pair of spaces.)

DEFINITION 10.2. Two elliptic operators

$$D_0 : C^\infty(M, E_0) \to C^\infty(M, F_0) \quad \text{and} \quad D_1 : C^\infty(M, E_1) \to C^\infty(M, F_1)$$

are said to be *stably homotopic* if there exist vector bundles $E, F \in \operatorname{Vect}(M)$ and bundle isomorphisms

$$e : E_0 \oplus E \longrightarrow E_1 \oplus F, \qquad f : F_1 \oplus F \longrightarrow F_0 \oplus E$$

such that the operators

$$D_0 \oplus \mathbf{1}_E \quad \text{and} \quad f(D_1 \oplus \mathbf{1}_F)e$$

are homotopic. Here $\mathbf{1}_E$ is the identity operator in a vector bundle $E$.

In other words, operators are stably homotopic if they become homotopic (modulo vector bundle isomorphisms) after stabilization by identity operators.

EXERCISE 10.3. Check that stable homotopy is indeed an equivalence relation.

***The group*** $\operatorname{Ell}(M)$**.** Let $\operatorname{Ell}(M)$ be the set of stable homotopy classes of elliptic operators on $M$. Elements of $\operatorname{Ell}(M)$ are sometimes referred to as "elliptic operators modulo stable homotopies." For an elliptic operator $D$ on $M$, by $[D]$ we denote the equivalence class of $D$ in $\operatorname{Ell}(M)$.

PROPOSITION 10.4. *The direct sum of elliptic operators descends to a well-defined operation*

$$[D_1] + [D_2] \overset{\text{def}}{=} [D_1 \oplus D_2]$$

*on* $\operatorname{Ell}(M)$, *which makes the set* $\operatorname{Ell}(M)$ *an abelian group. Moreover, the opposite element of* $[D]$ *is the class of an almost inverse*[1] *operator* $D^{-1}$,

$$-[D] = [D^{-1}],$$

*and the neutral element* $0 \in \operatorname{Ell}(M)$ *is the equivalence class of bundle isomorphisms.*

*Proof.* All assertions are obvious except for the statement about the opposite element. To prove the latter, one should construct a homotopy

$$\begin{pmatrix} D & 0 \\ 0 & D^{-1} \end{pmatrix} \sim \begin{pmatrix} 0 & 1 \\ 1 & 0 \end{pmatrix} : C^\infty(M, E \oplus F) \longrightarrow C^\infty(M, F \oplus E)$$

of $2 \times 2$ elliptic operator matrices. We leave this to the reader as an exercise. □

EXERCISE 10.5. Show that the index of elliptic operators induces an abelian group homomorphism $\operatorname{ind} : \operatorname{Ell}(M) \longrightarrow \mathbb{Z}$.

[1]That is, an inverse modulo compact operators.

***Statement of the problem.*** Now we are in a position to state the problem.

> *The homotopy classification problem for elliptic operators on a manifold $M$ is the problem of computing the group* $\mathrm{Ell}(M)$.

The following lemma shows that it suffices to classify symbols instead of operators.

LEMMA 10.6. *Elliptic operators are (stably) homotopic if and only if their principal symbols are (stably) homotopic in the class of elliptic symbols.*

*The proof* readily follows from the following facts. (a) Two operators having the same principal symbol are homotopic. (It suffices to take the linear homotopy.)

(b) A homotopy $\sigma_t$ of elliptic symbols can always be lifted to a homotopy of operators. The latter can be defined as $Q(\sigma_t)$, where $Q$ is a fixed quantization. □

***Homotopies and stable homotopies.*** Stable homotopy is coarser than homotopy. However, in applications to index theory it usually suffices to compute the stable homotopy classification, since the index is invariant under stable homotopy. Furthermore, stable homotopy has a simpler structure. To illustrate this, consider the simplest example. Let us try to classify elliptic symbols at some point of the manifold. An elliptic symbol at a point is a homogeneous function on the cotangent bundle with zero section deleted ranging in the set of invertible matrices: $\sigma : T^*_x M \setminus 0 \longrightarrow \mathbf{GL}(k,\mathbb{C})$. The homotopy classes of such mappings coincide with those of mappings $\mathbb{S}^{n-1} \to \mathbb{U}(k)$, $n = \dim M$, where $\mathbb{U}(k)$ is the unitary group. Hence the homotopy classification and the stable homotopy classification of symbols at a point have the form

$$[\mathbb{S}^{n-1} \longrightarrow \mathbb{U}(k)] \quad \text{and} \quad [\mathbb{S}^{n-1} \longrightarrow \mathbb{U}(\infty)],$$

respectively. Here the symbol $[X \to Y]$ stands for the set of homotopy classes of continuous mappings $X \to Y$. It is known in topology that the sets $[\mathbb{S}^{n-1} \longrightarrow \mathbb{U}(k)]$ are abelian groups that rather irregularly depend on $n$ and $k$ and are hard to compute for small $k$. However, under the stabilization $k \to \infty$, the corresponding groups of homotopy classes obey Bott periodicity:

$$[\mathbb{S}^{n-1} \longrightarrow \mathbb{U}(\infty)] \equiv \pi_{n-1}(\mathbb{U}(\infty)) = \begin{cases} 0 & \text{if } n \text{ is odd,} \\ \mathbb{Z} & \text{if } n \text{ is even.} \end{cases}$$

Hence throughout the following we deal only with stable homotopies.

## 10.2. Classification on Smooth Manifolds

For a smooth manifold $M$, the group $\mathrm{Ell}(M)$ can readily be computed in terms of topological $K$-theory. Namely, an elliptic operator $D$ on a manifold $M$ is determined modulo compact operators by its principal symbol

$$\sigma(D) : \pi^*E \longrightarrow \pi^*F, \quad \pi : T^*M \longrightarrow M,$$

which specifies an isomorphism of the bundles $\pi^*E$ and $\pi^*F$ over $T^*M$ outside the zero section of $T^*M$. In other words, we have a triple $(\pi^*E, \pi^*F, \sigma(D))$ consisting of two vector bundles over $T^*M$ and an isomorphism of these bundles outside a compact set. By definition, this triple defines a *difference element* in the $K$-group with compact supports of the space $T^*M$ (see Section 3.6). This difference element is denoted by

$$[\sigma(D)] \in K_c^0(T^*M).$$

THEOREM 10.7. *The mapping*

$$\begin{aligned} \chi : \operatorname{Ell}(M) &\longrightarrow K_c^0(T^*M), \\ [D] &\longmapsto [\sigma(D)], \end{aligned} \tag{10.1}$$

*is a well-defined group isomorphism.*

*The proof* (see Atiyah and Singer 1968) amounts to showing that triples $(\widetilde{E}, \widetilde{F}, \widetilde{\sigma})$ determining the $K$-group, where $\widetilde{E}$ and $\widetilde{F}$ are vector bundles over $T^*M$ and $\widetilde{\sigma}$ is an isomorphism of these bundles defined outside a compact subset of $T^*M$, are equivalent to triples of the special form $(\pi^*E, \pi^*F, \sigma)$, where $\sigma$ is an isomorphism on $T^*M \setminus 0$ homogeneous with respect to the covariables. The latter triples correspond to elliptic pseudodifferential operators. □

## 10.3. Atiyah–de Rham Duality

For operators on singular manifolds, the principal symbol contains not only the interior component but also a component responsible for the singularities. The latter does not occur in (10.1), and so one cannot hope that there is a generalization of (10.1) to singular manifolds. Hence we need a different approach to homotopy classification.

The alternative approach is based on the fundamental pairing

$$\langle \cdot , \cdot \rangle : \operatorname{Ell}(M) \times K^0(M) \longrightarrow \mathbb{Z}, \tag{10.2}$$

noticed by Atiyah, of the group of homotopy classes of elliptic operators and the topological $K$-group. Suppose that $M$ is a smooth closed manifold. The pairing (10.2) is determined in terms of the index: to an operator $D$ and a vector bundle $G$ on $M$, one assigns the index of the operator $D$ with coefficients in $G$:

$$\langle [D], [G] \rangle := \operatorname{ind} D \otimes \mathbf{1}_G, \qquad [D] \in \operatorname{Ell}(M), [G] \in K^0(M),$$

where the operator $D \otimes \mathbf{1}_G$ has the principal symbol $\sigma(D \otimes \mathbf{1}_G) := \sigma(D) \otimes 1_G$ and acts in the spaces

$$D \otimes \mathbf{1}_G : C^\infty(M, E \otimes G) \longrightarrow C^\infty(M, F \otimes G).$$

If $D$ is elliptic, then so is $D \otimes \mathbf{1}_G$, and hence the pairing is well defined.

The main property of the pairing (10.2) is given by the following assertion.

PROPOSITION 10.8. *The pairing* (10.2) *is nondegenerate on the free parts of the groups.*

◀ Recall that a pairing $\langle\,\cdot\,,\,\cdot\,\rangle : G \times H \to \mathbb{Z}$ is said to be *nondegenerate* on the free parts of the groups if the identity $\langle x, y\rangle = 0$ valid for all $y \in H$ implies that $x$ is a torsion element and if a similar assertion is valid with respect to the second argument. ▶

*Scheme of the proof.* Using the Atiyah–Singer formula and the Chern character

$$\mathrm{ch} : K_c^*(T^*M) \otimes \mathbb{Q} \simeq H_c^*(T^*M),$$

one reduces the problem to the nondegeneracy of the Poincaré duality

$$\langle\,\cdot\,,\,\cdot\,\rangle : H_c^*(T^*M) \times H^*(M) \longrightarrow \mathbb{R}, \qquad \langle \omega_1, \omega_2\rangle := \int_{T^*M} \omega_1 \wedge \omega_2,$$

which is well known. □

By way of corollary, the nondegeneracy of the pairing (10.2) implies the following expression for the groups $\mathrm{Ell}(M)$ modulo torsion as the dual group of $K^0(M)$:

$$\mathrm{Ell}(M) \otimes \mathbb{Q} \simeq \mathrm{Hom}(K^0(M), \mathbb{Q}). \tag{10.3}$$

Thus we obtain a classification modulo torsion. However, one can observe something more in these constructions.

***An analogy with the de Rham theory.*** The isomorphism (10.3) and the pairing (10.2) permit one to treat elliptic theory as an analog of the de Rham theory. Indeed, the key role in the latter is played by a pairing determined by integration of forms over cycles, while in elliptic theory we have a pairing between vector bundles and elliptic operators. For the reader's convenience, the following table lists some notions of the de Rham theory and their counterparts in elliptic theory.[2]

| de Rham theory | Elliptic theory |
|---|---|
| cocycle $\omega$, $d\omega = 0$ | vector bundle $E$ |
| cycle $\gamma$, $\partial\gamma = 0$ | elliptic operator $D$ |
| integral $\int_\gamma \omega$ | index $\mathrm{ind}(D \otimes \mathbf{1}_E)$ |
| cohomology $H^*(M)$ | topological $K$-theory $K^0(M)$ |
| homology $H_*(M)$ | the group $\mathrm{Ell}(M)$ |

[2]For brevity, we restrict ourselves to the even case, corresponding to the even $K$-group $K^0$.

***Elliptic theory as a $K$-homology theory.*** This correspondence between the theories suggests that the group $\mathrm{Ell}(M)$ of stable homotopy classes of elliptic operators can be viewed as a *realization* of the $K$-homology group $K_0(M)$ of the manifold $M$,

$$\mathrm{Ell}(M) \simeq K_0(M), \tag{10.4}$$

just as the usual cohomology $H^*(M)$ can be realized in terms of differential forms, Čech cochains, etc. Here the symbol $K_0$ stands for the generalized homology theory dual to topological $K$-theory. For the formal definition of duality of generalized (co)homology theories, we refer the reader to the literature (e.g., Baum and Douglas 1982, Whitehead 1962). Here we only note that there is a pairing between the dual theories, which in our case has the form

$$\langle\, \cdot\, , \cdot\, \rangle : K^0(X) \times K_0(X) \longrightarrow \mathbb{Z} \tag{10.5}$$

and is nondegenerate on the free parts of the groups. In particular, the nondegeneracy of the pairing (10.2) follows from (10.4) and (10.5).

It turns out that although the isomorphism (10.4) indeed holds, it is technically rather difficult to use elliptic operators in the construction of a generalized homology theory starting from the Ell-groups. The difficulty is in the definition of pseudodifferential operators on nonsmooth spaces as well as in the fact that it is completely unclear how to define the map

$$f_* : \mathrm{Ell}(M) \longrightarrow \mathrm{Ell}(N)$$

induced by a map $f : M \to N$ of manifolds (even smooth) in terms of elliptic operators.

In the following section, we introduce the notion of an abstract elliptic operator, which can be used to make elliptic theory a generalized homology theory (and in particular define the induced mappings $f_*$). Then we explain how one can use abstract elliptic operators and the related homology theory in the classification of operators on singular manifolds.

## 10.4. Abstract Elliptic Operators and Analytic $K$-Homology

***Abstract elliptic operators.*** Let $X$ be a compact space. Next, let $C(X)$ be the $C^*$-algebra of complex-valued continuous functions on $X$.

The following notion was introduced by Atiyah (1969).

DEFINITION 10.9. An *abstract elliptic operator* on $X$ is a Fredholm operator

$$D : H_1 \longrightarrow H_2$$

in Hilbert spaces $H_1$ and $H_2$ equipped with the structure of $*$-modules over $C(X)$ such that

$$[D, f] \in \mathcal{K}(H_1, H_2), \quad f \in C(X). \tag{10.6}$$

Here $\mathcal{K}$ is the space of compact operators from $H_1$ to $H_2$.

Thus $D$ almost commutes with the action of $C(X)$.

*Remark* 10.10. By Kasparov's lemma (Higson and Roe 2000), if we assume that the module structures are nondegenerate (i.e., $C(X)H_{1,2} = H_{1,2}$), then condition (10.6) is equivalent to *pseudolocality*: the operator $fDg$ is compact for any continuous functions $f, g$ on $X$ with disjoint supports.

The definition shows the main difference between abstract elliptic operators and usual pseudodifferential elliptic operators: the key role is played by pseudolocality, and the notion of symbol is not defined in general.

*Remark* 10.11. Abstract elliptic operators are covariant under continuous mappings of spaces. Indeed, consider a continuous mapping $f : X \to Y$. Then an abstract operator $D : H \to H$ over $X$ can also be viewed as an operator over $Y$ if the structure of a $C(Y)$-module on $H$ is defined as the composition of the induced mapping $f^* : C(Y) \to C(X)$ with the original $C(X)$-module structure on $H$.

***Analytic $K$-homology and Fredholm modules.*** A generalized homology theory based on abstract elliptic operators was constructed by Kasparov and independently by Brown, Douglas and Fillmore. It is known as the *theory of analytic $K$-homology*. It turned out that technically one works more conveniently with a modification of Atiyah's original definition, namely, with so-called *Fredholm modules*, which are a normalization of abstract elliptic operators.

DEFINITION 10.12. A *Fredholm module* on a space $X$ is a pair $(H, F)$ consisting of a $\mathbb{Z}_2$-graded Hilbert space $H = H_0 \oplus H_1$, where $H_0$ and $H_1$ are $*$-modules over $C(X)$, and a bounded self-adjoint operator

$$F : H \to H$$

such that $F$ is odd with respect to the gradation ($FH_{0,1} \subset H_{1,0}$) and

$$[F, f] \sim 0, \quad f(F^2 - 1) \sim 0, \tag{10.7}$$

for each function $f \in C(X)$, where $\sim$ stands for equality modulo compact operators.

The first condition in (10.7), already known to us, expresses pseudolocality, and the second is an analog of the ellipticity condition in this situation.

The *even analytic $K$-homology* group $K_0^a(X)$ is defined as the group generated by Fredholm modules modulo stable homotopy.

Stable homotopies for Fredholm modules (see Blackadar 1998, Chap. VIII) are defined by analogy with homotopies of elliptic operators. Namely, Fredholm modules are said to be *isomorphic* if the corresponding $C(X)$-modules $H$ are isomorphic and the operators $F$ are taken to one another by the isomorphism. Fredholm modules are said to be *homotopic* if they become isomorphic after some homotopy of the operators $F$. A *trivial module* is a module for which the congruences in (10.7) are equalities. Finally, two Fredholm modules are *stably homotopic* if they become homotopic after the addition of some trivial modules to each of them.

***Normalization of abstract elliptic operators.*** Usual pseudodifferential elliptic operators are obviously abstract elliptic operators. To obtain Fredholm modules from these, one needs a simple normalization based on the polar decomposition. Namely, if $D : H_0 \to H_1$ is an abstract elliptic operator, then one can construct the Fredholm module $(H, F)$ with components

$$H = H_0 \oplus H_1, \quad F = (P_{\operatorname{Ker}\mathcal{D}} + \mathcal{D}^*\mathcal{D})^{-1/2}\mathcal{D}, \tag{10.8}$$

where

$$\mathcal{D} = \begin{pmatrix} 0 & D^* \\ D & 0 \end{pmatrix} : H \longrightarrow H$$

and $P_{\operatorname{Ker}\mathcal{D}}$ is the orthogonal projection onto the kernel of the operator $\mathcal{D}$.

***Properties of the analytic $K$-homology.*** By analogy with the even group, one defines the odd analytic $K$-homology group $K_1^a(X)$. One simply does not require that the spaces involved are $\mathbb{Z}_2$-graded.

THEOREM 10.13 (Kasparov and Brown–Douglas–Fillmore).

1. *The groups $K_*^a(X)$ give rise to a generalized homology theory on the category of finite cell complexes.*

2. *There exists an isomorphism*

$$K_*^a(X) \simeq K_*^t(X),$$

*where by $K_*^t(X)$ we denote the homology theory dual to topological $K$-theory. In particular, there is a well-defined pairing*

$$K_*^a(M) \times K^*(M) \longrightarrow \mathbb{Z}$$

*nondegenerate on the free parts of the groups.*

3. *For a smooth closed manifold $M$, one has the isomorphism*

$$\begin{aligned} \operatorname{Ell}(M) &\longrightarrow K_0^a(M) \\ D &\longmapsto (H, F), \end{aligned}$$

*where the pair $(H, F)$ is defined as in* (10.8).

*The proof* can be found in Brown, Douglas and Fillmore (1977), Baum and Douglas (1982), and Higson and Roe (2000). □

In view of this theorem, we denote the analytic $K$-homology simply by $K_*(X)$ in what follows.

*Remark* 10.14. One can note that the $K$-homology groups $K_0(X)$ are defined in terms of the algebra $C(X)$ of functions on $X$. Hence these groups are well defined for an arbitrary $C^*$-algebra, and the theory of analytic $K$-homology has gone far beyond the purely topological framework and is widely used (e.g., Higson and Roe 2000) in the study of operator algebras, geometric analysis, etc.

## 10.5. Classification on Singular Manifolds

Here we apply the technique of analytic $K$-homology developed in the preceding section to the classification problem for elliptic operators on manifolds with conical points and manifold with edges. To be definite, we consider the case of manifolds with edges.

A zero-order elliptic pseudodifferential operator

$$D : \mathcal{W}^{0,\gamma}(\mathcal{M}, E) \longrightarrow \mathcal{W}^{0,\gamma}(\mathcal{M}, F)$$

on a manifold $\mathcal{M}$ with edges is an abstract elliptic operator on the manifold $\mathcal{M}$ viewed as a compact topological space. This follows from the pseudolocality

$$[D, f] \in \mathcal{K} \text{ for all } f \in C(\mathcal{M}).$$

Hence we have a well-defined mapping

$$i : \mathrm{Ell}(\mathcal{M}) \longrightarrow K_0(\mathcal{M}), \tag{10.9}$$

under which pseudodifferential elliptic operators are viewed as abstract elliptic operators and then the normalization (10.8) is used. The following theorem is proved in (Savin 2005, Nazaikinskii, Savin, Schulze and Sternin 2004*a*).

THEOREM 10.15 (on the homotopy classification). *The mapping* (10.9) *is an isomorphism for manifolds with conical points and manifolds with edges.*

COROLLARY 10.16. *The rank of the group* $\mathrm{Ell}(\mathcal{M})$ *is given by the formula*

$$\operatorname{rank} \mathrm{Ell}(\mathcal{M}) = \sum_j \beta_{2j}(\mathcal{M}),$$

*where the* $\beta_j(\mathcal{M})$ *are the Betti numbers of* $\mathcal{M}$.

One obtains the expression for the rank by applying the Chern character

$$\mathrm{ch} : K_0(\mathcal{M}) \otimes \mathbb{Q} \simeq H_{ev}(\mathcal{M}) \otimes \mathbb{Q}$$

to the isomorphism (10.9). Here by $H_{ev}(\mathcal{M})$ we denote the even-dimensional homology of the manifold.

*Idea of proof of Theorem* 10.15. This theorem is similar to the de Rham theorem, which establishes the equivalence of the two realizations of the homology groups (see the table in Section 10.3). Let us sketch the proof for the case of manifolds with edges by modeling the classical proof based on the reduction to the five lemma. Namely, for the pair $(\mathcal{M}, X)$, where $X \subset \mathcal{M}$ is the edge, one has the exact sequence

$$K_1(\mathcal{M}\backslash X) \longrightarrow K_0(X) \longrightarrow K_0(\mathcal{M}) \longrightarrow K_0(\mathcal{M}\backslash X) \longrightarrow K_1(X)$$

in $K$-homology. Note that here the $K$-homology groups of the noncompact complement $\mathcal{M} \setminus X$ are determined in terms of the algebra $C_0(\mathcal{M} \setminus X)$ of functions vanishing on the edge. A similar exact sequence

$$\mathrm{Ell}_1(\mathcal{M}\backslash X) \longrightarrow \mathrm{Ell}(X) \longrightarrow \mathrm{Ell}(\mathcal{M}) \xrightarrow{j} \mathrm{Ell}(\mathcal{M}\backslash X) \xrightarrow{\partial} \mathrm{Ell}_1(X)$$

in elliptic theory is constructed in Savin (2005) and Nazaikinskii, Savin, Schulze and Sternin (2004*a*). Here we clarify that to the odd $K$-homology there correspond odd elliptic groups $\mathrm{Ell}_1$ generated by *self-adjoint* elliptic operators and that the Ell-groups of the complement $\mathcal{M} \setminus X$ are defined in terms of operators subjected only to the condition of ellipticity of the interior symbol. Next, these exact rows can be embedded in the diagram

$$\begin{array}{ccccccccc}
\mathrm{Ell}_1(\mathcal{M}\backslash X) & \longrightarrow & \mathrm{Ell}(X) & \longrightarrow & \mathrm{Ell}(\mathcal{M}) & \xrightarrow{j} & \mathrm{Ell}(\mathcal{M}\backslash X) & \xrightarrow{\partial} & \mathrm{Ell}_1(X) \\
\downarrow & & \downarrow & & \downarrow{\scriptstyle i} & & \downarrow & & \downarrow \\
K_1(\mathcal{M}\backslash X) & \longrightarrow & K_0(X) & \longrightarrow & K_0(\mathcal{M}) & \longrightarrow & K_0(\mathcal{M}\backslash X) & \longrightarrow & K_1(X),
\end{array} \tag{10.10}$$

where the vertical arrows, by analogy with (10.9), are defined by normalization of elliptic operators.

Suppose that we have constructed the diagram (10.10) and proved its commutativity (see Nazaikinskii, Savin, Schulze and Sternin 2004*a*). Then the desired isomorphism, represented in (10.10) by the central vertical arrow, follows from the five lemma, since the other vertical mappings are isomorphisms. (They classify operators on manifolds without singularities, namely, on $X$ and $\mathcal{M} \setminus X$.)

## 10.6. Some Applications

Let us give some applications of the homotopy classification Theorem 10.15.

***Atiyah duality.*** Consider the pairing

$$\langle \cdot, \cdot \rangle : \mathrm{Ell}(\mathcal{M}) \times K^0(\mathcal{M}) \longrightarrow \mathbb{Z}$$

defined by the formula $\langle D, E\rangle := \operatorname{ind} D \otimes \mathbf{1}_E$ (see Section 10.3). It follows from the homotopy classification theorem and Theorem 10.13 that this pairing is nondegenerate on the free parts of the groups. In particular,

$$[D] = 0 \quad \text{in} \quad \mathrm{Ell}(\mathcal{M}) \otimes \mathbb{Q}$$

if and only if

$$\operatorname{ind} D \otimes \mathbf{1}_E = 0 \quad \text{for all bundles} \quad E \in \mathrm{Vect}(\mathcal{M}).$$

In other words, the indices with coefficients determine the homotopy class of an operator modulo torsion.

***Removable singularities and homotopy classification.*** Let $M$ be a smooth closed manifold and $N \subset M$ a smooth submanifold. Then a tubular neighborhood of $N$ is a bundle of cones (with spherical base); i.e., $M$ can be viewed as a manifold with edge $N$. Then, along with the usual pseudodifferential operators on $M$, one can also consider operators that are edge-degenerate on $N$. The classification theorem shows that the groups of stable homotopy classes in both cases coincide with $K_0(M)$ and are even independent of the choice of the smooth structure on the manifold.

***Obstruction to elliptic operators on manifolds with edges.*** In Chapter 6, we considered the problem of constructing an elliptic edge operator (or problem) with a given interior elliptic symbol. It turned out that the construction is not always possible, and the corresponding obstruction was indicated.

Here we show how the obstruction formula can readily be derived from the homotopy classification. Needless to say, the result is equivalent to the one obtained earlier.

Let $\mathcal{M}$ be a manifold with edge $X$, and let $M$ be the corresponding stretched manifold. Suppose that $\mathcal{M}$ is equipped with an operator $D$ whose interior symbol $\sigma(D)$ is elliptic. (The edge symbol is not supposed to be elliptic.) The operator defines an element

$$[D] \in \mathrm{Ell}(\mathcal{M} \setminus X)$$

in the Ell-group of the complement of the edge. Let us solve the problem of constructing an elliptic edge operator with interior symbol $\sigma(D)$ at the level of Ell-groups, i.e., let us find conditions under which $[D] \in \operatorname{Im} \beta$, where

$$\beta : \mathrm{Ell}(\mathcal{M}) \to \mathrm{Ell}(\mathcal{M} \setminus X)$$

is the mapping induced by the embedding of elliptic operators. We write out the following part of the diagram (10.10):

$$\begin{array}{ccccc} \mathrm{Ell}(\mathcal{M}) & \xrightarrow{\beta} & \mathrm{Ell}(\mathcal{M} \setminus X) & & \\ \simeq\downarrow & & \downarrow\simeq & & \\ K_0(\mathcal{M}) & \xrightarrow[i]{} & K_0(\mathcal{M} \setminus X) & \xrightarrow[\partial]{} & K_1(X). \end{array}$$

Since the vertical arrows are isomorphisms and the lower row is exact, it follows that $[D] \in \operatorname{Im} \beta$ if and only if

$$\partial[D] = 0 \in K_1(X). \tag{10.11}$$

This relation is just the desired obstruction to the construction of an elliptic operator on a manifold with edges.

Let us show that the condition $\partial[D] = 0$ is equivalent to the condition obtained in Chapter 6. Recall that earlier the obstruction was expressed in terms of topological $K$-theory. To translate relation (10.11) into topological $K$-theory, we first rewrite the boundary mapping $\partial$ in $K$-homology as the composition

$$K_0(\mathcal{M} \setminus X) \simeq K_0(M \setminus \partial M) \xrightarrow{\partial'} K_1(\partial M) \xrightarrow{\pi_*} K_1(X)$$

of the boundary mapping $\partial'$ corresponding to the pair $\partial M \subset M$ and the natural mapping $\pi_*$ induced in $K$-homology by the projection $\pi : \partial M \to X$. Then we supplement this sequence to obtain the diagram

$$\begin{array}{ccccc} K_0(\mathcal{M} \setminus X) & \longrightarrow & K_1(\partial M) & \xrightarrow{\pi_*} & K_1(X) \\ \simeq\uparrow & & \simeq\uparrow & & \updownarrow\simeq \\ K_c^0(T^*M) & \longrightarrow & K_c^0(T^*M|_{\partial M}) & \xrightarrow[\pi_!]{} & K_c^1(T^*X), \end{array}$$

where the lower row is the composition of the restriction to the boundary with the direct image map $\pi_!$ (see Atiyah 1989) induced in $K$-theory by the projection $\pi$.

The vertical arrows in this diagram are isomorphisms. The squares of the diagram commute, and hence the condition $\partial[D] = 0$ can be rewritten in the form

$$\pi_!(\sigma(D)|_{\partial T^*M}) = 0.$$

Just this obstruction formula was obtained in Chapter 6.

# Chapter 11

# Lefschetz Formulas

In this chapter, we extend the Atiyah–Bott–Lefschetz fixed point theorem to the case of manifolds with conical singularities. The classical theorem (Atiyah and Bott 1967) deals with a geometric endomorphism of an elliptic complex on a smooth closed manifold. It says that if all fixed points of the underlying self-mapping of the manifold are nondegenerate, then the Lefschetz number of the endomorphism is equal to the sum of their contributions. Moreover, these contributions are determined by the restrictions of the mappings comprising the endomorphism to the respective fixed points (and independent of the operators constituting the complex).

When generalizing this theorem to singular manifolds, one faces the following questions:

- Is it still true that the Lefschetz number is the sum of contributions of fixed points?
- If this is the case, how should the contributions of the fixed points in the new setting be computed?

It turns out that the answer to the first question is "yes" provided that we give an appropriate definition of nondegeneracy for conical fixed points. Furthermore, very predictably, the contributions of fixed points in the smooth part of the manifold have just the same form as in the classical theorem. (However, the proof is not straightforward, and we have to resort to a technique suggested by Sternin and Shatalov (1996*a*) in their semiclassical method in Lefschetz theory.) As to the contributions of conical fixed points, there is a novelty: they depend not only on the endomorphism but also on the operators occurring in the complex. More precisely, they are represented as integrals of meromorphic operator families involving the conormal symbols of the operators in the complex. Under additional conditions, these integrals are represented as sums of residues.

The structure of the chapter is clear from the table of contents. Here we only mention two Supplements, the first of which recalls the notion of Lefschetz number and the second gives a brief survey of the Sternin–Shatalov semiclassical method in Lefschetz theory.

## 11.1. Main Result

We consider only endomorphisms of short (two-term) complexes, having in mind that the case of general finite complexes can be analyzed in a similar manner.

Let $\mathcal{M}$ be a compact manifold with conical singular points $\alpha_1, \ldots, \alpha_N$. The bases of the corresponding cones will be denoted by $\Omega_{\alpha_1}, \ldots, \Omega_{\alpha_N}$, respectively. Next, let

$$D = H^{s,\gamma}(\mathcal{M}, E_1) \to H^{s-m,\gamma}(\mathcal{M}, E_2), \quad s \in \mathbb{R}, \quad \gamma = (\gamma_1, \ldots, \gamma_N),$$

be an elliptic differential operator of order $m$ in weighted Sobolev spaces of sections of finite-dimensional vector bundles $E_1$ and $E_2$ over $\mathcal{M}$.[1] Recall that ellipticity means that the interior symbol of $D$ is invertible everywhere outside the zero section of the cotangent bundle $T^*\mathcal{M}$ and the conormal symbols $\sigma_c(D)(\alpha_j, p)$ of $D$ at the conical points $\alpha_j$ are invertible on the corresponding weight lines $\mathcal{L}_j = \{\operatorname{Im} p = \gamma_j\}, \quad j = 1, \ldots, N$.

***Geometric endomorphisms.*** Let a commutative diagram

$$\begin{array}{ccccccc} 0 & \longrightarrow & H^{s,\gamma}(\mathcal{M}, E_1) & \xrightarrow{D} & H^{s-m,\gamma}(\mathcal{M}, E_2) & \longrightarrow & 0 \\ & & \downarrow{\scriptstyle T_1} & & \downarrow{\scriptstyle T_2} & & \\ 0 & \longrightarrow & H^{s,\gamma}(\mathcal{M}, E_1) & \xrightarrow[D]{} & H^{s-m,\gamma}(\mathcal{M}, E_2) & \longrightarrow & 0, \end{array} \tag{11.1}$$

i.e., an *endomorphism of the complex* specified by $D$, be given.

Here $T_j$ is a *geometric endomorphism*:

$$(T_j u)(x) = A_j(x)\, u(g(x)), \quad x \in \mathcal{M}, \qquad j = 1, 2, \tag{11.2}$$

where

$$g : \mathcal{M} \longrightarrow \mathcal{M}$$

is a smooth self-mapping of $\mathcal{M}$ and the $A_j : g^*E_j \longrightarrow E_j$ are bundle homomorphisms.

◀ A continuous mapping $g : \mathcal{M} \longrightarrow \mathcal{M}$ is said to be *smooth* if it is obtained from a smooth mapping $g : M \longrightarrow M$ (which we denote by the same letter) of the stretched manifold $M$ such that $g(\partial M) \subset \partial M$. ▶

Furthermore, we assume that $g$ is a *diffeomorphism* in the category of manifolds with singularities, that is, the inverse mapping $g^{-1}$ exists and is also a smooth self-mapping of $\mathcal{M}$. Apparently, just as in usual Lefschetz theory, this restriction can be dropped, but this leads to some technical complications, which we wish to avoid.

[1] In this chapter, we shall represent operators in neighborhoods of conical points in the form (1.29) omitting the factor $r^{-m}$ (which amounts to multiplying the operator by a nonzero smooth function equal to $r^m$ in a neighborhood of each conical point); hence the weight exponent $\gamma$ on the source and target spaces is the same.

By definition, $g$ preserves the sets of conical points and interior points and has a special structure near the conical points. Namely, let $g(\alpha) = \alpha'$, where $\alpha$ and $\alpha'$ are conical points. Then in conical coordinate neighborhoods of $\alpha$ and $\alpha'$ the mapping $g$ is given by the formulas

$$r' = rB(r, \omega), \qquad \omega' = C(r, \omega), \tag{11.3}$$

where $\omega \in \Omega$ and $\omega' \in \Omega'$ are points on the bases of the corresponding model cones, $r$ and $r'$ are radial variables, and the functions $B(r, \omega) \geq 0$ and $C(r, \omega)$ are smooth up to $r = 0$; moreover, $B(r, \omega) > 0$ for $r > 0$. Since $g$ is a diffeomorphism, it follows that $B(r, \omega)$ is positive also for $r = 0$. Furthermore, the mapping

$$g_\alpha \equiv C(0, \cdot) : \Omega \longrightarrow \Omega',$$

which will be referred to as the *boundary mapping* of $g$ at $\alpha$, is also a diffeomorphism.

***Conormal symbols.*** We define the conormal symbols of $T_j$, $j = 1, 2$, at a conical point $\alpha$ as follows. We pass to the limit as $\lambda \to 0$ in the product $\varkappa_\lambda T_j \varkappa_\lambda^{-1}$, where $\varkappa_\lambda$ is the dilation group

$$\varkappa_\lambda v(r, \omega) = \lambda^{\dim M/2} v(\lambda r, \omega) \tag{11.4}$$

defined in neighborhoods of $\alpha$ and $g(\alpha)$, and then to the Mellin transform from the variable $r$ to the variable $p$ in the resulting operator commuting with dilations. We obtain the operator family

$$\sigma_c(T_j)(\alpha, p) = A_{j0}(\omega) e^{ib(0,\omega)p} g_\alpha^*, \tag{11.5}$$

where $A_{j0}(\omega) = A_j(x)|_{r=0}$ is the restriction of the bundle homomorphism $A_j$ to the singular point $\alpha$ and

$$b(r, \omega) = -\ln B(r, \omega). \tag{11.6}$$

This family depends on the parameter $p \in \mathbb{C}$ and takes functions on $\Omega_{g(\alpha)}$ to functions on $\Omega_\alpha$.

DEFINITION 11.1. The operator family (11.5) is called the *conormal symbol* of the endomorphism $T_j$.

***Main theorem.*** Let $x \in \operatorname{fix}(g)$ be a fixed point of $g$.

DEFINITION 11.2. We say that $x$ is a *nondegenerate fixed point* if one of the following conditions is satisfied:

1) $x$ is an interior point, and $\det(1 - g_*(x)) \neq 0$

2) $x$ is a conical point, and in formulas (11.3) describing $g$ in a neighborhood of $x$ either $B(0,\omega) > 1$ for all $\omega \in \Omega$ (a *repulsive* conical fixed point), or $B(0,\omega) < 1$ for all $\omega \in \Omega$ (an *attractive* conical fixed point)

◀ Here $g_*(x)$ is the differential of $g$ at $x$. Condition 1) is just the classical nondegeneracy condition in the smooth case. ▶

Consider the geometric endomorphism (11.1), where $T_{1,2}$ are operators of the form (11.2) and $D$ is an elliptic differential operator on a manifold $\mathcal{M}$ with conical singularities.

THEOREM 11.3 (Nazaikinskii, Schulze, Sternin and Shatalov 1999, Nazaikinskii and Sternin 2002). *Suppose that all fixed points of $g$ are nondegenerate. Then the Lefschetz number of the diagram* (11.1) *has the form*

$$\begin{aligned}\mathcal{L} &= \sum_{x^*} \mathcal{L}_{\rm int}(x^*) + \sum_{\alpha^*} \mathcal{L}_{\rm sing}(\alpha^*)\\ &\equiv \sum_{x^*} \frac{\operatorname{Trace} A_1(x^*) - \operatorname{Trace} A_2(x^*)}{\left|\det\left(1 - g_*(x^*)\right)\right|}\\ &\quad + \frac{1}{2\pi i} \sum_{\alpha^*} \operatorname{Trace} \int\limits_{\operatorname{Im} p = \gamma(\alpha^*)} \sigma_c(T_1)(\alpha^*, p)\sigma_c(D)^{-1}(\alpha^*, p)\frac{\partial \sigma_c(D)}{\partial p}(\alpha^*, p)\, dp.\end{aligned} \tag{11.7}$$

*Here the first term is the sum of contributions $\mathcal{L}_{\rm int}(x^*)$ of interior fixed points $x^*$, and the second term is the sum of contributions $\mathcal{L}_{\rm sing}(\alpha^*)$ of conical fixed points $\alpha^*$. The integration in each term of the second sum is over the weight line determined by the weight exponent $\gamma(\alpha^*)$ of the corresponding conical point and the integral is understood as an oscillatory integral (see below) and gives a trace class operator in $L^2(\Omega)$, where $\Omega$ is the base of the cone at $\alpha^*$, so that the trace is well defined.*

***Oscillatory integrals.*** Let us explain the meaning of the integral representing the contribution of a conical fixed point.

For brevity, in what follows we often omit the argument $\alpha^*$. (For example, we write $\sigma_c(D)(p)$ instead of $\sigma_c(D)(\alpha^*, p)$ etc.) Furthermore, without loss of generality we assume that all weight exponents are zero, and so the weight line is the real axis.

Consider any of the integral terms on the right-hand side in (11.7). Let us analyze the integrand. The operator family

$$H(p) = \sigma_c(D)^{-1}(p)\frac{\partial \sigma_c(D)}{\partial p}(p)$$

is defined everywhere on the real line (by the ellipticity condition, there are no poles of $\sigma_c(D)^{-1}(p)$ there) and is a pseudodifferential operator of order $-1$ with parameter $p$ on the real line. Next, since the conical fixed point is nondegenerate, it follows that the function $b(0,\omega)$ defined in (11.6) is nonzero for all $\omega \in \Omega$. Consequently, we can transform the expression (11.5) for the conormal symbol of $T_1$ as follows:

$$\sigma_c(T_1)(p) = \left(\frac{1}{ib(0,\omega)}\right)^l \frac{\partial^l \sigma_c(T_1)}{\partial p^l}(p).$$

Now we can regularize the integral by formally integrating by parts $l$ times:

$$\frac{1}{2\pi i}\int_{-\infty}^{\infty} \sigma_c(T_1)(p)\sigma_c(D)^{-1}(p)\frac{\partial \sigma_c(D)}{\partial p}(p)\,dp$$
$$= \frac{1}{2\pi i}\int_{-\infty}^{\infty} \left(\frac{i}{b(0,\omega)}\right)^l \sigma_c(T_1)(p)\frac{\partial^l H(p)}{\partial p^l}\,dp. \quad (11.8)$$

For sufficiently large $l$, the pseudodifferential operator $\partial^l H(p)/\partial p^l$ with parameter $p$ of order $-l-1$ is trace class, and the integral on the right-hand side in (11.8) converges in the usual operator norm as well as in the trace norm in every $H^s(\Omega)$, the result being independent of $l$ and consistent for various $s$. This provides the desired regularization. Simultaneously, we have shown that this integral specifies a trace class operator.

*Remark* 11.4. The asymmetry in the expression for $\mathcal{L}_{\text{sing}}(\alpha^*)$ in (11.7) is only apparent. It follows from the relation

$$DT_1 = T_2 D$$

that

$$\sigma_c(D)(p)\sigma_c(T_1)(p) = \sigma_c(T_2)(p)\sigma_c(D)(p)$$

and consequently,

$$\sigma_c(D)^{-1}(p)\sigma_c(T_2)(p) = \sigma_c(T_1)(p)\sigma_c(D)^{-1}(p).$$

Cyclically permuting the factors, we can reduce the integrand to the form

$$\frac{\partial \sigma_c(D)}{\partial p}(p)\sigma_c(D)^{-1}(p)\sigma_c(T_2)(p).$$

(Although taking the trace cannot be permuted with integration or computing the residue, for the integrand is not a trace class operator, we can justify the validity of this transformation.)

## 11.2. Proof of the Theorem

The proof consists of three steps. At the first step, we localize the task. At the second and third steps, we compute the contributions of interior fixed points and conical fixed points, respectively.

### 11.2.1. Localization

Our first task is to show that the Lefschetz number $\mathcal{L}$ can be represented as the sum of contributions corresponding to connected components of the set $\operatorname{fix}(g)$ of fixed points of $g$. Although these contributions will be defined here as integrals over some neighborhoods of these components with integrands ambiguous to a certain extent, the values of the integrals are independent of the freedom in the choice of the integrand as well as of the structure of the operators $D$, $T_1$, and $T_2$ outside an arbitrarily small neighborhood of the set $\operatorname{fix}(g)$. Formulas explicitly depending on the values at fixed points alone will be given later.

Our starting point is the trace formula (e.g., Fedosov 1993)

$$\mathcal{L} = \operatorname{Trace} T_1(1 - RD) - \operatorname{Trace} T_2(1 - DR). \tag{11.9}$$

Here $R$ is an arbitrary almost inverse of $D$ modulo trace class operators. (In other words, the operators $1 - RD$ and $1 - DR$ are trace class.)

***A special almost inverse.*** In turns out that for a special choice of $R$ the right-hand side of (11.9) splits into a sum of integrals over arbitrarily small neighborhoods of the components of $\operatorname{fix}(g)$. Let us equip $\mathcal{M}$ with a natural Riemannian metric $d\rho^2$ that has the product structure

$$d\rho^2 = dr^2 + r^2 d\omega^2$$

near the conical points in some conical coordinates. (Here $d\omega^2$ is some Riemannian metric on the base $\Omega$ of the cone.) Next, we introduce a measure $d\mu(x)$ that is smooth outside the conical points and has the form

$$d\mu(x) = \frac{dr}{r} \wedge d\omega$$

in the same coordinates near the conical points, where $d\omega$ is a smooth measure on $\Omega$. Using this measure, one can treat the kernels of pseudodifferential operators as (generalized) functions (or sections of vector bundles) on $M^\circ \times M^\circ$.

Let

$$\operatorname{fix}(g) = K_1 \cup \ldots \cup K_k,$$

where the $K_j$, $j = 1, \ldots, k$, are disjoint compact sets.

Next, let $V_1, \ldots, V_k$ be sufficiently small neighborhoods of these sets. The function $\rho(x, g(x))$ is positive and continuous on the compact set $\mathcal{M} \setminus (V_1 \cup \ldots \cup V_k)$ and hence has a nonzero minimum $2\varepsilon$. Since $D$ is an elliptic differential operator on $\mathcal{M}$, it follows that there exists an almost inverse $R$ of $D$ modulo trace class operators such that the kernel $R(x, y)$ of $R$ has the property

$$R(x, y) = 0 \text{ for } \rho(x, y) > \varepsilon. \tag{11.10}$$

(An operator $R$ with this property is said to be $\varepsilon$-*narrow*.) Then the kernels $K_1(x, y)$ and $K_2(x, y)$ of the operators $1 - RD$ and $1 - DR$ have the same property, since $D$ is a differential operator and does not enlarge supports.

***Localization of the Lefschetz number.*** In terms of these kernels, formula (11.9) becomes

$$\mathcal{L} = \int_M \left[\operatorname{Trace} A_1(x)K_1(g(x),x) - \operatorname{Trace} A_2(x)K_2(g(x),x)\right] d\mu(x), \quad (11.11)$$

where Trace in the integrand is the matrix trace. With regard to (11.10), we can rewrite this expression in the form

$$\mathcal{L} = \sum_{j=1}^{k} \int_{V_j} \left[\operatorname{Trace} A_1(x)K_1(g(x),x) - \operatorname{Trace} A_2(x)K_2(g(x),x)\right] d\mu(x), \quad (11.12)$$

since the integrand is identically zero outside the union of $V_j$.

It turns out that not only the sum as a whole but also each separate term is independent of the choice of an $\varepsilon$-narrow almost inverse operator (at least if the neighborhoods $V_j$ are sufficiently small). This readily follows from the *locality* of the construction of an almost inverse operator: if there are two $\varepsilon$-narrow almost inverse operators, then one can construct a new $\varepsilon$-narrow almost inverse operator coinciding with the first operator in some $V_{j_0}$ and with the second operator in all $V_j$ with $j \neq j_0$. It follows that one can separately analyze each of the integrals specifying the contributions of the components of the set of fixed points and choose an $\varepsilon$-narrow almost inverse operator as is convenient near the component whose contribution is to be analyzed.

***Localization of the integrand.*** The expression (11.12) is not very convenient for the subsequent analysis, and we shall obtain a different, more convenient expression. One can assume that the operator $R$ is sufficiently narrow and the neighborhoods $V_j$ are so small that not only $V_j$ but also the sets

$$W_j = V_j \cup g(V_j)$$

are disjoint. Under these conditions, consider the $j$th term

$$\mathcal{L}_j = \int_{V_j} \left[\operatorname{Trace} A_1(x)K_1\big(g(x),x\big) - \operatorname{Trace} A_2(x)K_2\big(g(x),x\big)\right] d\mu(x) \quad (11.13)$$

in (11.12). It will be referred to as the *contribution of the component $K_j$ of the set of fixed points* to the Lefschetz number. To compute (11.13), we can modify the operator $R$ arbitrarily provided that the integrand in (11.13) remains unchanged. Let $f_j$ be a continuous function on $\mathcal{M}$ that is smooth in $M^\circ$ and constant near each singular point and satisfies $f_j\big|_{V_j} \equiv 1$. Next, let $\psi_j$ have similar properties and satisfy $\psi_j\big|_{g(V_j)} \equiv 1$. We replace the operator $R$ by $\psi_j R f_j$ and the identity operator by $f_j$ (that is, assume that $K_1$ and $K_2$ are the kernels of the operators $f_j - \psi_j R f_j D$ and $f_j - \widehat{D}\psi_j R f_j$, respectively.) Then the integral (11.13) remains unchanged.

Moreover, if $\operatorname{supp} f_j$ and $\operatorname{supp} \psi_j$ do not meet $V_k$ and $g(V_k)$ for $k \neq j$, then one can extend the integration in (11.13) to the entire manifold $M^\circ$. (The integrand vanishes identically on $M^\circ \setminus V_j$.)

*Remark* 11.5. *Formally*, $\mathcal{L}_j$ can be rewritten in the form

$$\mathcal{L}_j = \operatorname{Trace} T_1(f_j - \psi_j R f_j D) - \operatorname{Trace} T_2(f_j - D\psi_j R f_j). \qquad (11.14)$$

Although the operators $T_1(f_j - \psi_j R f_j D)$ and $T_2(f_j - D\psi_j R f_j)$ are not trace class in general and do not have traces, the integrals of their kernels over the diagonal are well defined, and we use notation (11.14) for the corresponding integral.

Let us find the contributions of fixed points. We carry out the computations for interior fixed points in Section 11.2.2 and for conical fixed points in 11.2.3.

### 11.2.2. Contributions of interior fixed points

Let $x^*$ be an interior fixed point. Let us show how one can prove that its contribution has the same form as in the classical Atiyah–Bott–Lefschetz theorem. (Intuitively, this seems to be rather obvious in view of the localization property proved above. However, we have no surgery that would permit us to replace the endomorphism of the elliptic complex on the singular manifold by an endomorphism of an elliptic complex on a smooth manifold preserving the structure of the problem near a given interior fixed point and thus reduce the problem to the case considered by Atiyah and Bott. Thus we have to give a direct proof.)

The proof closely follows the scheme suggested for the case of smooth manifolds by Sternin and Shatalov (1996*a*); see also Supplement 11.5. In a neighborhood of an interior fixed point $x^*$, consider local coordinates $x_1, \dots, x_n$. The endomorphisms $T_{1,2}$ can be locally represented as the integral operators with kernels

$$K_j(x, y) = \frac{1}{(2\pi)^n} \int_{\mathbb{R}^n} A_j(x) e^{ip(g(x)-y)}\, dp, \qquad j = 1, 2. \qquad (11.15)$$

The main idea of the method suggested in (Sternin and Shatalov 1996*a*) is to proceed to the "semiclassical" situation by making the regularizer depend on the small parameter $h \to 0$. Since the Lefschetz number is actually independent of this parameter, it suffices to compute its *asymptotics* as $h \to 0$; this asymptotics gives the exact answer. The same argument is used in the proof of our theorem. The only difference is that instead of the Lefschetz number itself we deal with the contribution (11.14) of a single fixed point. The independence of this integral of $h$ follows from the localization principle proved in the preceding subsection. First, note that we can introduce a small parameter in the formula specifying the endomorphism by changing the integration variable as follows:

$$K_j(x, y) = \frac{1}{(2\pi h)^n} \int_{\mathbb{R}^n} A_j(x) e^{ip(g(x)-y)/h}\, dp, \qquad j = 1, 2. \qquad (11.16)$$

Next, let $m$ be the order of our differential operator. The Lefschetz number does not change if we multiply $D$ by $(-ih)^m$. The product $(-ih)^m D$ can obviously be represented in the form of a semiclassical differential operator (see Maslov 1972, Mishchenko, Shatalov and Sternin 1990):

$$h^m D\left(x, -i\frac{\partial}{\partial x}\right) = \widetilde{D}\left(x, -ih\frac{\partial}{\partial x}, h\right),$$

where $\widetilde{D}(x,p,h)$ regularly depends on $h$ (and is even a polynomial in $h$). Moreover,

$$\sigma(D) = \tilde{D}(x,p,0);$$

i.e., the interior symbol of $D$ viewed as a usual differential operator coincides with the principal symbol of $\widetilde{D}$ viewed as a semiclassical pseudodifferential operator.

From now on, we deal only with the new operator and omit the tilde. Sternin and Shatalov (1996*a*) showed that $D$ has an almost inverse in the form of a semiclassical pseudodifferential operator,

$$R = R\left(x, -ih\frac{\partial}{\partial x}\right).$$

(Their assertion pertains to smooth manifolds; however, it is local and hence we can use it in a neighborhood of the interior fixed point, where the manifold is smooth.) Moreover, one can always ensure that modulo $O(h^l)$, where $l$ is arbitrarily large, the symbols of the operators $1 - RD$ and $1 - DR$ are equal to unity on a compact set of values of $p$ containing the zero section. Now the formula for the contribution of the Lefschetz number acquires the form

$$\begin{aligned}\mathcal{L}_{\text{int}}(x^*) = &\frac{1}{(2\pi h)^n}\int_{\mathbb{R}^{2n}} A_1(x)e^{ip(g(x)-x)/h}\phi_1(x,p,h)\,dp\,dx \\ &- \frac{1}{(2\pi h)^n}\int_{\mathbb{R}^{2n}} A_2(x)e^{ip(g(x)-x)/h}\phi_2(x,p,h)\,dp\,dx,\end{aligned} \tag{11.17}$$

where the functions $\phi_{1,2}$ smoothly depend on $h$, are compactly supported in $x$, decay as $|p| \to \infty$, and are equal, modulo $O(h)^N$ for large $N$, to unity in a neighborhood of the section $\{p = 0\}$ near the point $x = x^*$. We compute the asymptotics of the integrals (11.17) by the stationary phase method. The phase function in both integrals is the same and has the form

$$\Phi(x,p) = p(g(x) - x). \tag{11.18}$$

Let us write out the stationary point equations. They have the form

$$\frac{\partial\Phi}{\partial p} \equiv g(x) - x = 0; \tag{11.19}$$

$$\frac{\partial\Phi}{\partial x} \equiv p(g_*(x) - 1) = 0. \tag{11.20}$$

It follows from (11.19) that $x = g(x)$, i.e., $x = x^*$ is a fixed point; now it follows from (11.20), since the fixed point is nondegenerate, that $p = 0$. The Hessian matrix of the phase function at the stationary point has the form

$$\operatorname{Hess}(\Phi) = \begin{pmatrix} 0 & g_*(x^*) - 1 \\ {}^t g_*(x^*) - 1 & 0 \end{pmatrix}, \tag{11.21}$$

and hence

$$\det \operatorname{Hess}(-\Phi) = (-1)^n |\det(1 - g_*(x^*))|^2. \tag{11.22}$$

By applying the formulas of the stationary phase method (e.g., Mishchenko, Shatalov and Sternin 1990), we find that the contribution of the fixed point is given by the formula

$$\mathcal{L}_{\text{int}}(x^*) = \frac{\operatorname{Trace} A_1(x^*) - \operatorname{Trace} A_2(x^*)}{|\det(1 - g_*(x^*))|} + O(h). \tag{11.23}$$

Since this contribution is actually independent of $h$, we can omit the term $O(h)$ and obtain the desired expression.

### 11.2.3. Contribution of conical fixed points

Let $\alpha^*$ be a conical fixed point. We use the conical coordinates $(r, \omega)$ in a neighborhood of $\alpha^*$. Let us introduce a function $f(r)$ supported in a neighborhood of zero and equal to 1 for small $r$ and a function $\psi(r)$, also supported in a neighborhood of zero, such that $\psi(r)f(r) = f(r)$. One can choose these functions in such a way that

$$\psi(\lambda^{-1}r) f\big(\lambda^{-1} g_1(r, \omega)\big) = f\big(\lambda^{-1} g_1(r, \omega)\big)$$

for all $\lambda < 1$, where $g_1$ is the $r$-component of $g$. Then it follows from the results of Section 11.2.1 that for all sufficiently small $\lambda$ the contribution of $\alpha^*$ is given by the expression

$$\begin{aligned} \mathcal{L}_{\text{sing}} &= \operatorname{Trace} T_1 \big[\varkappa_\lambda^{-1} f(r) \varkappa_\lambda - \varkappa_\lambda^{-1} \psi(r) \varkappa_\lambda R \varkappa_\lambda^{-1} f(r) \varkappa_\lambda D\big] \\ &\quad - \operatorname{Trace} T_2 \big[\varkappa_\lambda^{-1} f(r) \varkappa_\lambda - D U_\lambda \psi(r) \varkappa_\lambda R \varkappa_\lambda^{-1} f(r) \varkappa_\lambda\big], \end{aligned} \tag{11.24}$$

where, as before, Trace is the trace integral (see Remark 11.5) and $\varkappa_\lambda$ is the dilation group (11.4).

In particular, the right-hand side of (11.24) is independent of $\lambda$. We use the identity

$$\operatorname{Trace} \varkappa_\lambda W \varkappa_\lambda^{-1} = \operatorname{Trace} W,$$

pass to the limit as $\lambda \to 0$, and obtain

$$\begin{aligned} \mathcal{L}_{\text{sing}} &= \operatorname{Trace} T_{10} \big[f(r) - \psi(r) R_0 f(r) D_0\big] \\ &\quad - \operatorname{Trace} T_{20} \big[f(r) - D_0 \psi(r) R_0 f(r)\big], \end{aligned} \tag{11.25}$$

where the operators $D_0$, $T_{j0}$, and $R_0$ on the infinite cone are given by the expressions

$$D_0 = \sigma_c(D)\left(ir\frac{\partial}{\partial r}\right), \quad T_{j0} = \sigma_c(T_j)\left(ir\frac{\partial}{\partial r}\right), \quad R_0 = \sigma_c(R)\left(ir\frac{\partial}{\partial r}\right).$$

LEMMA 11.6. *The expression* (11.25) *is equal to the corresponding term* $\mathcal{L}_{\text{sing}}(\alpha^*)$ *on the right-hand side in* (11.7).

This completes the proof of Theorem 11.3. It remains to give the computations that prove the lemma.

*Proof of the lemma.* Note that $D_0R_0 = 1$ and $R_0D_0 = 1$, and so we can simplify (11.25) by commuting the operator $D_0$ with the functions $f(r)$ and $\psi(r)$ in the first and second term, respectively. Since $f\psi = f$, we see that only terms containing commutators remain in the expression:

$$\mathcal{L}_{\text{sing}} = \operatorname{Trace} T_{10}\psi(r)R_0[D_0, f(r)] - \operatorname{Trace} T_{20}[D_0, \psi(r)]R_0 f(r).$$

Since $D_0$ is a differential operator and does not enlarge supports, it follows from the construction of $f$ and $\psi$ that the integrand in the second integral vanishes identically, while in the first integral the factor $\psi$ is equal to unity on the support of the integrand and hence can be omitted. Consequently, we obtain

$$\mathcal{L}_{\text{sing}} = \operatorname{Trace} T_{10}R_0[D_0, f(r)]. \tag{11.26}$$

A straightforward computation shows that the commutator $[D_0, f(r)]$ can be represented in the form

$$[D_0, f(r)] = i\frac{\partial \sigma_c(D)}{\partial p}\left(ir\frac{\partial}{\partial r}\right) r\frac{\partial f}{\partial r}(r) + \sum_{l=2}^{m} B_l\left(ir\frac{\partial}{\partial r}\right)\left[\left(r\frac{\partial}{\partial r}\right)^l f(r)\right],$$

where the $B_l(p)$ are some conormal symbols. By substituting this into (11.26), we obtain

$$\begin{aligned}\mathcal{L}_{\text{sing}} = i \operatorname{Trace} T_{10}R_0\frac{\partial \sigma_c(D)}{\partial p}\left(ir\frac{\partial}{\partial r}\right) r\frac{\partial f}{\partial r}(r)\\ + \sum_{l=2}^{m} \operatorname{Trace} T_{10}R_0B_l\left(ir\frac{\partial}{\partial r}\right)\left[\left(ir\frac{\partial}{\partial r}\right)^l f(r)\right].\end{aligned} \tag{11.27}$$

Let us compute the first term. We pass to the cylindrical coordinates ($r = e^{-t}$). Let $Z(\omega, \omega', t - t')$ be the kernel of the operator

$$\sigma_c(D)^{-1}\left(-i\frac{\partial}{\partial t}\right)\frac{\partial \sigma_c(D)}{\partial p}\left(-i\frac{\partial}{\partial t}\right).$$

Then

$$\operatorname{Trace} T_{10} R_0 \frac{\partial \sigma_c(D)}{\partial p} \left(-i\frac{\partial}{\partial t}\right) \frac{\partial}{\partial t} f(e^{-t}) = \int_{\Omega\times\mathbb{R}} A_0(\omega) Z(\omega, g_{\alpha^*}(\omega), b(0,\omega)) \frac{\partial}{\partial t} f(e^{-t})\, d\omega\, dt. \quad (11.28)$$

The integral with respect to $t$ is equal to $-1$ by the definition of $f(r)$, and we obtain

$$\operatorname{Trace} T_{10} R_0 \frac{\partial \sigma_c(D)}{\partial p} \left(-i\frac{\partial}{\partial t}\right) \frac{\partial}{\partial t} f(e^{-t}) = \int_{\Omega} A_0(\omega) Z(\omega, g_{\alpha^*}(\omega), b(0,\omega))\, d\omega. \quad (11.29)$$

Since $b(0,\omega) \neq 0$, it follows that $Z(\omega, g_{\alpha^*}(\omega), b(0,\omega))$ is infinitely differentiable with respect to $\omega$. (This is the kernel of a pseudodifferential operator away from the diagonal.) We multiply and divide the integrand by $b(0,\omega)^l$, where $l$ is sufficiently large, and use the properties of the Fourier transform to obtain

$$i \operatorname{Trace} T_{10} R_0 \frac{\partial \sigma_c(D)}{\partial p} \left(-i\frac{\partial}{\partial t}\right) \frac{\partial}{\partial t} f(e^{-t}) = \frac{1}{2\pi i} \operatorname{Trace} \int_{-\infty}^{\infty} \left(\frac{i}{b(0,\omega)}\right)^l \sigma_c(T_1)(p) \left(\frac{\partial}{\partial p}\right)^l \left(\sigma_c(D)^{-1}(p) \frac{\partial \sigma_c(D)(p)}{\partial p}\right) dp, \quad (11.30)$$

which is exactly the trace of the desired regularization (11.8). It remains to note that all other terms in (11.27) are zero, since, computing them in a similar way, we can use the fact that

$$\int_0^{\infty} \left(\frac{\partial}{\partial t}\right)^l f(e^{-t})\, dt = 0$$

for $l \geq 2$ by our assumptions. □

## 11.3. Contributions of Conical Points as Sums of Residues

Under certain conditions on the conormal symbol of $D$, the expression for the contribution of conical fixed points can be computed in terms of residues.

Since $D$ is a *differential* operator, we see that its conormal symbol at each singular point is just a polynomial in $p$ (a polynomial operator pencil on the corresponding manifold $\Omega_{\alpha_j}$). By the ellipticity, it is Agranovich–Vishik elliptic with parameter $p$ (Agranovich and Vishik 1964) in some double sector of nonzero opening angle containing the real axis and is finite-meromorphically invertible in the entire complex plane. All but finitely many poles of the inverse family lie in the complement of the above-mentioned sector.

It follows that each strip $\{|\operatorname{Im} p| < R\}$ contains only finitely many poles of the family $\sigma_c(D)^{-1}(\alpha_j, p)$ and the principal part of the Laurent series at each pole is of finite rank. We impose some technical conditions on the behavior of the number of these poles and some of their characteristics as $R \to \infty$.

DEFINITION 11.7. We say that a conormal symbol $B(p)$ of order $m$ is of *power type* if the following conditions are satisfied.

1) The orders of the poles of $B^{-1}(p)$ are bounded by some constant.

2) The number of the poles of $B^{-1}(p)$ in the strip $\{|\operatorname{Im} p| < R\}$ grows as $R \to \infty$ no faster than some power of $R$.

3) The ranks and the operator norms in $L^2(\Omega_{\alpha_j})$ of the coefficients in the principal parts of the Laurent series for $B^{-1}(p)$ around the poles in the strip $\{|\operatorname{Im} p| < R\}$ are bounded above by some power of $R$ as $R \to \infty$.

4) There exists a sequence of circles centered at the origin with radius tending to infinity such that $B^{-1}(p)$ has no poles on these circles and can be estimated for every $s$ in the operator norm $H^s(\Omega_{\alpha_j}) \to H^{s+m}(\Omega_{\alpha_j})$ by some power of the radius (which may depend on $s$).

*Remark* 11.8. In some cases (e.g., for the Beltrami–Laplace operator associated with a conical metric on $M$) the power type property of the conormal symbol can be established with the use of results on the spectral asymptotics for self-adjoint elliptic pseudodifferential operators on closed manifolds (e.g. Hörmander 1968, Shubin 1985, Martinez 1997).

THEOREM 11.9. *Suppose that the assumptions of Theorem* 11.3 *are satisfied and the conormal symbol* $\sigma_c(D)(\alpha^*, p)$ *of the operator* $D$ *at some conical fixed point* $\alpha^*$ *is of power type. Then the contribution of this conical point can be represented by the absolutely convergent series of traces of finite rank operators given by the residues of the integrand*:

$$\mathcal{L}_{\text{sing}}(\alpha^*) = \sum_{\operatorname{Im} p_j > \gamma(\alpha^*)} \operatorname{Trace} \operatorname*{Res}_{p=p_j} \left\{ \sigma_c(T_1)(\alpha^*, p)\sigma_c(D)^{-1}(\alpha^*, p) \frac{\partial \sigma_c(D)}{\partial p}(\alpha^*, p) \right\} \tag{11.31}$$

*if* $\alpha^*$ *is an attractive point, and*

$$\mathcal{L}_{\text{sing}}(\alpha^*) = -\sum_{\operatorname{Im} p_j < \gamma(\alpha^*)} \operatorname{Trace} \operatorname*{Res}_{p=p_j} \left\{ \sigma_c(T_1)(\alpha^*, p)\sigma_c(D)^{-1}(\alpha^*, p) \frac{\partial \sigma_c(D)}{\partial p}(\alpha^*, p) \right\} \tag{11.32}$$

*if* $\alpha^*$ *is a repulsive point. Here the* $p_j$ *are the poles of the family* $\sigma_c(D)^{-1}(\alpha^*, p)$ *and* $\gamma(\alpha^*)$ *is the weight exponent at the point* $\alpha^*$.

*Proof.* Without loss of generality, we can assume that $\gamma(\alpha^*) = 0$. Let $\sigma_c(D)(p)$ be the conormal symbol of $D$ at $\alpha^*$. Suppose that $\sigma_c(D)(p)$ is of power type.

Let $S_k$ be a family of circles of radii $R_k \to \infty$ on which $\sigma_c(D)^{-1}(p)$ satisfies a power-law estimate of the norm according to item 4) of Definition 11.7.

Since $\sigma_c(D)(p)$ is a polynomial of $p$, one can readily prove by induction that all derivatives $\partial^l H(p)/\partial p^l$ satisfy power-law estimates of the norm on these circles in appropriate pairs of Sobolev spaces.

Since the norm of the operator $\partial^l H(p)/\partial p^l$ decays polynomially as $p \to \infty$ in the above-mentioned double sector of nonzero angle containing the real axis, we see that the integration contour in the integral on the right-hand side of (11.8) can be closed in the half-plane where the exponential $e^{ib(0,\omega)p}$ decays. (This is the upper half-plane for attracting points and the lower half-plane for repulsive points.) More precisely, consider the sequence of contours consisting of the segments $[-R_k, R_k]$ of the real axis and the corresponding half-circles $S_k$ in the upper or lower half-plane. The integral over the half-circle tends to zero as $k \to \infty$ by virtue of the above considerations, and in the limit we find that the integral over the real axis is equal to the limit of the sums of residues in the half-disks bounded by these half-circles and the real axis. It follows from items 1)–3) of Definition 11.7 that this limit is just equal to the sum of residues in the corresponding half-plane. (This sum converges absolutely.) These residues coincide with the residues of the integrand on the left-hand side, since formal integration by parts does not affect the residues at the poles. (This is a trivial consequence of the Cauchy integral formula applied to the integral over a small circle centered at a pole.)

It remains to prove that one can transpose summation and trace computation. (Recall that so far we have established the convergence of the series in the operator rather than the trace norm.) This is however trivial. The terms of the series are finite rank operators, and for such operators the trace norm does not exceed the operator norm times the rank. It follows from items 1)–3) that the total rank and the total operator norm of the coefficients of the principal parts of the Laurent series at the poles of the family $H(p)$ in the strip $|\operatorname{Im} p| < R$ grow at most polynomially in $R$, and the presence of the exponential factor $e^{ib(0,\omega)p}$ provides the absolute convergence of the series in the trace norm as well. □

## 11.4. Supplement. The Lefschetz Number

Let

$$0 \longrightarrow E_0 \xrightarrow{D_0} E_1 \xrightarrow{D_1} \cdots \xrightarrow{D_{m-1}} E_m \longrightarrow 0 \qquad (11.33)$$

be a complex of vector spaces over $\mathbb{C}$ with finite-dimensional cohomology, and let

$$T = \{T_j : E_j \to E_j\}\big|_{j=0,\ldots,m} \qquad (11.34)$$

be an endomorphism of the complex (11.33), that is, a collection of linear mappings such that the diagram

$$\begin{array}{ccccccccccc} 0 & \longrightarrow & E_0 & \xrightarrow{D_0} & E_1 & \xrightarrow{D_1} & \cdots & \xrightarrow{D_{m-1}} & E_m & \longrightarrow & 0 \\ & & \Big\downarrow T_0 & & \Big\downarrow T_1 & & & & \Big\downarrow T_m & & \\ 0 & \longrightarrow & E_0 & \xrightarrow[D_0]{} & E_1 & \xrightarrow[D_1]{} & \cdots & \xrightarrow[D_{m-1}]{} & E_m & \longrightarrow & 0 \end{array} \tag{11.35}$$

commutes. It follows from the commutativity of this diagram that

$$T_j(\operatorname{Ker} D_j) \subset \operatorname{Ker} D_j, \quad T_j(\operatorname{Im} D_{j-1}) \subset \operatorname{Im} D_{j-1}, \quad j = 0, \ldots, m,$$

where $\operatorname{Ker} A$ and $\operatorname{Im} A$ are the kernel and range of the operator $A$, respectively, and by definition $D_{-1}$ and $D_m$ are zero maps. Hence the endomorphism (11.34) induces well-defined operators

$$\widetilde{T}_j : H^j(E) \to H^j(E), \quad j = 0, \ldots, m, \tag{11.36}$$

on the cohomology spaces $H^j(E) = \operatorname{Ker} D_j / \operatorname{Im} D_{j-1}$ of the complex (11.33), and the *Lefschetz number* of the endomorphism (11.34) is defined as the alternating sum

$$\mathcal{L} \equiv \mathcal{L}(D, T) = \sum_{j=0}^{m} (-1)^j \operatorname{Trace} \widetilde{T}_j \tag{11.37}$$

of traces of the finite-dimensional operators (11.36). For $m = 1$, the complex consists of the single operator $D = D_0$. In this case, the Lefschetz number has the form

$$\mathcal{L} = \operatorname{Trace} T_0 \big|_{\operatorname{Ker} D} - \operatorname{Trace} T_1 \big|_{\operatorname{Coker} D}. \tag{11.38}$$

## 11.5. Supplement. The Sternin–Shatalov Method

The semiclassical technique that we have used in the proof of the theorem arises from the semiclassical method in Lefschetz theory due to Sternin and Shatalov, and here we give a brief account of their ideas.

The class of geometric endomorphisms is not a natural framework for the problem on the Lefschetz number if one does not restrict oneself to complexes of differential operators but has in mind also pseudodifferential operators. Indeed, as differential operators form a subclass of the more general class of pseudodifferential operators, so geometric endomorphisms form a subclass of the class of Fourier integral operators, and hence one can naturally try to obtain a Lefschetz type formula for the case in which (11.33) is an elliptic complex of *pseudodifferential* operators and the endomorphism (11.34) is given by a set of Fourier integral operators. It turns out that once we pass to endomorphisms associated with mappings of the phase space, it is more convenient to deal with *asymptotic* theory. To obtain meaningful formulas, one should introduce a small parameter $h \in (0, 1]$ and consider semiclassical (or $1/h$-)pseudodifferential operators (e.g., Maslov 1973, Mishchenko, Shatalov and Sternin 1990) and Fourier–Maslov integral operators associated with a canonical transformation

$$g : T^*M \longrightarrow T^*M.$$

Thus, it is *symplectic* rather than contact geometry that underlies the Lefschetz formula. Then the Lefschetz number depends on $h$, and under appropriate assumptions about the fixed points of $g$ one obtains an expression for the asymptotics of the Lefschetz number as $h \to 0$ by applying the stationary phase method to the trace integrals representing this number.

Namely, the Lefschetz number of the endomorphism (11.34) (with $m = 1$ for simplicity) is given by the formula (11.9). Now if $D$ and $R$ are pseudodifferential operators and $T_0$ and $T_1$ are Fourier–Maslov integral operators, then $T_0(1 - RD)$ and $T_1(1 - DR)$ are also Fourier–Maslov integral operators associated with the same canonical transformation as $T_0$ and $T_1$. Hence, the problem is reduced to the evaluation of *traces of Fourier–Maslov integral operators*. This is carried out with the help of the stationary phase method; only fixed points of $g$ give a nonzero contribution to the asymptotics of these traces as $h \to 0$. As usual in the stationary phase method, the contribution of each isolated component of the set of fixed points can be treated separately; in other words, the *microlocalization principle* holds in this case for the Lefschetz number.

We see that the semiclassical method provides a straightforward computation of the Lefschetz number. Let us now write out the Lefschetz formula for semiclassical endomorphisms of elliptic complexes on smooth manifolds following Sternin and Shatalov (1998).

Let $M$ be a smooth closed manifold of dimension $n$. We suppose that $M$ is oriented and is equipped with a volume form $dx$. For a canonical transformation

$$g : T^*M \to T^*M \tag{11.39}$$

and a smooth function $\varphi$ on $T^*M$ satisfying appropriate conditions at infinity in the fibers, by $T(g, \varphi)$ we denote the semiclassical Fourier–Maslov integral operator (i.e., a Fourier–Maslov integral operator with small parameter $h \in (0, 1]$) with amplitude $\varphi$ associated with the graph $\operatorname{graph} g$ of the transformation $g$. We assume that the graph of $g$ is a quantized Lagrangian submanifold in $T^*M \times T^*M$. (A detailed definition of Fourier–Maslov integral operators can be found in Mishchenko, Shatalov and Sternin (1990), and precise conditions ensuring that $T(g, \varphi)$ is well defined are given in Sternin and Shatalov (1998).)

Consider the commutative diagram

$$\begin{array}{ccccccc} 0 & \longrightarrow & C^\infty(M,F_1) & \xrightarrow{D} & C^\infty(M,F_2) & \longrightarrow & 0 \\ & & \downarrow{\scriptstyle T_1} & & \downarrow{\scriptstyle T_2} & & \\ 0 & \longrightarrow & C^\infty(M,F_1) & \xrightarrow[D]{} & C^\infty(M,F_2) & \longrightarrow & 0, \end{array} \tag{11.40}$$

where $F_1$ and $F_2$ are vector bundles over $M$, $T_j = T(g,\varphi_j)$, $j=1,2$, are Fourier–Maslov integral operators associated with some canonical transformation (11.39), and $D$ is an elliptic operator.[2]

THEOREM 11.10 (Sternin and Shatalov 1998). *Let the transformation $g$ have only isolated nondegenerate fixed points $\alpha_1,\dots,\alpha_N$. (The nondegeneracy condition means that* $\det(1-g_*(\alpha_k))\neq 0,\quad k=1,\dots,N$, *where $g_*(\alpha_k)$ is the differential of $g$ at the point $\alpha_k$.) Then the Lefschetz number of the diagram* (11.40) *has the asymptotics*

$$\mathcal{L}=\sum_{k=1}^{N}\exp\left(\frac{i}{h}S_k\right)\frac{\operatorname{Trace}\varphi_1(\alpha_k)-\operatorname{Trace}\varphi_2(\alpha_k)}{\sqrt{\det\left(1-g_*(\alpha_k)\right)}}+O(h), \tag{11.41}$$

*where $\varphi_1$ and $\varphi_2$ are the amplitudes of $T_1$ and $T_2$, $S_k$ is the value of the generating function of $g$ (the choice of which is fixed in the definition of the Fourier–Maslov integral operators $T_1$ and $T_2$) at the point $\alpha_k$, and the branch of the square root is chosen according to the stationary phase method.*[3]

### 11.5.1. Application to the classical Atiyah–Bott–Lefschetz theorem

The theorem due to Atiyah and Bott (1967) deals with the case in which (11.33) is an elliptic complex of differential operators on a smooth closed manifold $M$ and (11.34) is a geometric endomorphism associated with a smooth mapping $f: M\longrightarrow M$. Suppose that the fixed points of $f$ are nondegenerate in the sense that

$$\det\left(1-\frac{\partial f}{\partial x}(x)\right)\neq 0,\quad x\in\operatorname{fix}(f). \tag{11.42}$$

(Here $\operatorname{fix}(f)$ is the set of fixed points of $f$.) Then they are isolated, and the Atiyah–Bott–Lefschetz theorem states that the Lefschetz number $\mathcal{L}$ can be expressed by the formula

$$\mathcal{L}=\sum_{x\in\operatorname{fix}(f)}\frac{\sum_{j=0}^{m}(-1)^j\operatorname{Trace}A_j(x)}{\left|\det\left(1-\frac{\partial f}{\partial x}(x)\right)\right|}. \tag{11.43}$$

[2]By definition, this means that there exists a $1/h$-pseudodifferential operator $R$ on $M$ such that the symbols of the operators

$$1-DR \quad\text{and}\quad 1-RD$$

belong to the Hörmander class $S^{-\infty}(T^*M)$ uniformly with respect to $h\in(0,1]$.

[3]See the explicit formulas in (Sternin and Shatalov 1998).

Thus the Lefschetz number of the endomorphism (11.34) is expressed in classical terms. Note also that the operators $D_j$ themselves do not occur in (11.43); only the *existence* of a complex (11.33) that makes the diagram (11.35) commute is relevant to the theorem.

In fact, this classical result follows from Theorem 11.10 (strictly speaking, for short complexes, but the general case can be handled in a similar manner). The computations are identical to those used in Section 11.2.2, and hence we omit them.

# Appendix A
# Spectral Flow

In this appendix, we recall the notion of spectral flow for families of self-adjoint operators (Atiyah, Patodi and Singer 1976) and families of parameter-dependent operators (Nazaikinskii and Sternin 1999). We also discuss higher spectral flows (Dai and Zhang 1998, Melrose and Piazza 1997).

## A.1. The Classical Spectral Flow

### A.1.1. Definition and main properties

***The finite-dimensional case.*** Consider a continuous family $A_t$, $t \in [0,1]$, of Hermitian matrices. Informally, the *spectral flow* of this family is the net number of eigenvalues passing through zero as $t$ varies from 0 to 1. Needless to say, one can easily give a rigorous definition in the finite-dimensional case: the spectral flow is just half the difference of signatures of the corresponding quadratic forms at the beginning and end of the homotopy.

In what follows, we show that although the number of eigenvalues in the case of operators can be infinite (so that the above definition in terms of signatures becomes meaningless), one still can give a well-defined notion of spectral flow at least for *elliptic* self-adjoint operators.

***The infinite-dimensional case.*** Let $\{A_\tau\}_{\tau\in[0,1]}$ be a continuous family of elliptic self-adjoint operators on a closed compact manifold $M$. We assume that the operators have positive order.

DEFINITION A.1. The *spectral flow* of the family $\{A_\tau\}_{\tau\in[0,1]}$ is the integer equal to the net number of eigenvalues of $A_\tau$ that change their sign as $\tau$ varies, the change of sign from minus to plus being counted as $+1$ (see Fig. A.1). The spectral flow is denoted by $\operatorname{sf}\{A_\tau\}_{\tau\in[0,1]}$.

We point out that this definition makes sense only in general position, where the graph of the spectrum of the family and the line $\lambda = 0$ intersect transversally. Otherwise, one obtains a well-defined formula for the spectral flow by bringing the intersection to general position. The simplest way to do this is to perturb the line $\lambda = 0$ slightly and replace it by a broken line (see Fig. A.2) with alternating horizontal and vertical segments such that the horizontal segments avoid the spectrum of the family. Such broken lines are said to be *admissible*.

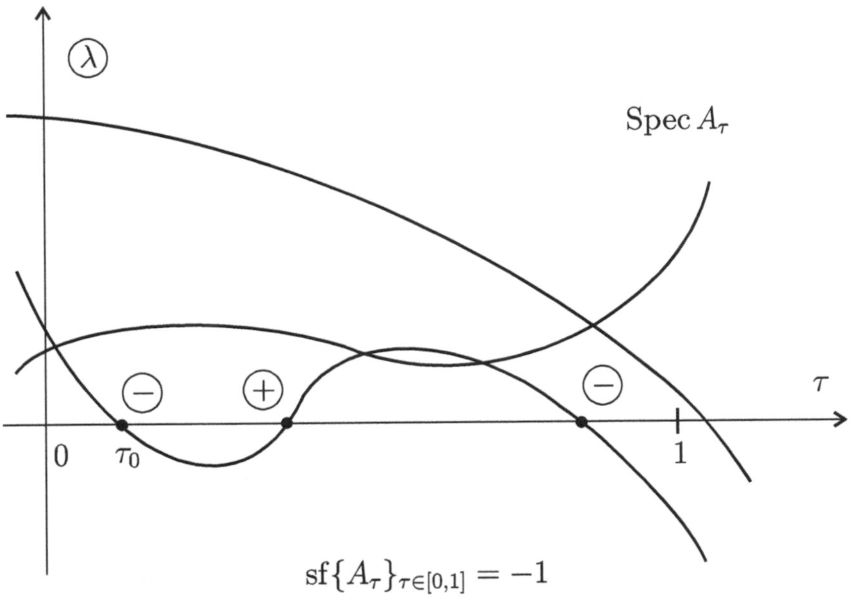


Figure A.1. The spectral flow.

LEMMA A.2. *Admissible broken lines always exist.*

*Proof.* This follows from the fact that the spectrum of an elliptic self-adjoint operator is discrete. □

Using an admissible broken line with vertices $\{(\tau_i, \lambda_i)\}_{i=0,N}$, one can compute[1] the spectral flow as the net sum over all vertical segments of the numbers of eigenvalues (counted according to their multiplicities) that leave the region under the broken line:

$$\mathrm{sf}\,\{A_\tau\}_{\tau\in[0,1]} = \sum_{i=1}^{N-1} (-1)^{\mathrm{sgn}(\lambda_{i-1}-\lambda_i)} N(\lambda_i, \lambda_{i-1}), \tag{A.1}$$

where

$$N(\lambda_i, \lambda_{i-1}) = \#\left\{\lambda \in \mathrm{spec} A_{\tau_i} \middle| \lambda \text{ lies between } \lambda_i \text{ and } \lambda_{i-1}\right\}.$$

(Here # stands for the cardinality of a set.)

The boundary points $\tau = 0$ and $\tau = 1$ deserve special attention. Here the choice of the horizontal segment is ambiguous. We adopt the following convention: if the family is invertible at these points, then the first (and, accordingly, the last) segment

[1] Or, more precisely, one can define.

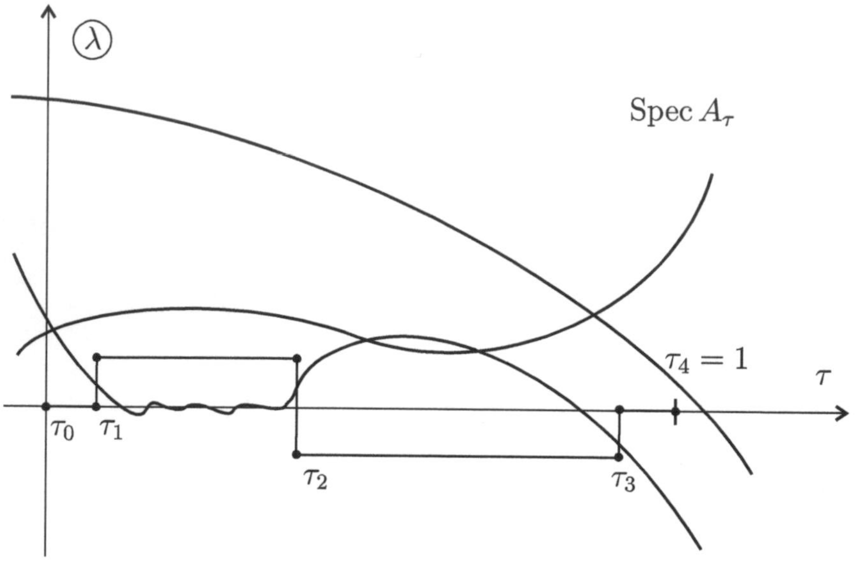


Figure A.2. The spectral flow. Computation via a broken line.

lies on the abscissa axis $\lambda = 0$. In the general case, these segments are taken below the abscissa axis so close to it that no eigenvalues of $A_0$ and $A_1$ lie between the abscissa axis and the corresponding horizontal segment.

The main properties of the spectral flow are given by the following theorem.

THEOREM A.3. *The spectral flow* (A.1) *is well defined, i.e., independent of the choice of an admissible broken line, and has the following properties:*

1) (*Additivity*) *The spectral flow of the concatenation of two homotopies is the sum of the spectral flows of the two parts*:

$$\mathrm{sf}\{A_t\}_{t\in[0,2]} = \mathrm{sf}\{A_t\}_{t\in[0,1]} + \mathrm{sf}\{A_t\}_{t\in[1,2]}.$$

2) (*Homotopy invariance*) *The spectral flow is invariant under deformations of the family* $\{A_t\}_{t\in[0,1]}$ *such that the operators* $A_0$ *and* $A_1$ *either do not change at all or change in such a way that the multiplicity of the zero eigenvalue remains constant.*

*Proof.* To prove that the notion of spectral flow is well defined, it suffices to verify that the expression (A.1) is invariant under the following transformations of the admissible broken line:

- subdivision of segments or taking the union of neighboring adjacent horizontal segments

- a vertical displacement of a horizontal segment

Indeed, two arbitrary admissible broken lines can be transformed into each other by a sequence of such transformations via a sequence of admissible broken lines.

The invariance of the spectral flow with respect to the first transformation is fairly obvious. As to the second transformation, one should exhaust all possible mutual arrangements of the segment to be moved and its nearest neighbors. This is an easy but somewhat painstaking task, for there are eight distinct possibilities. However, one can avoid the exhaustion by rewriting the expression for the spectral flow in terms of the index of Fredholm operators as follows:

$$\operatorname{sf}\{A_\tau\}_{\tau\in[0,1]} = \sum_{i=1}^{N-1} \operatorname{ind}(\Pi_{\lambda_i}(A_{\tau_i}), \Pi_{\lambda_{i-1}}(A_{\tau_i})). \tag{A.2}$$

Here $\Pi_\lambda(A)$ is the *spectral projection* onto the subspace spanned by the eigenvectors of a self-adjoint operator $A$ with eigenvalues $\geq \lambda$, and $\operatorname{ind}(P, Q)$ is the *relative index of projections* $P$ and $Q$ with compact difference, defined as the index of the following Fredholm operator:

$$\operatorname{ind}(P, Q) \stackrel{\text{def}}{=} \operatorname{ind}(Q : \operatorname{Im} P \longrightarrow \operatorname{Im} Q).$$

EXERCISE A.4. Check that the expressions (A.1) and (A.2) for the spectral flow are equivalent.

Now if one broken line is obtained from another by the replacement $\lambda_i \mapsto \lambda_i'$ in a single segment, then the difference of the spectral flows computed via these broken lines is equal to the difference

$$\operatorname{ind}(\Pi_{\lambda_i}(A_{\tau_i}), \Pi_{\lambda_i'}(A_{\tau_i})) - \operatorname{ind}(\Pi_{\lambda_i}(A_{\tau_{i+1}}), \Pi_{\lambda_i'}(A_{\tau_{i+1}})) \tag{A.3}$$

of relative indices. To obtain this relation, it is convenient to use the cyclic property

$$\operatorname{ind}(P, Q) + \operatorname{ind}(Q, R) = \operatorname{ind}(P, R)$$

of the relative index of projections.

Now the difference (A.3) is zero by virtue of the following fact of functional analysis: if a continuous family of self-adjoint operators does not have eigenvalues $\lambda$ and $\mu$, then the number of eigenvalues (counted according to their multiplicities) between $\lambda$ and $\mu$ is constant. Thus we have proved that the spectral flow is well defined. Let us prove properties 1) and 2).

1) Additivity now follows by construction.

2) Homotopy invariance. Formula (A.2) shows that the spectral flow is invariant under small deformations of the family $\{A_t\}_{t\in[0,1]}$ as long as the broken line remains admissible. Now the invariance of the spectral flow under large deformations follows with regard to the fact that the definition is independent of the choice of an admissible broken line. □

*Remark* A.5. The reader has possibly noticed that the definition of the spectral flow essentially uses only the self-adjointness of operators and the discreteness of the spectrum in the vicinity of zero. Thus the definition is actually given in the framework of functional analysis. A detailed discussion of the spectral flow from the viewpoint of functional analysis can be found in the literature.

### A.1.2. The spectral flow formula for periodic families

First, note that the knowledge of the principal symbol is not sufficient for computing the spectral flow; this is already seen in the matrix case, where the symbol is zero identically. However, the spectral flow is completely determined by the principal symbol in the class of *periodic* families ($A_0 = A_1$). In this subsection, we compute the spectral flow in terms of the principal symbol.[2]

In fact, the spectral flow formula is written out in terms of some vector bundle. To construct this bundle, recall that the principal symbol of an elliptic self-adjoint operator $A$ on a manifold $M$ determines the vector bundle[3]

$$\operatorname{Im}\Pi_+\sigma(A) \in \operatorname{Vect}(S^*M)$$

over the bundle $S^*M$ of unit spheres (with respect to some Riemannian metric) in the cotangent bundle $T^*M$. The vector bundle is constructed as follows. The symbol of an elliptic self-adjoint operator is Hermitian and invertible, and so its positive spectral projection $\Pi_+\sigma(A)$ is a smooth function on the spheres. The desired vector bundle is just the range of this projection.

THEOREM A.6. *The spectral flow of a periodic family $\mathcal{A} = \{A_t\}_{t\in\mathbb{S}^1}$ of elliptic self-adjoint operators is given by the formula*

$$\operatorname{sf}\{A_t\}_{t\in\mathbb{S}^1} = \left\langle \operatorname{ch}\left[\operatorname{Im}\Pi_+\sigma(\mathcal{A})\right]\operatorname{Td}\left(T^*M\otimes\mathbb{C}\right), \left[S^*M\times\mathbb{S}^1\right]\right\rangle. \tag{A.4}$$

*Here* $\operatorname{ch}\left[\operatorname{Im}\Pi_+\sigma(\mathcal{A})\right] \in H^{ev}\left(S^*M\times\mathbb{S}^1\right)$ *is the Chern character of the bundle*

$$\operatorname{Im}\Pi_+\sigma(\mathcal{A}) \in \operatorname{Vect}\left(S^*M\times\mathbb{S}^1\right)$$

*determined by the principal symbol of the family,* Td *stands for the Todd class, and* $\langle\cdot,[S^*M\times\mathbb{S}^1]\rangle$ *is the value of a cohomology class of the top degree on the fundamental cycle.*

The idea of derivation of the formula (A.4) is to express the spectral flow as the index of some elliptic operator on a closed manifold and then apply the Atiyah–Singer formula. We state these two parts (analytical and topological) of the proof in the form of the following two propositions.

[2] And hence in topological terms.

[3] Here and in the following, $\operatorname{Vect}(N)$ is the set of vector bundles on a manifold $N$.

PROPOSITION A.7. *The spectral flow of a periodic family $\{A_t\}$ is equal to the index of the elliptic operator*

$$D = \frac{\partial}{\partial t} + A_t, \quad t \in \mathbb{S}^1, \tag{A.5}$$

*on the torus $M \times \mathbb{S}^1$.*

This is a restatement of Proposition 5.6 for first-order operators. □

PROPOSITION A.8. *One has*

$$\operatorname{ind} D = \left\langle \operatorname{ch}\left[\operatorname{Im}\Pi_+\sigma(\mathcal{A})\right] \operatorname{Td}\left(T^*M \otimes \mathbb{C}\right), \left[S^*M \times \mathbb{S}^1\right]\right\rangle . \tag{A.6}$$

*Sketch of proof.* The Atiyah–Singer formula

$$\operatorname{ind} D = \left\langle \operatorname{ch}[\sigma(D)]\operatorname{Td}(T^*M \otimes \mathbb{C}), [T^*(M \times \mathbb{S}^1)]\right\rangle \tag{A.7}$$

expresses the index of the operator $D$ via the Chern character of the difference element $[\sigma(D)] \in K_c^0(T^*(M \times \mathbb{S}^1))$ in $K$-theory of the cotangent space. This element is determined by the principal symbol of the operator. To rewrite the formula (A.7) in terms of the original family $A_t$, we note that the element $[\sigma(D)]$ can be expressed via the symbol of the original family by the formula

$$[\sigma(D)] = \partial[\operatorname{Im}\Pi_+\sigma(\mathcal{A})] \tag{A.8}$$

as the image of the element $[\operatorname{Im}\Pi_+\sigma(\mathcal{A})] \in K(S^*M \times \mathbb{S}^1)$ under the coboundary mapping

$$\partial : K(S^*M \times \mathbb{S}^1) \longrightarrow K_c^1(T^*M \times \mathbb{S}^1) \simeq K_c(T^*(M \times \mathbb{S}^1))$$

corresponding to the pair $\mathbb{S}^1 \times S^*M \subset \mathbb{S}^1 \times B^*M$. (Here $B^*M$ is the unit ball bundle.) Transferring the computation of the pairing (A.6) from the cotangent bundle $T^*(M \times \mathbb{S}^1)$ to the product $S^*M \times \mathbb{S}^1$, one can show that the Atiyah–Singer formula acquires the desired form (A.6). □

EXAMPLE A.9. Let us give an example of a periodic family with nonzero spectral flow. Such a family exists even on the circle. Indeed, consider the family of first-order operators

$$A_\tau = (2P_{\varphi,\tau} - 1)\left(-i\frac{d}{d\varphi}\right) - i\frac{dP_{\varphi,\tau}}{d\varphi},$$

where $P_{\varphi,\tau}$ in $\mathbb{C}^N$ is a periodic family of orthogonal projections. (The second term ensures that the operators are self-adjoint.) The cosphere bundle of the circle is the disjoint union of two circles. Moreover, the positive spectral bundle of the principal symbol $\sigma(A_\tau)$ coincides with the range of $P_{\varphi,\tau}$ over one circle and with the complement of this range over the other circle:

$$\operatorname{Im} \Pi_+\sigma\left(A_\tau\right)(\varphi, \xi) = \begin{cases} \operatorname{Im} P_{\varphi,\tau} & \text{for } \xi > 0, \\ \operatorname{Im}\left(1 - P_{\varphi,\tau}\right) & \text{for } \xi < 0. \end{cases}$$

Recall that the two connected components of $S^*\mathbb{S}^1$ have opposite orientations. Hence we derive an expression for the spectral flow via the Chern character of the range of the family of projections from the spectral flow formula:

$$\operatorname{sf}\left\{A_\tau\right\}_{\tau\in\mathbb{S}^1} = -2\left\langle \operatorname{ch}\left(\operatorname{Im} P_{\varphi,\tau}\right), \left[\mathbb{S}^1 \times \mathbb{S}^1\right]\right\rangle.$$

If $P_{\varphi,\tau}$ is a projection associated with a nontrivial vector bundle over the torus, then we obtain the desired family with nonzero spectral flow. For example, one can take the trivial one-dimensional bundle on the cylinder $\mathbb{S}^1 \times [0,1]$ and glue the fibers over the bases of the cylinder, twisting them by a nonzero complex function of nonzero degree on the circle. We invite the reader to write out the corresponding projection matrices in closed form and thus obtain a specific family $A_\tau$.

## A.2. The Spectral Flow of Families of Parameter-Dependent Operators

In this section, we define the spectral flow for families of elliptic parameter-dependent operators. Informally speaking, the spectral flow is defined as the net number of zeros of the family crossing some fixed weight line. This notion is similar to the spectral flow of self-adjoint operators. Actually, the spectral flow for self-adjoint operators turns out to be a special case of the spectral flow that we define in this section. Indeed, the self-adjoint theory can be embedded in the theory of parameter-dependent operators by assigning the family $p - iA$ to a self-adjoint operator $A$. This family is analytic, and the spectral flow is just the number of zeros of the family (counted according to their multiplicities) crossing the real axis.

Thus, to define the spectral flow for general families, we should introduce the notion of "multiplicity of zero" for analytic operator functions. From the viewpoint of complex analysis, it is more natural to deal not only with *holomorphic* but also with *meromorphic* operator functions. In this case, one should define the multiplicity of a singular point of a meromorphic operator function.

### A.2.1. Meromorphic operator functions and multiplicities of singular points

DEFINITION A.10. 1. An operator function $D(p) : H \longrightarrow G$ (where $H$ and $G$ are Hilbert spaces) in a domain $U \subset \mathbb{C}$ is said to be *finitely meromorphic* if it is meromorphic and the coefficients of the principal parts of its Laurent series expansions at all poles are finite rank operators.

2. An operator function $D(p)$ is said to be *strongly finitely meromorphic* in $U$ if both $D(p)$ and $D^{-1}(p)$ is finitely meromorphic in $U$.

In a sense, strongly finitely meromorphic operator functions are invertible neglecting a finite-dimensional subspace.

PROPOSITION A.11. *Let $D(p)$ be a strongly finitely meromorphic operator function in a neighborhood of a point $p_0$. Then there exist direct sum decompositions*

$$H = H_0 \oplus H_0^{\perp}, \quad G = G_0 \oplus G_0^{\perp}$$

*with finite-dimensional subspaces $H_0$ and $G_0$ such that in the corresponding block matrix representation*

$$D(p) = \begin{pmatrix} D_{11}(p) & D_{12}(p) \\ D_{21}(p) & D_{22}(p) \end{pmatrix} \tag{A.9}$$

*of $D(p)$ the only infinite-dimensional component $D_{22}(p)$ is holomorphic and invertible*[4] *in a neighborhood of $p_0$.*

*Remark* A.12. This is an operator function counterpart of the following well-known fact: *a Fredholm operator becomes invertible when restricted to appropriate subspaces of finite codimension.*

*Proof.* 1. First, let us ensure the holomorphy. For brevity, we set $p_0 = 0$.

Consider the decompositions $H = H_1 \oplus H_1^{\perp}$ and $G = G_1 \oplus G_1^{\perp}$, where $G_1$ is the (finite-dimensional) linear span of ranges of the coefficients in the principal part of the Laurent series

$$D(p) = \frac{D_{-N}}{p^N} + \ldots + \frac{D_{-1}}{p} + D_0 + D_1 p + \ldots$$

and $H_1$ is the corresponding subspace for the inverse function $D^{-1}(p)$.

In this decomposition, our operator functions have the representations

$$D(p) = \begin{pmatrix} D_{11}(p) & D_{12}(p) \\ D_{21}(p) & D_{22}(p) \end{pmatrix}, \quad D^{-1}(p) = \begin{pmatrix} C_{11}(p) & C_{12}(p) \\ C_{21}(p) & C_{22}(p) \end{pmatrix},$$

where only the blocks in the first row may be nonholomorphic at $p_0$. Let us show that $D_{22}(p)$ is Fredholm up to $p = p_0$. This easily follows from the obvious identities

$$D_{22}(p)C_{22}(p) = 1 - D_{21}(p)C_{12}(p), \quad C_{22}(p)D_{22}(p) = 1 - C_{21}(p)D_{12}(p).$$

Indeed, here the left-hand sides are holomorphic. Hence so are the products of finite rank operator functions on the right-hand sides. Letting $p \to p_0$, we find that $C_{22}(p_0)$ is an almost inverse of $D_{22}(p_0)$.

[4]Note that if an operator function is holomorphic and invertible, then the inverse is also holomorphic.

2. To ensure invertibility, from the subspaces $H_1^\perp$ and $G_1^\perp$ we subtract the kernel and cokernel, respectively, of the Fredholm operator $D_{22}(p_0)$. Thus we finally set $H_0 = H_1 \oplus \operatorname{Ker} D_{22}(p_0)$ and $G_0 = G_1 \oplus \operatorname{Coker} D_{22}(p_0)$. □

Using this proposition, we can prove the following proposition.

PROPOSITION A.13. *A strongly finitely meromorphic operator function $D(p)$ can be represented in a neighborhood of an arbitrary point (even a pole!) as*

$$D(p) = (1 + K(p))Q(p), \tag{A.10}$$

*where $Q(p)$ is holomorphic and invertible and $K(p)$ has a uniformly finite rank:* $\operatorname{rank} K(p) \le N < \infty$.

*Proof.* Using the invertibility of $D_{22}$, we can factorize the operator function (A.9) as the product of upper- and lower-triangular matrices. This product has the desired form. □

The poles of $D(p)$ and $D^{-1}(p)$ will be called *singular points* of $D(p)$. Now let us define the notion of multiplicity for singular points.

DEFINITION A.14. The *multiplicity of a singular point* $p = p_0$ of a strongly finitely meromorphic operator function $D(p)$ is the integer

$$m_D(p_0) = \operatorname{Tr} \operatorname{Res}_{p=p_0} \left( D^{-1} \frac{dD}{dp} \right).$$

It is also sometimes useful to represent the multiplicity by the Cauchy type integral

$$m_D(p_0) = \frac{1}{2\pi i} \operatorname{Tr} \left( \int_{|p-p_0|=\varepsilon} D^{-1} \frac{dD}{dp} dp \right) \tag{A.11}$$

over a circle of small radius $\varepsilon$ centered at $p_0$.

These expressions are obviously well defined (since the coefficients of the Laurent series are of finite rank). It is also clear that for scalar functions they indeed give the multiplicity of the singular point. However, the fact that the multiplicity is always *integer* requires a subtler proof. The following theorem comprises the main properties of multiplicities, including the fact that they are integer.

THEOREM A.15. *The multiplicities of singular points have the following properties.*

1) *They satisfy the logarithmic property:*

$$m_{D_1 D_2}(p_0) = m_{D_1}(p_0) + m_{D_2}(p_0).$$

2) *For a strongly finitely meromorphic operator function with a triangular block matrix*[5]

$$D(p) = \begin{pmatrix} D_1(p) & * \\ 0 & D_2(p) \end{pmatrix}, \tag{A.12}$$

*one has* $m_D(p_j) = m_{D_1}(p_j) + m_{D_2}(p_j)$.

3) *The multiplicities are integers*: $m_D(p_0) \in \mathbb{Z}$.

*Proof.* First, note that the functional $\operatorname{Tr}\operatorname{Res}_{p=p_0}$ on the set of finitely meromorphic operator functions is a trace, i.e., satisfies

$$\operatorname{Tr}\operatorname{Res}_{p=p_0}(AB) = \operatorname{Tr}\operatorname{Res}_{p=p_0}(BA).$$

This property permits one to exchange the operator functions $D$ and $D^{-1}$ in the expression for the multiplicity (the sign of the expression then changes to the opposite) and change the order of factors (without change in the sign).

1. The logarithmic property can be verified by a straightforward algebraic computation with the use of the fact that $\operatorname{Tr}\operatorname{Res}_{p=p_0}$ is a trace.

2. The formula (A.12) for the multiplicity of a singular point of a triangular operator function can also be obtained by a straightforward computation. On the one hand, the trace of an operator with a triangular matrix can be computed via the diagonal entries. On the other hand, the diagonal entries in the product $D^{-1}dD/dp$ are just $D_{1,2}^{-1}dD_{1,2}/dp$.

3. Now let us prove that the multiplicities are integer. To this end, we use the factorization (A.10). By the logarithmic property, it suffices to prove the desired assertion for the multiplicity $m_{1+K}(p_0)$ of the strongly finitely meromorphic operator function $1 + K(p)$ with meromorphic function $K(p)$ whose rank is bounded by some number $N$.

In this finite-dimensional case, the multiplicity is given by the formula

$$m_{1+K}(p_0) = w_{p_0}(\det(1 + K(p))) \in \mathbb{Z}, \tag{A.13}$$

where $w_{p_0}(f)$ is the degree of the singular point $p = p_0$ of a function $f$. To prove this, it suffices to note that the integrand in the expression of the multiplicity by the Cauchy integral (A.11) is just the logarithmic derivative of the determinant:

$$\operatorname{Tr}\left((1 + K(p))^{-1}\frac{d}{dp}K(p)\right) = \frac{d}{dp}\ln\det(1 + K(p)).$$

By integrating this relation, we obtain (A.13).

Thus we have proved that the multiplicity is an integer. □

[5]With respect to some direct sum decompositions $H = H_1 \oplus H_2$ and $G = G_1 \oplus G_2$.

***Homotopy invariance of the multiplicities.*** The multiplicities possess an important property: they are homotopy invariant (and hence constant) under continuous deformations.

Consider a family $D_t(p)$, $t \in [0,1]$, of strongly finitely meromorphic operator functions in a domain $U \subset \mathbb{C}$.

PROPOSITION A.16. *The sum of multiplicities of singular points of $D_t(p)$ in a subdomain $V \Subset U$ with smooth boundary $\partial\overline{V}$ is independent of the parameter $t$ provided that the following conditions are satisfied.*

1) *The family $D_t(p)$ is continuous in $(t,p) \in [0,1] \times \partial\overline{V}$.*

2) *No singular point of $D_t(p)$ lies on the boundary $\partial\overline{V}$ for any $t$.*

3) *The dimension of the linear span of the ranges of the coefficients (and their adjoints) in the principal parts of the Laurent series of $D_t(p)$ and $D_t^{-1}(p)$ at the singular points lying in $V$ does not exceed some finite number $N$ independent of $t$.*

*Proof.* The sum of multiplicities over $V$ is an integer, and so to establish the desired invariance property it suffices to show that the sum continuously depends on $t$.

Indeed, the sum of multiplicities is given by the Cauchy integral (A.11) over $\partial V$. By assumptions 1) and 2) of the theorem, the integrand continuously depends on the parameter $t$ in the *operator norm*. However, to prove the continuous dependence of the multiplicity on $t$, we should prove that the integral itself is continuous in the *trace norm*. To this end, we shall prove that the rank of the integral is bounded uniformly in $t$. Since the operator and trace norms are equivalent on the set of operators with rank uniformly bounded by some number $M$, this will imply the desired assertion.

Let us establish the uniform boundedness of the rank of the integral

$$\frac{1}{2\pi i}\int_{\partial\overline{V}} D^{-1} dD.$$

Consider some parameter value $t = t_0$. Then the family $D_{t_0}(p)$ has finitely many singular points $p_1, \ldots, p_M$ in $V$. The integral in question is the sum of residues at these points. Consider the Laurent series of our operator functions at a singular point (for brevity, we assume that this singular point is zero):

$$\frac{d}{dp}D(p) = \frac{D_{-N}}{p^N} + \ldots + \frac{D_{-1}}{p} + D_0 + D_1 p + \ldots,$$
$$D^{-1}(p) = \frac{C_{-N}}{p^N} + \ldots + \frac{C_{-1}}{p} + C_0 + C_1 p + \ldots$$

Then the contribution of this point to the integral is the sum

$$\sum_{i+j=-1} C_i D_j.$$

We split the sum into two parts, one with negative $i$ and the other with negative $j$. The rank of the sum of the first parts over all singular points is bounded by $N$ by virtue of the third assumption of the theorem. (The dimension of the linear span of the ranges of all $C_i$, $i < 0$, is bounded by $N$.) The rank of the sum of second parts can be estimated in a similar way. (Here one uses the fact that the linear span of the ranges of all $D_i^*$, $i < 0$, has a dimension than does not exceed $N$.)

Thus the rank of the integral is uniformly bounded. The proof is complete. □

*Remark* A.17. Note that the last assumption of the theorem is automatically satisfied if the operator function in question is finite-dimensional or holomorphic in the entire $V$.

◀ One can prove that this condition holds for a holomorphic operator function by analyzing the proof of Gokhberg's theorem (e.g., Egorov and Schulze 1997) saying that a Fredholm holomorphic family invertible at least at one point is strongly finitely meromorphic. More precisely, one observes that if the operator function continuously depends on a parameter, then the number of poles of the inverse function, their orders, and their ranks are bounded locally uniformly with respect to the parameter. ▶

### A.2.2. Definition of the spectral flow

Now that we have studied the multiplicities of singular points of general strongly finitely meromorphic operator functions, we can define the spectral flow of families of such operator functions.

Let $M$ be a smooth closed manifold. We consider holomorphic parameter-dependent operators $D(p)$ defined in a neighborhood of a sector $|\arg p| \leq c < \pi/2$ with nonzero angle and elliptic with parameter $p$ in the sector.

It is well known (e.g., Egorov and Schulze 1997) that such an operator function is finitely meromorphic and each strip $|\operatorname{Im} p| < \lambda$ of finite width (lying in the domain where the operator function is defined) contains only finitely many singular points of the operator function. Thus the multiplicities of singular points considered earlier in this subsection are well defined for parameter-dependent elliptic operators.

The *spectral flow* of a continuous homotopy $D_t(p), t \in [0, 1]$, of parameter-dependent elliptic operators is the integer equal to the net number of singular points of the family crossing the line $\operatorname{Im} p = 0$ upwards in the course of the homotopy.

One can give a rigorous definition of this notion. Namely, we take an admissible broken line with vertices $(t_i, \gamma_i)$,

$$0 = t_0 < t_1 < \cdots < t_N = 1,$$

i.e., a line such that the operator function $D_t(p)$ is holomorphic and invertible on the line $\operatorname{Im} p = \gamma_i$ for all $t \in [t_{i-1}, t_i]$.

DEFINITION A.18. The *spectral flow* of the homotopy $\{D_t\}_{t\in[0,1]}$ is the number

$$\operatorname{sf}\{D_t\}_{t\in[0,1]} = \sum_{i=1}^{N-1} N_i, \tag{A.14}$$

where

$$N_i = \begin{cases} -\sum\limits_{\operatorname{Im} p_j \in (\gamma_i, \gamma_{i+1})} m_{D_{t_i}}(p_j) & \text{if } \gamma_i < \gamma_{i+1}, \\ \sum\limits_{\operatorname{Im} p_j \in (\gamma_{i+1}, \gamma_i)} m_{D_{t_i}}(p_j) & \text{if } \gamma_i \geq \gamma_{i+1}. \end{cases}$$

Here the $p_j$ are the singular points of the conormal symbol $D_{t_i}$ in the strip between the weight lines $\operatorname{Im} p = \gamma_i$ and $\operatorname{Im} p = \gamma_{i+1}$.

THEOREM A.19. *The spectral flow has the following properties.*

1) *It is independent of the choice of an admissible broken line.*

2) *The spectral flow depends only on the homotopy class of a path $D_t$ with fixed beginning and end. It also remains unchanged under deformations of $D_t$ such that the beginning and the end are not fixed but the families $D_0$ and $D_1$ have no singular points on the real axis for any values of the homotopy parameter.*

*The proof* is similar to that of Theorem A.3 on the ordinary spectral flow. □

EXERCISE A.20. Check that the spectral flow of a homotopy $\{A_t\}_{t\in[0,1]}$ of elliptic self-adjoint operators coincides with the spectral flow of the corresponding homotopy of operator functions $p - iA_t$:

$$\operatorname{sf}\{A_t\}_{t\in[0,1]} = \operatorname{sf}\{p - iA_t\}_{t\in[0,1]}.$$

*Remark* A.21. The reader possibly noticed that the definition of the spectral flow in this section, as well as Theorem A.19, remains valid for abstract strongly finitely meromorphic operator functions. One has only to require that these operator functions have at most finitely many singular points in a neighborhood of the real axis $\mathbb{R} \subset \mathbb{C}$.

## A.3. Higher Spectral Flows

The spectral flow and the $\eta$-invariant are nearly as universal as the index itself, which occurs in various disguises depending on the geometric context: for a single operator, it is an integer; for group-invariant operators, it is a virtual representation; for operator families, it is an element of a $K$-group, and so on. Similar generalizations also exist for the spectral flow. By way of example, in this section we consider the notion of *higher spectral flow*, i.e., the spectral flow ranging in a $K$-group. The reader can find other generalizations in special literature.

There is a more powerful approach to the definition of spectral flow than that in the preceding subsection. The notion of a spectral section is a milestone of this approach.

### A.3.1. Spectral sections

Let $A = \{A_x\}_{x \in X}$ be a continuous family of elliptic self-adjoint operators on a closed manifold with compact parameter space $X$. For some parameter values, the operators in this family may not be invertible. Accordingly, the corresponding family of spectral projections is not continuous in the general case.

DEFINITION A.22. A *spectral section* of a family $\{A_x\}_{x \in X}$ is a continuous family $\{P_x\}_{x \in X}$ of projections that differ from the spectral projections by compact operators:

$$P_x - \Pi_+(A_x) \in \mathcal{K}.$$

EXERCISE A.23. Prove that there exists a spectral section if and only if there exists a perturbation of the family $\{A_x\}_{x \in X}$ by a family of compact operators such that the perturbed family is invertible.

It turns out that there exists a *topological obstruction* to the existence of spectral sections. Indeed, according to Atiyah and Singer (1969), a family of elliptic self-adjoint operators has a well-defined "index," which is an element of the $K^1$-group of the parameter space. In the simplest way, this index is defined by the formula

$$\operatorname{ind} A \stackrel{\text{def}}{=} \operatorname{ind}(p - iA) \in K_c(X \times \mathbb{R}) \simeq K^1(X)$$

as the ordinary index of the family $p - iA$. This family of Fredholm operators is parametrized by the product $X \times \mathbb{R}$ and invertible outside a compact subset of this space. Consequently, its index indeed belongs to the $K$-group with compact supports (see Section 3.6.2).

This index is zero for an invertible family and is preserved under perturbations of the family by compact operators. Thus we obtain an obstruction to the existence of spectral sections. It turns out that this is thc only obstruction.

THEOREM A.24. *There exists a spectral section of a family $A$ if and only if the index* $\operatorname{ind} A \in K^1(X)$ *is zero.*

*The proof* can be found in Melrose (1995). □

EXAMPLE A.25. If the parameter space is the interval $[0, 1]$, then there always exists a spectral section.

EXAMPLE A.26. If the parameter space is the circle $\mathbb{S}^1$, i.e., we deal with a periodic family, then there exists a single integer-valued obstruction to the existence of a spectral section. By way of example, the reader can verify that the obstruction coincides in this case with the spectral flow of the periodic family.

Let us use the notion of spectral section to give a new definition of the spectral flow.

***Definition of the spectral flow via spectral sections.*** Let $A = \{A_t\}_{t\in[0,1]}$ be a homotopy of elliptic self-adjoint operators. By $P_t$ we denote some spectral section of this homotopy.

DEFINITION A.27. The *spectral flow* of the homotopy $\{A_t\}_{t\in[0,1]}$ is the number

$$\operatorname{sf}\{A_t\}_{t\in[0,1]} = -\operatorname{ind}(\Pi_+(A_0), P_0) + \operatorname{ind}(\Pi_+(A_1), P_1).$$

### A.3.2. Spectral flow of homotopies of families of self-adjoint operators

Using the notion of a spectral section, one can readily give the definition of the spectral flow for the case of families.

Let $\{A_{t,x}\}$, $(t, x) \in [0, 1] \times X$, be a homotopy of families of elliptic self-adjoint operators. Suppose that for $t = 0$ and $t = 1$ the families are invertible. Then by Theorem A.24 there exists a spectral section $P_{t,x}$ on the product $[0, 1] \times X$.

DEFINITION A.28. The *spectral flow* of the family $A = \{A_{t,x}\}$, $(t, x) \in [0, 1] \times X$, is the element

$$\operatorname{sf}\{A_t\}_{t\in[0,1]} = -\operatorname{ind}(\Pi_+(A_0), P_0) + \operatorname{ind}(\Pi_+(A_1), P_1) \in K(X)$$

of the even $K$-group.

The spectral flow thus defined possesses all properties of the ordinary integer spectral flow (see Theorem A.3). In other words, the spectral flow is independent of the choice of a spectral section, additive, and homotopy invariant.

Note that in this case one can also give a definition in terms of admissible broken lines (Nazaikinskii, Savin, Schulze and Sternin 2002*a*). Using broken lines whose links depend on the parameters, one can quite explicitly construct a virtual bundle representing the spectral flow. Unfortunately, the construction is complicated and will not be presented here. We restrict ourselves to the following simplest case.

EXERCISE A.29. Show that if a family admits a broken line with vertices $(\tau_i, \lambda_i)$ *independent* of the parameters $x \in X$, then the spectral flow can be computed by the well-known formula (A.2):

$$\operatorname{sf}\{A_\tau\}_{\tau\in[0,1]} = \sum_{i=1}^{N-1} \operatorname{ind}(\Pi_{\lambda_i}(A_{\tau_i}), \Pi_{\lambda_{i-1}}(A_{\tau_i})) \in K(X).$$

Here ind is the relative index of families of projections.

### A.3.3. Spectral flow of homotopies of families of parameter-dependent operators

One can introduce and successfully use a notion similar to that of a spectral section for families of parameter-dependent operators.

***Spectral sections.*** Consider a family $D_x(p)$ of operators elliptic with parameter $p$ in a sector of nonzero angle and holomorphic in $p$ in a neighborhood of the sector. We assume that the variable $x$ ranges over a compact set $X$.

We define a *spectral section* of the family as an invertible family $\widetilde{D}_x(p)$ that differs from $D_x(p)$ by a family of smoothing operators. By analogy with the self-adjoint case, there exists a topological obstruction to the existence of spectral sections. It is given by the index

$$\operatorname{ind} D \in K_c(X \times \mathbb{R}) \simeq K^1(X).$$

THEOREM A.30. *A family $D_x(p)$ admits a spectral section if and only if*

$$\operatorname{ind} D = 0 \in K^1(X).$$

*Proof.* The necessity is obvious. The sufficiency was proved by Nistor (2003). □

***Definition of the spectral flow.*** Let $D_{t,x}(p)$ be a homotopy of families of parameter-dependent elliptic operators. We require the family to be invertible for $t = 0$ and $t = 1$. Then the index of the family is zero, and by Theorem A.30 there exists a spectral section, which we denote by $\widetilde{D}_{t,x}(p)$.

DEFINITION A.31. The *spectral flow* of the homotopy $D_{t,x}(p)$ of families of parameter-dependent operators is the element

$$\operatorname{sf} D = -[D_0\widetilde{D}_0^{-1}] + [D_1\widetilde{D}_1^{-1}] \in K_c^1(X \times \mathbb{R}) \simeq K(X)$$

of the even $K$-group.

Let us comment on the last formula. The families $D_t\widetilde{D}_t^{-1}$ have the form "1 plus a compact-valued family." Recall that the space of invertible operators of the form $1+K$ has the homotopy type of the infinite-dimensional unitary group[6] $\mathbb{U}(\infty)$, i.e., is the classifying space of the $K^1$-group (see Atiyah 1989). Since the ratio $D_t\widetilde{D}_t^{-1}$ tends to unity at infinity, it follows that it defines an element of the $K$-group with compact supports.

The reader may state and prove the properties of the spectral flow in this case himself or herself. In particular, in the absence of additional parameters ($X = pt$) the spectral flow coincides with the similar notion defined in Section A.2.2.

[6]One can readily prove this by approximating compact operators by finite-dimensional ones.

# Appendix B

# Eta Invariants

This appendix deals with the spectral $\eta$-invariant of Atiyah, Patodi and Singer (1975) and its generalization given by Melrose (1995). This invariant is almost as fundamental in elliptic theory as the index. It appears in many index formulas, while its applications range from algebraic geometry (Atiyah, Donnelly and Singer 1983) to quantum field theory (Witten 1985). The study of the $\eta$-invariant in this appendix is based on the paper by Seeley (1967) (see also Grubb and Seeley 1996). A detailed exposition of the $\eta$-invariant can be found in the monographs by Gilkey (1995, 1989*a*) and Booß-Bavnbek and Wojciechowski (1993). The $\eta$-invariant of parameter-dependent elliptic operators was subsequently studied and generalized by Lesch and Pflaum (2000), Nistor (2002), and Moroianu (2003).

## B.1. The Atiyah–Patodi–Singer Eta Invariant

For a quadratic form defined on a finite-dimensional real space, there is a well-defined notion of *signature*, which is the difference between the numbers of positive and negative squares in the canonical representation of the form. Note that the signature can be treated as a spectral invariant of the symmetric operator defining the quadratic form. Moreover, the signature is homotopy invariant in the class of invertible operators (nondegenerate forms).

In the infinite-dimensional case, we replace symmetric matrices that define quadratic forms by elliptic self-adjoint operators. The number of eigenvalues is now infinite, and the signature in the usual sense is meaningless.

### B.1.1. Definition of the eta invariant

Now let us view the signature as the sum of the series whose terms are the signs of eigenvalues of an elliptic self-adjoint operator $A$ on a manifold $M$. In the infinite-dimensional case, we regularize this series by analogy with the $\zeta$-function. Namely, we define the spectral *$\eta$-function* of $A$ by the formula

$$\eta\,(A, s) = \sum_{\lambda_j \in \operatorname{Spec} A, \lambda_j \neq 0} \frac{\operatorname{sgn}\lambda_i}{|\lambda_i|^s} \equiv \operatorname{Tr}\left[A\left(A^2\right)^{-s/2-1/2}\right], \tag{B.1}$$

where the $\lambda_j$ are the nonzero eigenvalues of $A$. The function (B.1) is analytic in the half-plane $\operatorname{Re} s > \dim M/\operatorname{ord} A$ (where the series converges absolutely).

DEFINITION B.1 (Atiyah, Patodi and Singer 1975). The *$\eta$-invariant* of the operator $A$ is the real number

$$\eta(A) = \frac{1}{2}(\eta(A,0) + \dim \operatorname{Ker} A) \in \mathbb{R}. \tag{B.2}$$

Clearly, the $\eta$-invariant of an invertible operator in the finite-dimensional case is equal to the signature of the corresponding quadratic form up to the factor $1/2$. However, to make Definition B.1 meaningful in the general case, we have to continue the $\eta$-function analytically to the point $s = 0$.

THEOREM B.2 (Atiyah, Patodi and Singer 1976, Gilkey 1981). *The $\eta$-function of an elliptic operator on a smooth closed manifold has a meromorphic continuation into the entire complex plane, possibly, with poles at the points*

$$s_j = \frac{\operatorname{ord} A - j}{\dim M}, \quad j \in \mathbb{Z}_+.$$

*There is no pole at the point $s = 0$.*

A detailed proof of this difficult theorem can be found in the cited papers (see also Gilkey 1989*a*). Hence we only describe the key points of the proof.

***1. Meromorphic continuation.*** For positive definite operators, the $\eta$-function $\eta(A,s)$ coincides with the $\zeta$-function $\zeta(A,s)$, which is known to be meromorphic (e.g., Seeley 1967). If the operator is not sign definite, then one can use the expression

$$\eta(A,s) = \frac{\zeta(A_+,s) - \zeta(A_-,s)}{2^s - 1} \tag{B.3}$$

of the $\eta$-function via the $\zeta$-functions of the positive definite operators

$$A_\pm = (3|A| \pm A)/2.$$

***2. Holomorphy at zero.*** It follows from the formula (B.3) that the $\eta$-function may have at most a simple pole at zero, since the $\zeta$-function is holomorphic there. The residue at this point is equal to

$$\operatorname{Res}_{s=0} \eta(A,s) = \frac{\zeta(A_+,0) - \zeta(A_-,0)}{\ln 2}. \tag{B.4}$$

Seeley (1967) showed that the $\zeta$-invariants can be represented as integrals over the manifold of some expressions determined by the complete symbol of $A$. The corresponding integrand is in general nonzero. However, it was shown by Atiyah, Patodi and Singer (1976) for odd-dimensional manifolds and Gilkey (1981) for even-dimensional manifolds that the residue is still zero! Thus the $\eta$-function is holomorphic at zero and the $\eta$-invariant is well defined.

*Remark* B.3. In contrast to the spectral flow, which can be defined in a purely functional-analytic situation, the definition of the $\eta$-invariant heavily relies on the fact that the operator is defined on a closed manifold. For example, the $\eta$-function of self-adjoint boundary value problems on a manifold with boundary may have a pole at zero (Grubb and Seeley 1996). It is interesting that so far there is no *analytic* proof of the holomorphy of the $\eta$-function at zero. (The proofs cited above use global topological methods.) The vanishing of the residue can be verified by a straightforward analytic computation only for Dirac type operators (e.g., Bismut and Freed 1986).

EXAMPLE B.4. Consider the operator family

$$A_t = -i\frac{d}{d\varphi} + t$$

on a circle of length $2\pi$ with coordinate $\varphi$, where $t$ is a real constant. Let us compute the $\eta$-invariant of this operator. The spectrum of $A_t$ is the lattice $t + \mathbb{Z}$. Hence the $\eta$-invariant is a 1-periodic function of the parameter $t$. For $0 < t < 1$, we can arrange the eigenvalues into pairs and transform the $\eta$-function to the form

$$\eta\left(A_t, s\right) = \sum_{n\geq 1} \left[(n+t)^{-s} - (n-t)^{-s}\right] + t^{-s}.$$

This series absolutely converges on the half-line $s > 0$, and the limit as $s \to +0$ is equal to $-2t + 1$. One can readily establish the latter by substituting the Taylor expansion

$$\left[(n+t)^{-s} - (n-t)^{-s}\right] = -2tsn^{-s} + O\left(\frac{s}{n^{2+s}}\right)$$

into the series. Thus the $\eta$-invariant is equal to

$$\eta\left(A_t\right) = \frac{\eta\left(A_t, 0\right) + \dim \operatorname{Ker} A_t}{2} = \frac{1}{2} - \{t\},$$

where $\{t\} \in [0, 1)$ is the fractional part of $t$.

### B.1.2. Variation under deformations of the operator

In the preceding example, the $\eta$-invariant of a smooth elliptic family proves to be piecewise smooth. Moreover, the (integer) jumps occur at the parameter values where some eigenvalue changes its sign. It turns out that the $\eta$-invariant is a piecewise smooth function of the parameter in the general case as well. More precisely, the following result is valid.

THEOREM B.5 (Atiyah, Patodi and Singer 1976). *Let $A_t$, $t \in [0,1]$, be a smooth family of elliptic self-adjoint operators. Then the function $\eta(A_t)$ is piecewise smooth and admits the expansion*

$$\eta(A_{t'}) - \eta(A_0) = \mathrm{sf}\,(A_t)_{t\in[0,t']} + \int_0^{t'} \omega(t_0)\,dt_0 \tag{B.5}$$

*into a piecewise constant function (the spectral flow) and a smooth function of the parameter, which can be expressed as*

$$\omega(t_0) = \left[\frac{d}{dt}\zeta(B_{t,t_0})\right]_{t=t_0} \in C^\infty[0,1]$$

*via the derivative of the $\zeta$-invariant of the auxiliary family*

$$B_{t,t_0} = |A_{t_0}| + P_{\mathrm{Ker}\,A_{t_0}} + (t-t_0)\left.\frac{dA_t}{dt}\right|_{t=t_0}.$$

*Here $P_{\mathrm{Ker}\,A}$ is the projection onto the kernel of $A$ and the $\zeta$-invariant is given by the expression*

$$\zeta(A) \overset{\mathrm{def}}{=} \zeta(A,0)/2.$$

*Sketch of proof.* If the family is invertible, then one can express the derivatives of the $\eta$- and $\zeta$-functions by the formulas

$$\frac{d}{dt}\zeta(B_t,s) = -s\mathrm{Tr}(\dot{B}_t B_t^{-s-1}), \qquad \frac{d}{dt}\eta(A_t,s) = -s\mathrm{Tr}(\dot{A}_t(A_t^2)^{-\frac{1}{2}(s+1)}).$$

This readily implies (B.5) for $s=0$ and $t=t_0$.

To prove (B.5) in the general case, one should use the broken line occurring in the definition of spectral flow (see Fig. A.2), which permits one to deal with invertible operators alone. □

This result is of practical importance owing to the following *Seeley formula* for the $\zeta$-invariant. If $A$ is a positive semidefinite operator whose complete symbol has the asymptotic expansion

$$\sigma(A) \sim a_m + a_{m-1} + a_{m-2} + \cdots, \qquad m = \mathrm{ord}\,A,$$

then the $\zeta$-invariant can be computed according to the formula

$$2\zeta(A) = \frac{1}{(2\pi)^{\dim M}\,\mathrm{ord}A}\int_{S^*M} dx\,d\xi \int_0^\infty b_{-\dim M-\mathrm{ord}A}(x,\xi,-\lambda)\,d\lambda, \tag{B.6}$$

where the coefficient $b_{-\dim M-\operatorname{ord}A}(x,\xi,-\lambda)$ is determined by the recursion relation

$$b_{-m-j}(x,\xi,\lambda)\,(a_m(x,\xi)-\lambda) + \sum_{\substack{k+l+|\alpha|=j\\ l>0}} \frac{1}{\alpha!}(-i\partial_\xi)^\alpha\, b_{-m-k}(x,\xi,\lambda)\,(-i\partial_x)^\alpha\, a_{m-l}(x,\xi)=0 \quad \text{(B.7)}$$

and $b_{-m}(x,\xi,\lambda)=(a_m(x,\xi)-\lambda)^{-1}$.

Thus by substituting the Seeley formula (B.7) into Theorem B.5 we obtain a closed-form local expression for the derivative of the $\eta$-invariant in terms of the homogeneous components of the complete symbol of the operator.

*Remark* B.6. If we transpose the integral in (B.5) to the left-hand side, then we obtain the new definition

$$\operatorname{sf}\{A_t\}_{t\in[0,1]}=\eta(A_1)-\eta(A_0)-\int_0^1\left(\frac{d}{dt}\eta(A_t)\right)dt$$

of the spectral flow via the $\eta$-invariant. This is an analog of the expression of the spectral flow via the signature in the finite-dimensional case. Although at first glance this seems to be an expression of a simpler object via a more complicated one, it often proves useful (e.g., Zhang 2000) if there are independent methods for the computation of the $\eta$-invariant.

### B.1.3. Homotopy invariance. Example

The cases in which the $\eta$-invariant is homotopy invariant are especially important in applications in geometry and topology. Unfortunately, in contrast to the signature, the $\eta$-invariant is not homotopy invariant even in the class of invertible operators. More precisely, for deformations in this class the spectral flow is always zero and only the second, smooth term in formula (B.5) survives. To verify that this component is also zero in some narrower operator classes, it is useful to analyze symmetries of the Seeley formula for the $\zeta$-invariant. An example in which such symmetries can be used is considered in the remaining part of this subsection.

***Eta invariant and the Gilkey conditions.*** Consider the formula (B.5). It follows from this formula that although the $\eta$-invariant itself is not local, its variation modulo the spectral flow already enjoys this property. More precisely, the $\zeta$-invariant is obtained by integration over the cosphere bundle. The integral is obviously zero if the integrand is an odd function on the sphere, for the contributions of the antipodal points $\xi$ and $-\xi$ cancel each other. This idea is realized for differential operators. Namely, the components of the complete symbol of a differential operator are homogeneous polynomials and have the symmetry

$$a_k(x,-\xi)=(-1)^k a_k(x,\xi).$$

By substituting these relations into (B.6) and (B.5), we obtain the following result.

THEOREM B.7 (Gilkey 1989*b*). *The fractional part of the spectral $\eta$-invariant of an elliptic self-adjoint differential operator $A$ on a manifold $M$ is homotopy invariant if the parity condition*

$$\operatorname{ord} A + \dim M \equiv 1 (\operatorname{mod} 2) \tag{B.8}$$

*is satisfied.*

*Sketch of proof.* First, consider the case of even-order operators. By induction, we establish the following homogeneity of the coefficients determined by the operators $B_t$ (see the recursion relations (B.7)):

$$b_j(x, -\xi, \lambda, t) = (-1)^j b_j(x, \xi, \lambda, t). \tag{B.9}$$

Hence we obtain

$$\frac{d}{dt}\eta(A_t) = \frac{d}{dt}\zeta(B_t) = \text{const}\,\frac{d}{dt}\left[\int\limits_{S^*M} dx d\xi \int\limits_0^\infty b_{-\dim M - \operatorname{ord} A}(x, \xi, -\lambda, t) d\lambda\right].$$

It follows from the homogeneity (B.9) and the condition that the manifold is odd-dimensional that the integrand $b_{-\dim M - \operatorname{ord} A}(x, \xi, -\lambda, t)$ is an odd function on the sphere $S_x^*M$ under the parity condition (B.8). Hence the integral is zero and we obtain the desired relation

$$\frac{d}{dt}\eta(A_t) = 0.$$

For odd-order operators, we have the opposite homogeneity

$$b_j(x, -\xi, \lambda, -t) = (-1)^{j+1} b_j(x, \xi, \lambda, t).$$

By substituting these homogeneous functions into the formula for the $\zeta$-invariant, we obtain

$$\frac{d}{dt}\zeta(B_t) = \frac{d}{dt}\zeta(B_{-t}).$$

Thus the derivative is also zero. □

Note that as a by-product we have even proved that the $\zeta$-invariant of even-order differential operators on odd-dimensional manifolds is zero.

## B.2. The Eta Invariant of Families with Parameter (Melrose's theory)

It turns out that one can also arrive at the $\eta$-invariant starting from another finite-dimensional invariant, the winding number of a nonzero function, rather than the signature as in Section B.1. In this section, we show how this invariant can be transferred to the infinite-dimensional case.

Recall that if $f(p)$ is a nonvanishing complex-valued function on the real line with coordinate $p \in \mathbb{R}$, then the winding number $w(f) \in \mathbb{Z}$ of $f$ around zero is defined by the formula

$$w(f) = \frac{1}{2\pi i} \int_{\mathbb{R}} f^{-1} df.$$

The winding number is well defined if, say, the function is constant at infinity. Then it is well known that the integral is equal to the number of revolutions of the point $f(p)$ around zero $\mathbb{C} \setminus \{0\}$ as the parameter runs over the real line. A similar formula is valid for invertible matrix functions:

$$w(F) = \frac{1}{2\pi i} \int_{\mathbb{R}} \operatorname{tr}(F^{-1} dF). \tag{B.10}$$

(In this case, one speaks of the winding number of the determinant of the family.)

Let us describe the relationship, which we wish to transfer to the infinite-dimensional case, between the signature and the winding number. If $A$ is an invertible Hermitian matrix, then the family $F = p - iA$ is invertible on the real line. An easy computation shows that the winding number of this function is equal to the signature of $A$ divided by 2.

We are interested in an infinite-dimensional generalization of the winding number (B.10) to operator-valued functions. In this case, the expression $F^{-1}dF$ is still well-defined (provided that the family is sufficiently smooth). However, the remaining part of the formula may well be meaningless: the trace may fail to exist, and the integral may diverge. Thus, to define an analog of the winding number, we should generalize these two notions.

It turns out that one can make a beautiful generalization of the trace and integral by considering *families of parameter-dependent operators*, well known in analysis (e.g., Shubin 1985), rather than arbitrary families. This construction will be carried out in the next subsection.

### B.2.1. A trace on the algebra of parameter-dependent operators

Let $X$ be a closed manifold, and let $\Psi_p(X)$ be the algebra of classical parameter-dependent pseudodifferential operators on $X$ with parameter $p \in \mathbb{R}$ (see Section 3.2.5). By $\Psi_p^m(X)$ we denote the subspace of operators of order $\leq m$.

The formula (B.10) for the winding number contains the functional

$$\int_{\mathbb{R}} \operatorname{tr} A(p) dp.$$

In the remaining part of this section, we show how one can extend this functional, originally defined only for families $A(p)$ lying in the subalgebra $\Psi_p^{-n-2}(X)$, where

$n = \dim X$, to families of arbitrary order. This continuation permits us to define the $\eta$-invariant of families in the next subsection.

Obviously, we obtain the desired continuation of the functional once we are able to define the notions of *trace* and *integral* for families of arbitrary order. Let us perform these tasks successively.

***Regularization of the trace.*** The regularization procedure for the trace is based on the following two facts: a) the trace of a family is well defined if the family has a sufficiently large negative order; b) the order of the family can be diminished by differentiation with respect to the parameter. Thus the *regularized trace* can be defined as

$$(\mathrm{TR}\,A)(p) \stackrel{\text{def}}{=} \int_0^p \int_0^{p_1} \ldots \int_0^{p_{k-1}} \operatorname{tr}\left[\left(\frac{\partial}{\partial q}\right)^k A(q)\right] dq dp_{k-1} \ldots dp_1. \tag{B.11}$$

The resulting function of a real variable is well defined for $k > m + n$ and has asymptotics of a special form at infinity:

LEMMA B.8. *One has the asymptotic expansion*

$$(\mathrm{TR}\,A)(p) \sim \sum_{i \le k} p^i c_i^{\pm} + \sum_{j=0}^{k} p^j d_j^{\pm} \ln |p| \tag{B.12}$$

*as* $p \to \pm\infty$.

The proof of the expansion (B.12) is based on the fact that the trace takes a classical parameter-dependent operator to a classical symbol depending on the covariable $p$ and raises the order by the dimension $n$ of the manifold $X$. One can verify this using the local expression for the trace. Now the relation (B.12) is obtained by $k$-fold integration of the classical symbol.

One can readily verify that the traces $\mathrm{TR}\,A$ corresponding to various numbers $k$ differ by polynomials. Thus if by $S_{log}(\mathbb{R})$ we denote the space of smooth functions possessing asymptotic expansions of the form (B.12) for some $k$, then Lemma B.8 shows that there is a well-defined mapping

$$\mathrm{TR} : \Psi_p(X) \longrightarrow S_{log}(\mathbb{R})/\mathcal{P},$$

where $\mathcal{P}$ is the subalgebra of polynomials. Note that the space $S_{log}(\mathbb{R})$ is invariant with respect to differentiation and integration.

***Regularization of the integral.*** The regularized trace may prove to be a function growing at infinity. To integrate such functions, we use the Eisenstein and Hadamard rules.

The integral of a function belonging to the space $S_{log}(\mathbb{R})$ over the interval $(-T, T)$ viewed as a function of the parameter $T$ has an asymptotic expansion of the form (B.12). We define the *regularized integral* as the constant term in this asymptotic expansion as $T \to +\infty$:

$$\fint_{\mathbb{R}} f(p)dp \stackrel{\text{def}}{=} c_0, \text{ where } \int_{-T}^{T} f(p)dp \sim \sum_{i\le k} T^i c_i + \sum_{j=0}^{k} T^j d_j \ln T. \qquad \text{(B.13)}$$

It follows from the definition that the regularized integral coincides with the ordinary integral on functions that decay at infinity at least as $p^{-2}$. Note the following simple properties of the regularized integral: it vanishes on all odd functions as well as all polynomials.

By combining (B.12) with (B.13), we arrive at the following assertion, which is the main result of this subsection.

PROPOSITION B.9. *The expression*

$$\overline{\mathrm{Tr}} A \stackrel{\text{def}}{=} \fint_{\mathbb{R}} (\mathrm{TR}\, A)\, dp$$

*is a trace (i.e., satisfies* $\overline{\mathrm{Tr}}AB = \overline{\mathrm{Tr}}BA$*) on the algebra of parameter-dependent operators and an extension of the trace*

$$\int_{\mathbb{R}} \mathrm{tr}\, A(p)dp$$

*from the subalgebra* $\Psi_p^{-n-2}(X)$.

### B.2.2. Definition of the Melrose eta invariant

By analogy with ordinary operators, we introduce the notion of ellipticity for parameter-dependent operators.

DEFINITION B.10. A parameter-dependent operator $A(p)$ is said to be *elliptic* if its principal symbol $A(x,\xi,p)$ is invertible for $|\xi|^2 + p^2 \neq 0$.

The ellipticity of a parameter-dependent operator implies not only that the operators in the family are Fredholm but also that they are invertible for sufficiently large parameter values (e.g., Shubin 1985).

Let $A(p)$ be an invertible elliptic parameter-dependent operator. Then the *Melrose $\eta$-invariant* of $A$ is defined as

$$\eta(A) = \frac{1}{2\pi i} \overline{\mathrm{Tr}}(A^{-1}\dot{A}).$$

PROPOSITION B.11. 1) *The $\eta$-invariant has the logarithmic property*

$$\eta(AB) = \eta(A) + \eta(B).$$

2) *If* $1+S(p)$ *is an invertible family where* $S(p)$ *is a family of finite-dimensional operators vanishing at infinity, then the* $\eta$*-invariant of the family is equal to the winding number*:

$$\eta(1+S) = \frac{1}{2\pi i}\int_{\mathbb{R}} \operatorname{tr}((1+S)^{-1}dS).$$

*Proof.* The logarithmic property is a special case of the following algebraic fact: if some algebra $\mathcal{A}$ is equipped with a trace $T : \mathcal{A} \to \mathbb{C}$ and a derivation $d$, then the functional $T(A^{-1}dA)$ defined on invertible elements possesses the logarithmic property. The second property readily follows from the definition. □

### B.2.3. Relationship with the Atiyah–Patodi–Singer eta invariant

The term "$\eta$-invariant" in the preceding subsection is not occasional: it turns out that the Atiyah–Patodi–Singer $\eta$-invariant is a special case of the Melrose $\eta$-invariant. Namely, if $A$ is an elliptic self-adjoint first-order differential operator, then $D(p) = p-iA$ is an elliptic parameter-dependent operator. This family is invertible if we additionally assume that $A$ is invertible.

THEOREM B.12.

$$\eta(A) = \eta(p-iA).$$

The proof is rather cumbersome (see Lesch and Pflaum 2000). Hence we restrict ourselves to heuristic computations showing why this is the case.

Consider the integral representation

$$\eta(A,s) = \frac{1}{\Gamma\left(\frac{s+1}{2}\right)}\int_0^\infty t^{(s-1)/2}\operatorname{tr}(Ae^{-tA^2})dt$$

of the $\eta$-function. In a number of cases (say, for the Dirac operator), the integral converges absolutely even for $s = 0$. By formally substituting the factor

$$1 = (\sqrt{\pi})^{-1}\int_{\mathbb{R}} t^{1/2}e^{-t\tau^2}d\tau$$

into the integral and by reversing the order of integration, we obtain the "expression"

$$\eta(A) = \frac{\eta(A,0)}{2} = \frac{1}{2\pi}\int_{\mathbb{R}} \operatorname{tr}\left(A(A^2+\tau^2)^{-1}\right)d\tau \qquad \text{(B.14)}$$

for the $\eta$-invariant. On the other hand, for the Melrose $\eta$-invariant we have

$$\eta(p-iA) = \frac{1}{2\pi i}⨍_{\mathbb{R}} \operatorname{TR}((p-iA)^{-1})dp = \frac{1}{2\pi}⨍_{\mathbb{R}} \operatorname{TR}(A(p^2+A^2)^{-1})dp. \quad \text{(B.15)}$$

In the last equation, we have used the fact that the regularized integral vanishes on odd functions.

The expressions (B.14) and (B.15) obviously coincide.

### B.2.4. Locality of the derivative of the eta invariant. Examples

Just as with the classical $\eta$-invariant, the derivative of the Melrose $\eta$-invariant also has the locality property. Let us obtain a formula for the derivative of the $\eta$-invariant in terms of symbols of the operators involved.

Let $A_t(p)$ be a smooth homotopy of invertible elliptic parameter-dependent operators. Then the derivative of the $\eta$-invariant with respect to $t$ is given by the expression

$$\frac{d}{dt}\eta(A_t) = \frac{1}{2\pi i}\overline{\mathrm{Tr}}\left(\frac{\partial}{\partial p}(A_t^{-1}\dot{A}_t)\right).$$

EXERCISE B.13. Derive this formula.

A straightforward computation shows that the composition of the trace $\overline{\mathrm{Tr}}$ with differentiation with respect to $p$ is also a trace, which will be denoted by

$$\widetilde{\mathrm{Tr}} \stackrel{\mathrm{def}}{=} \overline{\mathrm{Tr}} \circ \frac{\partial}{\partial p}.$$

This trace vanishes on operators of large negative order (where the regularization of the integral and of the trace is not needed). A convenient formula was obtained by Melrose.

THEOREM B.14. *Let $B$ be a parameter-dependent operator defined in some coordinate neighborhood on a manifold $X$. Then*

$$\widetilde{\mathrm{Tr}}B = \lim_{L\to\infty}\int_{|\eta|\le L}\left[\mathrm{tr}\, b_{-\dim X}(y,\eta,1) - \mathrm{tr}\, b_{-\dim X}(y,\eta,-1)\right]\frac{\omega^n}{n!},$$

*$(y,\eta)\in T^*X$, where $\omega$ is the symplectic form on the cotangent bundle and $b_k$ is the degree $k$ component of the complete symbol of $B$.*

*Sketch of proof.* 1. Since the trace is local, it suffices to prove the formula for homogeneous symbols. It also suffices to consider components of homogeneity degree $\ge -\dim X$. Such components $b(y,\eta,p)$ are locally integrable on $T^*X\times\mathbb{R}$ for all $p$, and so in the construction of the operator from the symbol they require no smoothing at $\eta = p = 0$.

2. Consider a homogeneous component $b(x,\eta,p)$ of the symbol. Then the value of the trace $\widetilde{\mathrm{Tr}}$ coincides with the constant term in the asymptotic expansion as $p\to\infty$ of the integral

$$\int_{-p}^{p}\int_{-1}^{p_{k-1}}\cdots\int_{-1}^{p_1}\int_{\mathbb{R}^{2n}}\frac{\partial^k}{\partial q^k}b(y,\eta,q)\,dyd\eta\,dq\,dp_1\ldots dp_{k-1}.$$

The inner integral over $(y,\eta)$ is a homogeneous distribution of degree $\mathrm{ord}\, b + \dim X - k$. Hence the remaining integration gives a homogeneous distribution of degree $\operatorname{ord} b + \dim X$ plus a polynomial. The polynomial does not contribute to the regularized integral, while the homogeneous function contributes if and only if $\operatorname{ord} b = -\dim X$. Thus it remains to consider the latter case.

3. In the computation of the contribution of the homogeneous component of degree $-\dim X$, it suffices to take $k = 1$. Then, by Newton–Leibniz,

$$\fint_{\mathbb{R}} \int_{\mathbb{R}^{2n}} \frac{\partial b}{\partial q}\, dy d\eta\, dq = \operatorname{reg-}\lim_{p\to\infty} \int_{\mathbb{R}^{2n}} [b(y,\eta,p) - b(y,\eta,-p)]\, dy d\eta,$$

where reg- lim is the regularized limit (the constant term in the asymptotic expansion). After the change of variables $\eta = p\eta'$, we see that this limit is equal to the integral

$$\int_{\mathbb{R}^{2n}} [b(y,\eta,1) - b(y,\eta,-1)]\, dy d\eta.$$

The proof of the formula is complete. □

Using the formula for the derivative of the $\eta$-invariant, one can construct classes of operators in which the $\eta$-invariant is homotopy invariant. These results are parallel to those in Section B.1.3. Let us state them in the form of exercises.

EXERCISE B.15. The fractional part of the Melrose $\eta$-invariant is homotopy invariant in the class of operators whose complete symbols $b \sim b_m + b_{m-1} + \ldots$ possess the symmetry

$$b_k(-\xi, -p) = (-1)^k b_k(\xi, p)$$

provided that the manifold on which the family is defined is even-dimensional.

EXERCISE B.16. The fractional part of the Melrose $\eta$-invariant is homotopy invariant in the class of operators whose complete symbols $b \sim b_m + b_{m-1} + \ldots$ possess the symmetry

$$b_k(-\xi, p) = (-1)^k b_k(\xi, p),$$

provided that the manifold on which the family is defined is odd-dimensional.

◄ In these exercises, one should note first that these symmetries single out subalgebras in the algebra of classical parameter-dependent operators. Then, using Theorem B.14, one verifies that the symbolic trace $\widetilde{\operatorname{Tr}}$ vanishes on these subalgebras. ►

# Appendix C

# Index of Parameter-Dependent Elliptic Families

In this appendix, we discuss the index theorem for parameter-dependent elliptic families due to Nistor (2003).

Let $D(v) : C^\infty(M, E) \longrightarrow C^\infty(M, F)$ be a parameter-dependent elliptic family with parameters $v \in V$, where $V$ is a finite-dimensional vector space. Recall that the principal symbol

$$\sigma(D) : \pi^* E \longrightarrow \pi^* F, \qquad \pi : (T^* M \oplus V) \setminus 0 \longrightarrow M,$$

of a parameter-dependent elliptic family (in the sense of Agranovich–Vishik) is invertible.

It follows from ellipticity that the family $D(v)$ is everywhere Fredholm and invertible for large parameter values. Hence the index of such a family is an element of the $K$-group with compact supports of the parameter space (see Section 3.6.2):

$$\operatorname{ind} D \in K_c(V).$$

The stability properties of the index imply the following lemma.

LEMMA C.1. *The index* $\operatorname{ind} D$ *is determined by the principal symbol of the family.*

Thus we arrive at the problem of computing the index via the principal symbol of the family.

THEOREM C.2 (Nistor 2003). *The index of the parameter-dependent elliptic family is given by the relation*

$$\operatorname{ind} D = \pi_! [\sigma(D)] \in K_c(V) \tag{C.1}$$

*where* $[\sigma(D)] \in K_c(T^* M \times V)$ *is the difference element,* $\pi : T^* M \times V \to V$ *is the projection, and* $\pi_! : K_c(T^* M \times V) \to K_c(V)$ *is the direct image map in* $K$*-theory.*

*Remark* C.3. Note that although there is a standard Atiyah–Singer index theorem for families of elliptic operators (Atiyah and Singer 1971), the family $D(v)$ does not satisfy the assumptions of this theorem: Atiyah and Singer assume that the parameter space is compact, or, in the case of a noncompact parameter space, the operators in the family are induced by *vector bundle isomorphisms* for large parameter values. We shall prove the index formula (C.1) by a reduction to the Atiyah–Singer formula: we construct a deformation to a family consisting of vector bundle isomorphisms for large parameter values.

*Proof.* 1. First, note that it suffices to prove the formula only for zero-order families. Indeed, composition with the invertible family

$$\left(1+|v|^2-\Delta_M\right)^{-m/2}$$

of order $-m$ does not change the index and the difference element.

2. By virtue of the parameter ellipticity, $D(v)$ is invertible for all $v$ of sufficiently large absolute value. Let $V_R \subset V$ be the ball of radius $R$. Let us show that for sufficiently large $R$ the restriction of $D(v)$ to the boundary of $V_R$ is homotopic in the class of invertible families to a family of vector bundle isomorphisms.

Let $SV$ be the unit sphere in $V$. For each point $v_1 \in SV$, consider the hemisphere

$$S^+(T^*M \times V) = \{(q,v) \in S(T^*M \times V) \mid (v_1, v) \geq 0\}$$

of unit vectors having an acute angle with the fixed vector $(0, v_1)$. Obviously, the hemisphere retracts to the point $(0, v_1)$; see Fig. C.1. A retraction can be defined as

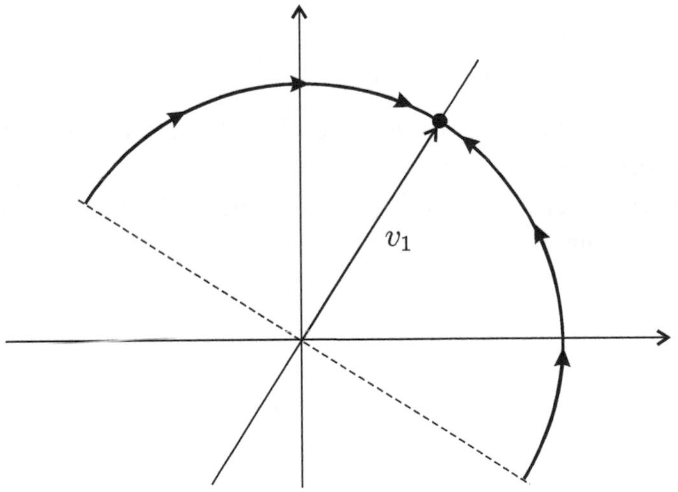


Figure C.1. Retraction of a hemisphere.

$$\gamma_{\varphi,v_1} : S^+(T^*M \times V) \longrightarrow S^+(T^*M \times V), \quad \varphi \in [0, \pi/2],$$

$$\gamma_{\varphi,v_1}(q,v) = \sqrt{\sin^2\varphi + \alpha^2\cos^2\varphi}(0,v_1) + \cos\varphi(q, v^{\perp}).$$

We use the orthogonal decomposition $v = \alpha v_1 + v^\perp$ ($v^\perp$ is orthogonal to $v_1$ and $\alpha \geq 0$).

For given $v_1$, this family is a smooth deformation of the identity (for $\varphi = 0$) to the mapping to the point $(0, v_1)$ (for $\varphi = \pi/2$).

Now for a parameter-elliptic symbol $\sigma = \sigma(D)$ we define the following family of symbols:

$$\sigma(\gamma_{\varphi,v_1}(q, v)). \tag{C.2}$$

This family of symbols with parameter $v$ is elliptic in a conical neighborhood of the ray $v = \alpha v_1, \alpha > 0$, for all $(\varphi, v_1) \in [0, \pi/2] \times SV$. Thus the corresponding family

$$\sigma\left(\gamma_{\varphi,v_1}\left(-i\frac{\partial}{\partial\omega}, v\right)\right)$$

of parameter-dependent elliptic operators with parameter $v$ is invertible on the ray $\{v = \alpha v_1, \alpha \geq R\}$ for all $(\varphi, v_1) \in [0, \pi/2] \times SV$ and sufficiently large $R$.

Now the desired homotopy of the family $D(v)$ on the sphere of radius $R$ is defined as

$$D_\varphi(v) = \sigma\left(\gamma_{\varphi,\frac{v}{|v|}}\left(-i\frac{\partial}{\partial\omega}, v\right)\right), \quad \varphi \in [0, \pi/2].$$

For $\varphi = \pi/2$, this family is induced by vector bundle isomorphisms, and for $\varphi = 0$ it is equal to the original family.

3. The original family, together with the homotopy thus constructed, can be viewed as a new family defined on the ball of radius $R + \pi/2$:

$$D'(v) = \begin{cases} D(v) & \text{if } |v| \leq R, \\ D_{|v|-R}(v) & \text{if } R + \pi/2 > |v| > R. \end{cases}$$

This family is extended as a constant in each radial direction to the remaining part of $V$. The resulting family $D'$ is a continuous family of elliptic pseudodifferential operators that are invertible outside $V_R$. Therefore, by the definition of the index for families we obtain

$$\text{ind} D = \text{ind} D' \in K_c(V). \tag{C.3}$$

4. Let us compute the latter index. The family $D'$ consists of elliptic operators induced by vector bundle isomorphisms for large parameter values. The index of such families is computed by the Atiyah–Singer index formula for families:

$$\text{ind}\, D' = \pi_![\sigma(D')],$$

where $[\sigma(D')] \in K_c(T^*M \times V)$ is the difference element of the principal symbol of $D'$. The proof of the index formula is finished once we note the equality

$$[\sigma(D')] = [\sigma(D)] \in K_c(T^*M \times V)$$

of the difference elements of the family $D'$ and of the family of parameter-elliptic operators $D$. □

EXERCISE C.4. Verify the last two formulas.

**A generalization.** The index formula remains valid if the family depends on additional parameters, more precisely, if $\pi_0 : X \to B$ is a fiber bundle ($M$ appears as a typical fiber) and $V$ is a vector bundle on $B$. In this case, the index of the family $D(v)$ of families with parameters in $V$ is again an element of $K_c(V)$. The index formula in this case is

$$\operatorname{ind} D = \pi_![\sigma(D)] \in K_c(V). \tag{C.4}$$

Here the principal symbol $\sigma(D)$ is defined on the sum $T^*M \oplus \pi_0^*V$, where $T^*M \in \operatorname{Vect}(X)$ is the cotangent bundle of the fibers of $\pi_0$, $[\sigma(D)] \in K_c(T^*M \oplus \pi_0^*V)$ is the usual difference element, and $\pi_!$ is the direct image map induced by the projection $\pi : T^*M \oplus \pi_0^*V \longrightarrow V$.

The proof of this index formula is the same as the proof of the special case described earlier.

# Bibliographic Remarks

To avoid any misinterpretation by the readers, we note that neither these bibliographic remarks nor the bibliography itself gives complete coverage of the literature on any of the diverse topics touched upon in this book or provides any information on priority issues.

While we always try to give proper credit to authors, this is by no means an essay on the history of the topic. We have included the literature that we had in mind when writing the book and/or that can, in our opinion, be useful for the readers who wish to gain further knowledge on the topics of the present book. That is why we have given preference to more accessible and readable sources (in particular, monographs and textbooks) rather than original papers whenever the former have been available. In many cases, extensive further bibliographies can be found in the sources cited here.

## Chapter 1

The idea of defining manifolds with singularities as ringed spaces (pairs $(M^\circ, \mathcal{D})$, where $M^\circ$ is a smooth open manifold and $\mathcal{D}$ an algebra of differential operators on $M^\circ$) goes back to Schulze, Sternin and Shatalov (1996, 1997, 1998*a*).

In the literature, there is another approach to the theory of manifolds with singularities, which was developed by Melrose (e.g., 1981, 1990) and his school. This approach is based on the consideration of operator algebras related to Lie algebras of degenerate vector fields.

Briefly speaking, Melrose's approach is as follows. Let $M$ be a smooth manifold with corners, $\operatorname{Vect}(M)$ the Lie algebra of smooth vector fields on $M$, and $\operatorname{Vect}_0(M)$ the Lie subalgebra of fields whose support does not contain boundary points. The idea is to describe singularities by specifying a Lie algebra $\mathcal{V}$ of vector fields such that

$$\operatorname{Vect}_0(M) \subset \mathcal{V} \subset \operatorname{Vect}(M).$$

Degenerate differential operators are then defined as elements of the algebra generated by $\mathcal{V}$ and $C^\infty(M)$.

For example, in Melrose's theory we obtain a manifold with conical singularities if we take $M$ to be a smooth manifold with boundary $\partial M = \Omega$ and $\mathcal{V} = \mathcal{V}_0$ to be the Lie algebra of vector fields everywhere tangent to the boundary $\partial M$ (the "$b$-differential calculus"). This algebra actually describes a manifold with cylindrical ends obtained from $M$ by stretching a collar neighborhood of the boundary $\partial M$ into an infinite cylinder and is associated with the singular metric

$$d\rho^2 = \frac{dr^2}{r^2} + d\omega^2$$

rather than our degenerating metric $ds^2 = dr^2 + r^2 d\omega^2$. Melrose's approach then produces operators of the form (1.30) *without* the factor $r^{-k}$.

Accordingly, instead of the *stretched cotangent bundle* Melrose defines the *compressed cotangent bundle* as follows. (We also use this notion in Chapter 7.)

The space $\mathcal{V}_0$ is a $C^\infty(M)$-module. (In other words, one can add vector fields tangent to the boundary and multiply them by smooth functions, the tangency being preserved.) Consider the dual module $\mathrm{Hom}_{C^\infty(M)}(\mathcal{V}_0, C^\infty(M))$. The elements of this module are $C^\infty(M)$-linear mappings of $\mathcal{V}_0$ into $C^\infty(M)$; they can be added and multiplied by functions in a natural way. This is a locally free $C^\infty(M)$-module and hence, by the Swan theorem, the module of sections of some vector bundle $\widetilde{T}^*\mathcal{M}$. This bundle is called the *compressed cotangent bundle* of $\mathcal{M}$ (Melrose 1981). The space $\widetilde{T}^*\mathcal{M}$ is a manifold with boundary and a vector bundle over $M$. Since in the interior of the manifold the set $\mathcal{V}_0$ coincides with the set of all vector fields, it follows that there is a canonical isomorphism

$$\mu : \widetilde{T}^*\mathcal{M} \backslash \partial \widetilde{T}^*\mathcal{M} \to T^*M^\circ.$$

Manifolds with edges are obtained by a similar pattern. One considers a manifold $M$ with boundary $\partial M$ fibered over a smooth manifold $X$ and takes $\mathcal{V}$ to be the Lie algebra of vector fields tangent to the fibers everywhere on the boundary.

## Chapter 2

The study of differential equations on manifolds with singularities has a long history. It was pioneered by Kondrat'ev (1967). It is virtually impossible to list all publications concerning this subject, and we refer the reader to Schulze, Sternin and Shatalov (1998*a*) and Kozlov, Maz'ya and Rossmann (2001), where one can find further references.

Edge-degenerate differential operators were considered in the framework of Melrose's approach by Mazzeo (1991). The main problem studied in this paper is finding conditions ensuring the Hausdorff normal solvability (i.e., conditions ensuring that the range is closed) for linear edge-degenerate operators in special function spaces.

Note that boundary value problems for elliptic equations in domains with edges were considered in spaces with weights given by powers of $r$ in numerous papers (e.g., Maz'ya and Rossmann 1988); a detailed bibliography can be found in the book by Kozlov, Maz'ya and Rossmann (2001). However, the methods used there are completely different (and in particular do not involve the construction of a pseudodifferential calculus).

There is a vast literature on semiclassical quantization. In the context of our problems, some references can be found in the monograph by Schulze, Sternin and Shatalov (1998*a*), which deals with the application of semiclassical methods in the theory of manifolds with singularities. Such methods were also applied in the geometric context, where they correspond to the adiabatic limit of Riemannian metrics; e.g., see Bismut and Cheeger (1990).

## Chapter 3

So far, there exists a wide literature on pseudodifferential calculus on singular manifolds (e.g., Plamenevskii 1989, Schulze 1991, Melrose 1996, Plamenevskii and Senichkin 1999, Egorov and Schulze 1997, Nistor 2004, and papers cited therein).

We follow the approach initiated by Schulze (1989) and developed in Schulze (1991). The calculus of pseudodifferential operators on singular manifolds with edges given in Chapter 3 is the simplest version of such a calculus, mainly adjusted for the study of *topological* aspects of elliptic theory on manifolds with edges. In particular, we do not require the conormal symbols of the edge symbols to be analytic: it suffices that they be defined only on the given weight line. A finer but more complicated theory can be found in Schulze (1991) (see also Egorov and Schulze 1997); one can also find further references in these books. Here we only note that under the condition of analyticity of symbols with respect to the conormal variable $p$ one can also obtain asymptotic expansions of solutions in a neighborhood of the singularities.

There also is an approach to pseudodifferential operators in the framework of $C^*$-algebras, based on the localization principle. The localization principle in $C^*$-algebras was developed by numerous authors (Simonenko 1965, Dauns and Hofmann 1968, Hofmann 1972, Douglas 1972, 1973, Varela 1974, Pedersen 1979, Gohberg and Krupnik 1992, Antonevich and Lebedev 1994, Plamenevskii and Senichkin 1994, Vasilevski 1989, 1994) and actively used in the study of pseudodifferential operators on singular manifolds (Plamenevskii and Senichkin 1994, 1996).

In particular, algebras of pseudodifferential operators on manifolds with edges were also constructed and studied in the framework of the $C^*$-algebras approach by Senichkin (1995, 1996).

The modern approach based on pseudodifferential operators on groupoids was applied to the analysis of pseudodifferential operators on manifolds with singularities in Nistor (2004), where one can also find further references.

A detailed construction (including composition formulas and changes of variables) of pseudodifferential operators in $\mathbb{R}^n$ or on a closed manifold can be found in the classical papers by Kohn and Nirenberg (1965) and Hörmander (1965) or the book by Hörmander (1985*a*). The composition and boundedness theorems 3.2, 3.4 without additional assumptions such as stabilization at infinity can be found in numerous places. In particular, they can be obtained by methods of noncommutative analysis (Maslov 1973; see also the elementary exposition in Nazaikinskii, Sternin and Shatalov 1995).

Our definitions of pseudodifferential operators with operator-valued symbols are a substantially improved version of the definitions apparently introduced for the first time by Luke (1972). The index theorem 3.89 was announced there, but the proof was only given for homogeneous symbols. We give the proof of this formula in the general case.

The reader can find the facts of $K$-theory required in this chapter, e.g., in Atiyah (1989) and also in Mishchenko (1984).

## Chapter 4

The procedure of cutting and pasting elliptic operators has a long history. Gluing a copy of the manifold followed by extending the operator or at least the symbol to the double (or, more generally, to a manifold containing the original manifold as a part) is in fact quite an old idea. Indeed, this technique was successfully applied to the index problem for boundary value problems in early Soviet papers on elliptic theory (e.g. Agranovich 1965, Dezin 1964); see also the survey by Agranovich (1997), where more detailed references can be found. Later, it was widely used by many authors. We mention the papers by Hsiung (1972, 1976) and Stong (1974), where gluing by an orientation-reversing automorphism of the boundary was used to obtain the index of the signature operator, as well as the papers by Gilkey and Smith (1983*a*, 1983*b*). Naturally, the list can be continued.

The fact that the extension to a wider manifold (in particular, to the double) proves to be fruitful in index problems is a consequence of the so-called *index locality principle* or *relative index theorem.*

For smooth manifolds, the index locality principle trivially follows from the so-called *local index formula* (e.g., Gilkey 1995). A method based on the use of trace integrals permitted Gromov and Lawson (1983) to prove relative index theorem for Dirac operators on complete noncompact Riemannian manifolds and later Anghel (1993) to generalize this theorem to arbitrary essentially self-adjoint supersymmetric Fredholm elliptic first-order operators. The heat equation method was used by Donnelly (1987) in the proof of a relative index theorem for the signature operator and also by Bunke (1992). We also mention the book by Booß-Bavnbek and Wojciechowski (1993). Teleman (1984) used a subtle homotopy technique to prove the relative index theorem for signature operators on Lipschitz manifolds. The well-known Agranovich and Agranovich–Dynin theorems (Dynin 1961, Agranovich and Dynin 1962, Agranovich 1965), which express the relative index of two boundary value problems with the same boundary conditions and different operators (coinciding on the boundary) or with the same operator but different boundary conditions, can essentially be interpreted as a statement of the relative index theorem for boundary value problems.

The technique of attaching a copy of the manifold and passing to the double has also been successfully used in the index theory of elliptic operators on manifolds with singularities. The index theorem for operators with a symmetric conormal symbol was obtained by this method by Schulze, Sternin and Shatalov (1998*b*). Later, these ideas were used in Fedosov, Schulze and Tarkhanov (1999).

Following Nazaikinskii and Sternin (2001), we establish the locality principle in a sufficiently general case so as to ensure that it can be applied to elliptic pseudodifferential operators as well as elliptic Fourier integral operators on manifolds with singularities. Thus we introduce a new class of Fredholm operators for which the locality principle holds and which includes both pseudodifferential operators and Fourier integral operators.

## Chapter 5

Stratified index formulas on manifolds with conical singularities were considered for the first time in Schulze, Sternin and Shatalov (1998*b*). This idea was further developed in Savin, Schulze and Sternin (1999), where similar formulas were obtained in the framework of spectral boundary value problems. (We refer the reader to Savin and Sternin (2004) for a detailed exposition and further references.) Moreover, a criterion for the existence of a homotopy invariant index splitting was stated. A similar criterion was given on manifolds with conical points and edges (see Nazaikinskii, Schulze and Sternin 1999, Nazaikinskii, Savin, Schulze and Sternin 2003).

There are a number of results concerning index formulas for manifolds with edges. First, we mention index formulas for model manifolds of infinite wedge type (Schrohe and Seiler 1996, Fedosov, Schulze and Tarkhanov 1998, Rozenblum 2002, 2003).

Second, for compatible Dirac type operators one can use the celebrated *heat equation* method (e.g., Gilkey 1995) to compute the index provided that the heat expansion is available. This approach was used by Vaillant (2001) to obtain index formulas on manifolds with fibered cusps. By analogy with manifolds with edges, manifolds with fibered cusps are also defined using manifolds with fibered boundary, but the metric has singularities (infinite growth) at the boundary. The index formula in this case involves the usual local contribution of the smooth part of the manifold and an additional term depending on the $\eta$-form of Bismut and Cheeger (1989). Lauter and Moroianu (2002) used regularized traces (as in Melrose and Nistor 1996) to obtain, in some cases, an index formula where the singularity set gives the contribution to the index in the form of the adiabatic limit of the $\eta$-invariant.

Third, one can try to find stratified index formulas with homotopy invariant contributions of the interior symbol and the edge symbol. The obvious benefit of this approach is that one can use homotopies of the operator (and other topological tools like surgeries, $K$-theory, etc.) to simplify the computations. We explain several index formulas of this type following Nazaikinskii, Savin, Schulze and Sternin (2003, 2004*b*). These index formulas apply to general edge-degenerate operators (with symmetry of boundary symbols). The results in Schulze, Sternin and Shatalov (1998*b*) and Savin and Sternin (1999, 2000) are precursors of such formulas.

## Chapter 6

The terms "boundary and coboundary operators" go back to Sternin (1967), who introduced these operators in the framework of Sobolev problems. They were then used by many people in various contexts.

The topological obstruction formulas given in this chapter were obtained in Nazaikinskii, Savin, Schulze and Sternin (2002*b*). Similar topological obstructions can be traced back to the paper by Atiyah and Bott (1964), where the classical Atiyah–Bott obstruction was introduced. It is the obstruction to the existence of elliptic classical boundary conditions on a smooth manifold with boundary.

The obstruction to the definition of elliptic operators on manifolds with singularities was also studied in Nistor (2003), where the index formula for elliptic parameter-dependent families was obtained.

## Chapter 7

The index problem for quantized contact transformations (Fourier integral operators) was posed by Weinstein (1976, 1997) and solved for the case of smooth manifolds by Epstein and Melrose (1998) and Leichtnam, Nest and Tsygan (2001). Here we deal with the same problem for manifolds with singularities following Nazaikinskii, Schulze and Sternin (2001).

## Chapter 8

This chapter is a survey of ideas, methods and results contained in Novikov and Sternin (1966*a*, 1966*b*), Sternin (1966, 1967, 1971, 1976), Sternin and Shatalov (1996*b*), and Nazaikinskii and Sternin (2004). In particular, it gives a new glance on classical elliptic Sobolev problems and relative elliptic theory from the viewpoint of modern theory of differential equations. For example, boundary and coboundary operators, as well as Green operators, are interpreted as Fourier integral operators.

The first papers on general Sobolev problems were written about forty years ago (see Sternin 1966). Since then, quite a few remarkable papers on relative elliptic theory have been published. The notion of a coboundary operator, originally introduced by Sternin (1967), is nowadays widely used in general theory of differential equations. In various situations, many authors independently developed constructions close in their spirit to the constructions of relative elliptic theory (e.g., Vishik and Eskin 1965, 1966, Boutet de Monvel 1971, Eskin 1973, Schulze 1991).

## Chapter 9

Numerous papers deal with index formulas on manifolds with conical singularities. It would be virtually impossible to name all the authors who contributed to index theory on manifolds with conical singularities. Let us only mention the early work of Cheeger (1979) linking the index of cone-degenerate operators with the Atiyah–Patodi–Singer index formula containing the $\eta$-invariant for operators on manifolds with cylindrical ends. The reader can find detailed exposition and further references in the books by Melrose (1993) and Lesch (1997).

A detailed exposition of the theory of spectral problems (together with the proof of the Fredholm property) can be found in the original paper by Atiyah, Patodi and Singer (1975) or in Booß-Bavnbek and Wojciechowski (1993). The stratified index formula for the case of the Dirac operator was obtained by Kreck and Stolz (1993).

## Chapter 10

The classification of elliptic operators on a smooth closed manifold in terms of topological $K$-theory was obtained by Atiyah and Singer (1968). The classification on a manifold with a conical point was obtained by Nazaikinskii, Schulze and Sternin (1999). In this case, the classification is also given by a topological $K$-group. The papers by Nazaikinskii, Savin, Schulze and Sternin (2004*a*) and Savin (2005) give an approach to the classification of operators on manifolds with singularities based on $K$-homology. The classification was obtained for isolated singularities, manifolds with fibered boundary, and manifolds with edges.

For the reader's convenience, we also give some references to papers concerning $K$-homology. First, these are the pioneering papers by Atiyah (1969), Kasparov (1973), and Brown, Douglas and Fillmore (1977), where the theory was constructed. A modern exposition can be found in Blackadar (1998) and Higson and Roe (2000). The topological aspects of duality are discussed in Baum and Douglas (1982).

## Chapter 11

Background of the investigation of Lefschetz formulas for singular manifolds is the classical paper by Atiyah and Bott (1967) for the case of smooth manifolds. Various alternative methods and generalizations were developed by Gilkey (1995) (heat equation approach), Bismut (1984, 1985) (probabilistic approach), Fedosov (1993, 1996), and Sternin and Shatalov (1994, 1996*a*, 1998) (semiclassical approach). The last of these papers provides an elementary proof of the semiclassical Lefschetz formula for endomorphisms of elliptic complexes given by arbitrary quantized canonical transformations. The classical formula due to Atiyah and Bott (1967), which deals with geometric endomorphisms, is covered as a special case. Note also the paper by Brenner and Shubin (1993) establishing a Lefschetz formula for manifolds with boundary. Our exposition is based on Nazaikinskii, Schulze, Sternin and Shatalov (1999) and Nazaikinskii and Sternin (2002).

# Bibliography

**Agranovich, M.** (1965), Elliptic singular integro-differential operators, *Uspekhi Matem. Nauk* **20**(5), 3–120.

**Agranovich, M.** (1997), Elliptic boundary problems, in *Partial Differential Equations* IX. *Elliptic Boundary Value Problems*, Agranovich, M., Egorov, Yu. and Shubin, M., Eds., Vol. 60 of *Encyclopaedia of Mathematical Sciences*, Springer-Verlag, Berlin–Heidelberg, 1–144.

**Agranovich, M. and Dynin, A.** (1962), General boundary value problems for elliptic systems in higher-dimensional regions, *Dokl. Akad. Nauk SSSR* **146**, 511–514.

**Agranovich, M. and Vishik, M.** (1964), Elliptic problems with parameter and parabolic problems of general type, *Uspekhi Matem. Nauk* **19**(3), 53–161.

**Anghel, N.** (1993), An abstract index theorem on non-compact Riemannian manifolds, *Houston J. Math.* **19**, 223–237.

**Antonevich, A. and Lebedev, A.** (1994), *Functional-Differential Equations.* I. *$C^*$-Theory*, Longman, Harlow, U.K.

**Arnold, V.** (1989), *Mathematical Methods of Classical Mechanics*, $2^d$ ed., Springer-Verlag, Berlin–Heidelberg–New York.

**Atiyah, M.** (1969), Global theory of elliptic operators, in *Proceedings of the International Symposium on Functional Analysis*, University of Tokyo Press, Tokyo, 21–30.

**Atiyah, M.** (1989), *K-Theory*, 2nd ed., Addison–Wesley, Redwood City, CA.

**Atiyah, M. and Bott, R.** (1964), The index problem for manifolds with boundary, in *Bombay Colloquium on Differential Analysis*, Oxford University Press, Oxford, U.K., 175–186.

**Atiyah, M. and Bott, R.** (1967), A Lefschetz fixed point formula for elliptic complexes. I, *Ann. Math.* **86**, 374–407.

**Atiyah, M., Bott, R. and Patodi, V.** (1973), On the heat equation and the index theorem, *Invent. Math.* **19**, 279–330.

**Atiyah, M., Donnelly, H. and Singer, I.** (1983), Eta invariants, signature defects of cusps and values of $L$-functions, *Ann. Math.* **118**, 131–177.

**Atiyah, M. and Singer, I.** (1968), The index of elliptic operators. I, *Ann. Math.* **87**, 484–530.

**Atiyah, M. and Singer, I.** (1969), Index theory for skew-adjoint Fredholm operators, *Publ. Math. IHES* **37**, 5–26.

**Atiyah, M. and Singer, I.** (1971), The index of elliptic operators. IV, *Ann. Math.* **93**, 119–138.

**Atiyah, M., Patodi, V. and Singer, I.** (1975), Spectral asymmetry and Riemannian geometry. I, *Math. Proc. Cambridge Philos. Soc.* **77**, 43–69.

**Atiyah, M., Patodi, V. and Singer, I.** (1976), Spectral asymmetry and Riemannian geometry. III, *Math. Proc. Cambridge Philos. Soc.* **79**, 71–99.

**Baum, P. and Douglas, R.** (1982), $K$-homology and index theory, in *Operator Algebras and Applications*, Kadison, R., Ed., Vol. 38 of *Proceedings of Symposia on Pure Mathematics*, American Mathematical Society, Providence, RI, 117–173.

**Bismut, J.-M.** (1984), The Atiyah–Singer theorems: a probabilistic approach. II. The Lefschetz fixed point formulas, *J. Funct. Anal.* **57**(3), 329–348.

**Bismut, J.-M.** (1985), The infinitesimal Lefschetz formulas: a heat equation proof, *J. Funct. Anal.* **62**(3), 437–457.

**Bismut, J.-M. and Cheeger, J.** (1989), $\eta$-invariants and their adiabatic limits, *J. Am. Math. Soc.* **2**(1), 33–70.

**Bismut, J.-M. and Cheeger, J.** (1990), Families index for manifolds with boundary, superconnections, and cones. I, *J. Funct. Anal.* **89**(2), 313–363.

**Bismut, J.-M. and Freed, D.** (1986), The analysis of elliptic families. I, *Comm. Math. Phys.* **106**(1), 159–176.

**Blackadar, B.** (1998), *$K$-Theory for Operator Algebras*, 2nd ed., Cambridge University Press, Cambridge, U.K.

**Booß-Bavnbek, B. and Bleecker, D. D.** (1985), *Topology and Analysis*, Springer-Verlag, New York–Berlin.

**Booß-Bavnbek, B. and Wojciechowski, K.** (1993), *Elliptic Boundary Problems for Dirac Operators*, Birkhäuser, Boston–Basel–Berlin.

**Bott, R. and Tu, L.** (1982), *Differential Forms in Algebraic Topology*, Springer-Verlag, Berlin–Heidelberg–New York.

**Boutet de Monvel, L.** (1971), Boundary problems for pseudodifferential operators, *Acta Math.* **126**(1–2), 11–51.

**Brenner, V. and Shubin, M.** (1993), Atiyah–Bott–Lefschetz formula for elliptic complexes on manifolds with boundary, *J. Sov. Math.* **64**, 1069–1111.

**Brown, L., Douglas, R. and Fillmore, P.** (1977), Extensions of $C^*$-algebras and $K$-homology, *Ann. Math.* **105**(2), 265–324.

**Brüning, J. and Seeley, R.** (1988), An index theorem for first order regular singular operators, *Am. J. of Math.* **110**(4), 659–714.

**Bunke, U.** (1992), Relative index theory, *J. Funct. Anal.* **105**, 63–76.

**Cheeger, J.** (1979), On the spectral geometry of spaces with cone-like singularities, *Proc. Nat. Acad. Sci. USA* **76**, 2103–2106.

**Connes, A., Sullivan, D. and Teleman, N.** (1994), Quasiconformal mappings, operators on Hilbert space, and local formulae for characteristic classes, *Topology* **33**(4), 663–681.

**Dai, X.** (1991), Adiabatic limits, nonmultiplicativity of signature and Leray spectral sequence, *J. Am. Math. Soc.* **4**(2), 265–321.

**Dai, X. and Zhang, W.** (1998), Higher spectral flow, *J. Funct. Anal.* **157**(2), 432–469.

**Dauns, J. and Hofmann, K.-H.** (1968), *Representation of Rings by Sections*, American Mathematical Society, Providence, RI.

**Dezin, A. A.** (1964), *Invariant Differential Operators and Boundary Value Problems*, American Mathematical Society, Providence, RI.

**Donnelly, H.** (1987), Essential spectrum and the heat kernel, *J. Funct. Anal.* **75**(2), 362–381.

**Douglas, R.** (1972), *Banach Algebra Techniques in Operator Theory*, 2nd ed., Springer-Verlag, New York.

**Douglas, R.** (1973), *Banach Algebra Techniques in the Theory of Toeplitz Operators*, American Mathematical Society, Providence, RI.

**Dynin, A.** (1961), $n$-Dimensional elliptic boundary problems with a single unknown function, *Dokl. Akad. Nauk SSSR* **141**, 285–287.

**Egorov, Yu. and Schulze, B.-W.** (1997), *Pseudo-Differential Operators, Singularities, Applications*, Birkhäuser, Boston–Basel–Berlin.

**Epstein, C. and Melrose, R.** (1998), Contact degree and the index of Fourier integral operators, *Math. Res. Lett.* **5**(3), 363–381.

**Eskin, G.** (1973), *Boundary Value Problems for Elliptic Pseudodifferential Equations*, American Mathematical Society, Providence, RI.

**Fedosov, B.** (1993), A trace formula for the Schrödinger operator, *Russian Journ. of Math. Phys.* **1**(4), 447–463.

**Fedosov, B.** (1996), *Deformation Quantization and Index Theory*, Akademie-Verlag, Berlin.

**Fedosov, B., Schulze, B.-W. and Tarkhanov, N.** (1998), On the index of elliptic operators on a wedge, *J. Funct. Anal.* **157**, 164–209.

**Fedosov, B., Schulze, B.-W. and Tarkhanov, N.** (1999), The index of elliptic operators on manifolds with conical points, *Selecta Math.* **5**(4), 467–506.

**Gilkey, P.** (1981), The residue of the global eta function at the origin, *Adv. Math.* **40**, 290–307.

**Gilkey, P.** (1989*a*), *The Geometry of Spherical Space Form Groups*, World Sientific, Singapore.

**Gilkey, P.** (1989*b*), The eta invariant of even order operators, *Lect. Notes Math.* **1410**, 202–211.

**Gilkey, P.** (1995), *Invariance Theory, the Heat Equation, and the Atiyah-Singer Index Theorem*, 2nd ed., CRC Press, Boca Raton, FL.

**Gilkey, P. and Smith, L.** (1983*a*), The eta invariant for a class of elliptic boundary value problems, *Comm. Pure Appl. Math.* **36**, 85–132.

**Gilkey, P. and Smith, L.** (1983*b*), The twisted index problem for manifolds with boundary, *J. Diff. Geometry* **18**(3), 393–444.

**Gohberg, I. and Krupnik, N.** (1992), *One-Dimensional Linear Singular Integral Equations. I*, Birkhäuser, Basel.

**Gromov, M. and Lawson, H.** (1983), Positive scalar curvature and the Dirac operator on complete Riemannian manifolds, *Publ. Math. IHES* **58**, 295–408.

**Grubb, G. and Seeley, R.** (1996), Zeta and eta functions for Atiyah–Patodi–Singer operators, *J. Geom. Anal.* **6**(1), 31–77.

**Higson, N. and Roe, J.** (2000), *Analytic $K$-Homology*, Oxford University Press, Oxford, U.K.

**Hirzebruch, F.** (1966), *Topological Methods in Algebraic Geometry*, Springer-Verlag, Berlin.

**Hofmann, K.** (1972), Representations of algebras by continuous sections, *Bull. Am. Math. Soc.* **78**, 291–373.

**Hörmander, L.** (1965), Pseudo-differential operators, *Comm. Pure Appl. Math.* **18**, 501–517.

**Hörmander, L.** (1968), The spectral function of an elliptic operator, *Acta Math.* **121**, 193–218.

**Hörmander, L.** (1985*a*), *The Analysis of Linear Partial Differential Operators.* III, Springer-Verlag, Berlin–Heidelberg–New York–Tokyo.

**Hörmander, L.** (1985*b*), *The Analysis of Linear Partial Differential Operators.* IV, Springer-Verlag, Berlin–Heidelberg–New York–Tokyo.

**Hsiung, C.-C.** (1972), The signature and $G$-signature of manifolds with boundary, *J. Diff. Geometry* **6**, 595–598.

**Hsiung, C.-C.** (1976), A remark on cobordism of manifolds with boundary, *Arch. Math.* **XXVII**, 551–555.

**Kasparov, G.** (1973), The generalized index of elliptic operators, *Funkts. Anal. Prilozhen.* **7**(3), 82–83.

**Kasparov, G.** (1988), Equivariant $KK$-theory and the Novikov conjecture, *Inv. Math.* **91**(1), 147–201.

**Kohn, J. and Nirenberg, L.** (1965), An algebra of pseudo-differential operators, *Comm. Pure Appl. Math.* **18**, 269–305.

**Kondrat'ev, V.** (1967), Boundary value problems for elliptic equations in domains with conical or angular points, *Trudy Mosk. Matem. Obshch.*, **16**, 209–292.

**Kozlov, V., Maz'ya, V. and Rossmann, J.** (2001), *Spectral Problems Associated with Corner Singularities of Solutions to Elliptic Equations*, American Mathematical Society, Providence, RI.

**Kreck, M. and Stolz, S.** (1993), Nonconnected moduli spaces of positive sectional curvature metrics, *J. Am. Math. Soc.* **6**(4), 825–850.

**Lauter, R. and Moroianu, S.** (2002), *An index formula on manifolds with fibered cusp ends*, Prépublication du Laboratoire Emile Picard n. 253.

**Lawson, H. and Michelsohn, M.** (1989), *Spin Geometry*, Princeton University Press, Princeton, NJ.

**Leichtnam, E., Nest, R. and Tsygan, B.** (2001), Local formula for the index of a Fourier integral operator, *J. Diff. Geometry* **59**(2), 269–300.

**Lesch, M.** (1997), *Differential Operators of Fuchs Type, Conical Singularities, and Asymptotic Methods*, Teubner Verlag, Stuttgart–Leipzig.

**Lesch, M. and Pflaum, M.** (2000), Traces on algebras of parameter dependent pseudodifferential operators and the eta-invariant, *Trans. Am. Math. Soc.* **352**(11), 4911–4936.

**Luke, G.** (1972), Pseudodifferential operators on Hilbert bundles, *J. Diff. Equations* **12**, 566–589.

**Luke, G. and Mishchenko, A.** (1998), *Vector Bundles and Their Applications*, Kluwer, Dordrecht.

**Martinez, A.** (1997), Microlocal exponential estimates and applications to tunneling, in *Microlocal Analysis and Spectral Theory*, Rodino, L., Ed., Kluwer, Dordrecht, 349–376.

**Maslov, V.** (1972), *Théorie des Perturbations et Méthod Asymptotiques*, Dunod, Paris.

**Maslov, V.** (1973), *Operator Methods*, Nauka, Moscow.

**Maz'ya, V. and Rossmann, J.** (1988), Über die Asymtotik der Lösungen elliptischer Randwertaufgaben in der Umbgebung von Kanten, *Math. Nachr.* **138**, 27–53.

**Mazzeo, R.** (1991), Elliptic theory of differential edge operators. I, *Comm. Partial Diff. Equations* **16**(10), 1615–1664.

**Melrose, R.** (1981), Transformation of boundary problems, *Acta Math.* **147**, 149–236.

**Melrose, R.** (1990), Pseudodifferential operators, corners, and singular limits, in *Proceedings of the International Congress of Mathematicians*, Kyoto, Springer-Verlag, Berlin–Heidelberg–New York, 217–234.

**Melrose, R.** (1993), *The Atiyah–Patodi–Singer Index Theorem*, A.K. Peters, Boston.

**Melrose, R.** (1995), The eta invariant and families of pseudodifferential operators, *Math. Res. Lett.* **2**(5), 541–561.

**Melrose, R.** (1996), Fibrations, compactifications and algebras of pseudodifferential operators, in *Partial Differential Equations and Mathematical Physics. The Danish–Swedish Analysis Seminar,* 1995, Hörmander, L., and Melin, A., Eds., Birkhäuser, Boston–Basel–Berlin, 246–261.

**Melrose, R. and Nistor, V.** (1996), Homology of pseudodifferential operators. I. Manifolds with boundary, http://arXiv.org/funct-an/9606005.

**Melrose, R. and Piazza, P.** (1997), Families of Dirac operators, boundaries and the $b$-calculus, *J. Diff. Geometry* **46**(1), 99–180.

**Mishchenko, A.** (1984), *Vector Bundles*, MGU, Moscow.

**Mishchenko, A., Shatalov, V. and Sternin, B.** (1990), *Lagrangian Manifolds and the Maslov Operator*, Springer-Verlag, Berlin–Heidelberg.

**Moroianu, S.** (2003), $K$-theory of suspended pseudo-differential operators, *$K$-Theory* **28**(2), 167–181.

**Nazaikinskii, V., Oshmyan, V., Sternin, B. and Shatalov, V.** (1981), Fourier integral operators and the canonical operator, *Uspekhi Matem. Nauk* **36**(2), 81–140.

**Nazaikinskii, V., Savin, A., Schulze, B.-W. and Sternin, B.** (2002*a*), *Elliptic Theory on Manifolds with Nonisolated Singularities. III. The Spectral Flow of Families of Conormal Symbols*, Preprint No. 2002/20, Universität Potsdam, Institut für Mathematik, Potsdam.

**Nazaikinskii, V., Savin, A., Schulze, B.-W. and Sternin, B.** (2002*b*), *Elliptic Theory on Manifolds with Nonisolated Singularities. IV. Obstructions to Elliptic Problems on Manifolds with Edges*, Preprint No. 2002/24, Universität Potsdam, Institut für Mathematik, Potsdam.

**Nazaikinskii, V., Savin, A., Schulze, B.-W. and Sternin, B.** (2003), *Elliptic Theory on Manifolds with Nonisolated Singularities. V. Index Formulas for Elliptic Problems on Manifolds with Edges*, Preprint No. 2003/02, Universität Potsdam, Institut für Mathematik, Potsdam.

**Nazaikinskii, V., Savin, A., Schulze, B.-W. and Sternin, B.** (2004*a*), *On the Homotopy Classification of Elliptic Operators on Manifolds with Edges*, Preprint No. 2004/16, Universität Potsdam, Institut für Mathematik, Potsdam.

**Nazaikinskii, V., Savin, A., Schulze, B.-W. and Sternin, B.** (2004*b*), On the index of differential operators on manifolds with edges, *Dokl. Ross. Akad. Nauk* **69**(1), 68–71.

**Nazaikinskii, V., Schulze, B.-W. and Sternin, B.** (1999), *On the Homotopy Classification of Elliptic Operators on Manifolds with Singularities*, Preprint No. 99/21, Universität Potsdam, Institut für Mathematik, Potsdam.

**Nazaikinskii, V., Schulze, B.-W. and Sternin, B.** (2001), The index of Fourier integral operators on manifolds with conical singularities, *Izv. Ross. Akad. Nauk, ser. Matem.* **65**(2), 329–355.

**Nazaikinskii, V., Schulze, B.-W. and Sternin, B.** (2002), *Quantization Methods in Differential Equations*, Taylor & Francis, London–New York.

**Nazaikinskii, V., Schulze, B.-W., Sternin, B. and Shatalov, V.** (1999), The Atiyah–Bott–Lefschetz fixed point theorem for manifolds with conical singularities, *Ann. Global Anal. Geometry* **17**(5), 409–439.

**Nazaikinskii, V. and Sternin, B.** (1999), *Surgery and the Relative Index in Elliptic Theory*, Preprint No. 99/17, Universität Potsdam, Institut für Mathematik, Potsdam.

**Nazaikinskii, V. and Sternin, B.** (2001), On the index locality principle in elliptic theory, *Funkts. Anal. Prilozhen.* **35**(2), 37–52.

**Nazaikinskii, V. and Sternin, B.** (2002), *Lefschetz Theory on Manifolds with Singularities*, Preprint No. 2002:31, Department of Mathemathics, Chalmers University of Technology and Göteborg University, Göteborg.

**Nazaikinskii, V. and Sternin, B.** (2004), Relative elliptic theory, in *Aspects of Boundary Problems in Analysis and Geometry*, Gil, J., Krainer, T. and Witt, I., Eds., Birkhäuser, Basel–Boston–Berlin, 495–560.

**Nazaikinskii, V., Sternin, B. and Shatalov, V.** (1995), *Methods of Noncommutative Analysis. Theory and Applications*, Walter de Gruyter, Berlin–New York.

**Nistor, V.** (2002), Asymptotics and index for families invariant with respect to a bundle of Lie groups, *Rev. Roum. Math. Pures Appl.* **47**(4), 451–483.

**Nistor, V.** (2003), An index theorem for gauge-invariant families: The case of solvable groups, *Acta Math. Hungarica* **99**(2), 155–183.

**Nistor, V.** (2004), Singular integral operators on non-compact manifolds and analysis on polyhedral domains, *Contemp. Math.* **366**, 307–328.

**Novikov, S. and Sternin, B.** (1966*a*), Traces of elliptic operators on submanifolds and $K$-theory, *Dokl. Akad. Nauk SSSR* **170**, 1265–1268.

**Novikov, S. and Sternin, B.** (1966*b*), Elliptic operators and submanifolds, *Dokl. Akad. Nauk SSSR* **171**, 525–528.

**Palais, R.** (1965), *Seminar on the Atiyah–Singer Index Theorem*, Princeton Univerity Press, Princeton, NJ.

**Pedersen, G.** (1979), *$C^*$-Algebras and Their Automorphism Groups*, Academic Press, London–New York.

**Plamenevskii, B.** (1989), *Algebras of Pseudodifferantial Operators*, Kluwer, Dordrecht.

**Plamenevskii, B. and Senichkin, V.** (1994), Solvable operator algebras, *Algebra i Analiz* **6**(5), 1–87.

**Plamenevskii, B. and Senichkin, V.** (1996), On composition series in algebras of pseudodifferential operators and in algebras of Wiener–Hopf operators, in *Schrödinger Operators, Markov Semigroups, Wavelet Analysis, Operator Algebras*, Akademie-Verlag, Berlin, 373–404.

**Plamenevskii, B. and Senichkin, V.** (1999), Pseudodifferential operators on manifolds with singularities, *Funkts. Anal. Prilozhen.* **33**(2), 88–91.

**Reed, M. and Simon, B.** (1975), *Fourier Analysis, Self-Adjointness*, Vol. II of *Methods of Modern Mathematical Physics*, Academic Press, San Diego.

**Rozenblum, G.** (2002), On some analytical index formulas related to operator-valued symbols, *Electron. J. Diff. Equations* **17**, 1–31.

**Rozenblum, G.** (2003), Regularisation of secondary characteristic classes and unusual index formulas for operator-valued symbols, in *Nonlinear Hyperbolic Equations, Spectral Theory, and Wavelet Transformations*, Birkhäuser, Basel, 419–437.

**Savin, A.** (2005), Elliptic Operators on Singular Manifolds and $K$-homology. E-print: *http://www.arxiv.org/math.KT/0403335*.

**Savin, A., Schulze, B.-W. and Sternin, B.** (1999), On invariant index formulas for spectral boundary value problems, *Differents. Uravn.* **35**(5), 709–718.

**Savin, A. and Sternin, B.** (1999), Elliptic operators in even subspaces, *Matem. Sbornik* **190**(8), 125–160. E-print: *http://www.arxiv.org/math.DG/9907027*.

**Savin, A. and Sternin, B.** (2000), Elliptic operators in odd subspaces, *Matem. Sbornik* **191**(8), 89–112. E-print: *http://www.arxiv.org/math.DG/9907039*.

**Savin, A. and Sternin, B.** (2004), Index defects in the theory of spectral boundary value problems, in *Aspects of Boundary Problems in Analysis and Geometry*, Gil, J., Krainer, T. and Witt, I., Eds., Birkhäuser, Basel–Boston–Berlin, 170–238.

**Schrohe, E.** (1992), Spectral invariance, ellipticity, and the Fredholm property for pseudodifferential operators on weighted Sobolev spaces, *Ann. Global Anal. Geometry* **10**(3), 237–254.

**Schrohe, E. and Seiler, J.** (1996), *An Analytic Index Formula for Pseudo-Differential Operators on Wedges*, Preprint No. MPI/96-172, Max-Planck-Institut für Mathematik, Bonn.

**Schulze, B.-W.** (1989), Pseudo-differential operators on manifolds with edges, in *Symposium "Partial Differential Equations,"* Holzhau, 1988, Teubner, Stuttgart–Leipzig, 259–288.

**Schulze, B.-W.** (1991), *Pseudodifferential Operators on Manifolds with Singularities*, North–Holland, Amsterdam.

**Schulze, B.-W., Sternin, B. and Shatalov, V.** (1996), Structure rings of singularities and differential equations, in *Differential Equations, Asymptotic Analysis, and Mathematical Physics*, Akademie-Verlag, Berlin, 325–347.

**Schulze, B.-W., Sternin, B. and Shatalov, V.** (1997), Operator algebras associated with resurgent transforms and differential equations on manifolds with singularities, in *Spectral Theory, Microlocal Analysis, Singular Manifolds*, Demuth, M., Schrohe, E., Schulze, B.-W., and Sjöstrand, J., Eds., Akademie-Verlag, Berlin, 300–333.

**Schulze, B.-W., Sternin, B. and Shatalov, V.** (1998*a*), *Differential Equations on Singular Manifolds. Semiclassical Theory and Operator Algebras*, Wiley–VCH, Berlin–New York.

**Schulze, B.-W., Sternin, B. and Shatalov, V.** (1998*b*), On the index of differential operators on manifolds with conical singularities, *Ann. Global Anal. Geometry* **16**(2), 141–172.

**Seeley, R.** (1967), Complex powers of an elliptic operator, *Proc. Sympos. Pure Math.* **10**, 288–307.

**Seeley, R.** (1990), Conic degeneration of the Dirac operator, *Colloquium Mathematicum* **LX/LXI**, 649–658.

**Senichkin, V.** (1995), Pseudodifferential operators on manifolds with edges, *J. Math. Sci.* **73**(6), 711–747.

**Senichkin, V.** (1996), The spectrum of the algebra of pseudodifferential operators on a manifold with smooth edges, *Algebra Analiz* **8**(6), 105–147.

**Shubin, M.** (1985), *Pseudodifferential Operators and Spectral Theory*, Springer-Verlag, Berlin–Heidelberg.

**Simonenko, I.** (1965), A new general method of investigating linear operator equations of singular integral equation type, *Izv. Akad. Nauk SSSR, ser. Matem.* **29**(3), 567–586, **29**(4), 757–782.

**Sternin, B.** (1966), Elliptic and parabolic problems on manifolds with boundary consisting of components of different dimension, *Trudy Mosk. Matem. Obshch.* **15**, 38–108.

**Sternin, B.** (1967), Elliptic (co)boundary morphisms, *Dokl. Akad. Nauk SSSR* **172**(1), 44–47.

**Sternin, B.** (1971), *Topological aspects of the Sobolev problem*, Moscow Institute of Electronic Engineering, Moscow.

**Sternin, B.** (1972), *Quasielliptic Equations on an Infinite Cylinder*, Moscow Institute of Electronic Engineering, Moscow.

**Sternin, B.** (1976), Relative elliptic theory and the Sobolev problem, *Dokl. Akad. Nauk SSSR* **230**(2), 287–290.

**Sternin, B. and Shatalov, V.** (1994), *Quantization of Symplectic Transformations and the Lefschetz Fixed Point Theorem*, Preprint No. MPI/94-92, Max-Planck-Institut für Mathematik, Bonn.

**Sternin, B. and Shatalov, V.** (1996*a*), Atiyah–Bott–Lefschetz fixed point theorem in symplectic geometry, *Dokl. Ross. Acad. Nauk* **348**(2), 165–168.

**Sternin, B. and Shatalov, V.** (1996*b*), Relative elliptic theory and the Sobolev problems, *Matem. Sbornik* **187**(11), 115–144.

**Sternin, B. and Shatalov, V.** (1998), The fixed point Lefschetz theorem for quantized canonical transformations, *Funkts. Anal. Prilozhen.* **32**(4), 35–48.

**Stong, R.** (1974), Manifolds with reflecting boundary, *J. Diff. Geometry* **9**, 465–474.

**Taylor, M.** (1981), *Pseudodifferential Operators*, Princeton University Press, Princeton, NJ.

**Teleman, N.** (1984), The index of signature operators on Lipschitz manifolds, *Publ. Math. IHES* **58**, 39–78.

**Vaillant, B.** (2001), *Index and Spectral Theory for Manifolds with Generalized Fibred Cusps*, Bonner Mathematische Schriften, 344, Universität Bonn, Mathematisches Institut, Bonn.

**Varela, J.** (1974), Duality of $C^*$-algebras, in *Recent Advances in the Representation Theory of Rings and $C^*$-Algebras by Continuous Sections* (Sem., Tulane Univ., New Orleans, LA, 1973), American Mathematical Society, Providence, RI, 97–108.

**Vasilevski, N.** (1989), Local principles in operator theory, in *Linear Operators in Function Spaces. Abstracts North Caucasus Regional Conference*, Groznyi, 32–33.

**Vasilevski, N.** (1994), Convolution operators on standard CR-manifolds. II. Algebras of convolution operators on the Heisenberg group, *Integr. Equat. Oper. Theory* **19**(3), 327–348.

**Vishik, M. and Eskin, G.** (1965), Convolution equations in a bounded domain, *Uspekhi Matem. Nauk* **20**(3), 89–152.

**Vishik, M. and Eskin, G.** (1966), Convolution equations in a bounded domain in spaces with weighted norms, *Matem. Sbornik* **69**, 65–110.

**Weinstein, A.** (1976), Fourier integral operators, quantization, and the spectra of riemannian manifolds, in *Géométrie symplectique et physique mathématique* (Aix-en-Provence, 1974), Vol. 237 of Colloque Internationale de CNRS, Paris, 289–298.

**Weinstein, A.** (1997), Some questions about the index of quantized contact transformations, *RIMS Kôkûryuku* **104**, 1–14.

**Whitehead, G.** (1962), Generalized homology theories, *Trans. Am. Math. Soc.* **102**, 227–283.

**Witten, E.** (1985), Global gravitational anomalies, *Comm. Math. Phys.* **100**, 197–229.

**Yosida, K.** (1968), *Functional Analysis*, Springer-Verlag, Berlin.

**Zhang, W.** (2000), $\eta$-invariants and the Poincaré-Hopf index formula, in *Geometry and topology of submanifolds, X* (Beijing/Berlin, 1999), World Scientific Publishing, Singapore, 336–345.

# Index

Milton Keynes UK
Ingram Content Group UK Ltd.
UKHW051946071024
449327UK00026B/2181